Probiotic Research in Therapeutics

Indu Pal Kaur
Editor-in-Chief

Parneet Kaur Deol • Simarjot Kaur Sandhu
Editors

Probiotic Research in Therapeutics

Volume 4: Probiotics in Neurodegenerative Disorders

 Springer

Editor-in-Chief
Indu Pal Kaur
University Institute of Pharmaceutical
Sciences, Panjab University
Chandigarh, Punjab, India

Editors
Parneet Kaur Deol
G.H.G. Khalsa College
of Pharmacy, Gurusar Sadha
Ludhiana, Punjab, India

Simarjot Kaur Sandhu
Taro Pharmaceutical Industries Ltd.
Brampton, ON, Canada

ISBN 978-981-16-6762-6   ISBN 978-981-16-6760-2   (eBook)
https://doi.org/10.1007/978-981-16-6760-2

This Springer imprint is published by the registered company Springer Nature Singapore Pte Ltd.
The registered company address is: 152 Beach Road, #21-01/04 Gateway East, Singapore 189721, Singapore

Foreword by J. V. Yakhmi

The saying attributed to Hippocrates, the Father of Medicine, that "Let food be thy medicine, and let medicine be thy food" never felt more valid than now when we are challenged by a variety of life-style diseases. The relevance of holistic healing has increasingly been related, in recent years, to the gut microbiome, composed of bacteria, archaea, viruses and eukaryotic microbes, all of which reside in our gut, and together have a strong potential to impact our physiology, both in health and in disease. When faced with a variety of diseases, our present-day knowledge lays emphasis on the importance of a healthy microbiome, not only limited to gut health but also to metabolic disorders, cancers, immunity, brain health and skin health. Can we manipulate the gut microbiota by probiotic intervention towards disease prevention and treatment? That is precisely what is receiving the attention of a large number of scientists engaged in research on human health. The growing market interest in health benefits of probiotics has intensified research and investments in this area. With an overwhelmingly large number of new products based on probiotics on the shelves of the supermarkets and pharmacies, it can be inferred that the research in this area is at a very exciting stage. Though the intricate mechanisms involved in the importance of gut flora may require some basic scientific expertise, surfing through scientific claims on usefulness of probiotic therapy can catch the fancy of even a general reader.

I have known Prof. Indu Pal Kaur, Chief Editor of this book series, for the past 12 years and have been closely following her research interests which essentially hover around being a formulation scientist, be it for small and large molecules, phytochemicals and probiotics. I have noticed her deep interest in trying to complement the observational data compiled in the traditional system of medicine with scientific rationale from currently available information. I have myself discussed with her, several times, the human microbiome and its manipulations for useful therapeutic options. She has been active in the topic of probiotics for a long time and had, in fact, published her first review on Potential Pharmaceutical Applications of Probiotics way back in 2002, which has been cited over 500 times to date. Her passion to bring probiotics into mainstream therapeutics is not limited only to the ailments of the gut, viz. inflammation, ulcers and cancers, but is also aimed to extend it to other life-style diseases, such as depression, chronic fatigue syndrome, vaginal candidiasis, wound healing and skin health.

The present ebook series, comprising five volumes, brings latest information and key insights on application of probiotics in cancer and immunological disorders, gut inflammation and infection, skin ailments, neurodegenerative disorders and metabolic disorders. The contributing authors are recognised experts which ensures that each chapter affords a critical insight into the topic covered, with a review of current research, and a discussion on future directions in order to stimulate interest. Each volume itself covers a broad theme in detail by including chapters disseminating basic information in the field in such a manner that it would attract the attention of even a stray reader or intending consumers. Of course, the whole series of five volumes is designed with care so as to not only ignite the minds of graduating students for future research but also boost the confidence of health professionals, physicians, dieticians, nutritionists and those practicing naturopathy by underlining the integrity of the data documented in the chapters of these volumes from well-established labs and groups. All in all, a very thoughtful compendium of probiotics research in therapeutics!

Asia Pacific Academy of Materials   J. V. Yakhmi
Mumbai, Maharashtra, India

Homi Bhabha National Institute
Mumbai, Maharashtra, India

Physics Group, Bhabha Atomic Research Centre
Mumbai, Maharashtra, India

Technical Physics and Prototype
Engineering Division, Bhabha Atomic Research Centre
Mumbai, Maharashtra, India

Spectroscopy, Synchrotron and Functional
Materials, Bhabha Atomic Research Centre
Mumbai, Maharashtra, India

Foreword by Pike-See Cheah

The brain is the body's most complex organ. There are around 86 billion neurons in the human brain, all of which are in use. Each neuron communicates with many other neurons to form circuits and share information. Any disruption in these neuronal communication circuits can contribute to the development of various neurological or psychiatric disorders.

The beneficial effects of gut microbiota are known for a long time. Initially, it seemed unlikely that these microorganisms could affect body organs other than the digestive system. However, it is now known that the gut microbiota composition affects a wide range of physiological processes, including mental health. In recent years evidence has shown that the resident microbes and brain communicate via a bidirectional axis. The traditional concept of the gut–brain axis has now broadened to the "microbiota–gut–brain axis", emphasising the prominence of the microbiome in the regulation of gut–brain communication.

The fourth volume of the ebook series *Probiotic Research in Therapeutics: Probiotics in Neurodegenerative Disorders* is an essential addition to our growing understanding of the "microbiota–gut–brain axis". In this volume, Prof. Indu Pal Kaur and her team of editors bolster the latest advances in brain research with a series of compelling discoveries about the underlying mechanism of the gut–brain axis and the current status of probiotic therapy in neurodegenerative disorders. In view of the therapeutic use of probiotics for neurodegenerative diseases, clear bright light of optimism shines through every chapter. As the editor in chief, Professor Kaur is genuinely the mastermind in providing an authoritative and timely overview of the field. This ebook should be one of the riveting reference materials of all time!

International Brain Research Organisation-Asia   Pike-See Cheah
Pacific Regional Committee (IBRO-APRC)
Paris, France

Universiti Putra Malaysia (UPM)
Selangor, Malaysia

Preface

It is only in the last decade or so that scientific community has actively deciphered the potential role of gut microbiome in modulating mental health. It is said that the next era therapeutics would involve manoeuvring the resident microbiome of human body to shift it from dysbiosis to symbiosis and hence from disease to health.

The multiplicity of bacteria in the gut to the count of a trillion or more has also shown profound effects on the brain and is now propounded to modulate an individual's mental health. Researchers from Massachusetts Institute of Technology, USA, recently validated the gut–brain axis via their organ-on-a-chip system and confirmed that the gut microflora communicates with brain cells. Latter occurs via a galaxy of pathways, viz. the hypothalamic–pituitary–adrenal axis, immune modulation, tryptophan and serotonin metabolism and production of short-chain fatty acids and other neuroactive compounds.

Thus the possibility to manipulate mental health by regulating gut microbiota through probiotic therapy is highly exciting and full of possibilities. Gut microbiome is suggested to be a neuronal health target, and the term "psychobiotic" has been coined to describe probiotic bacteria eliciting mental health benefits.

The fourth volume of the present eBook series entitled *Probiotic Research in Therapeutics: Probiotics in Neurodegenerative Disorders* is a sincere effort by us to compile scientific evidence on "Gut–Brain Axis" and the scope of probiotic therapy for improved mental health. The volume comprises ten chapters. The introductory chapter demonstrates the importance of gut microbiome in adult brain function and its impact on neurological disorders. Chapter 2 unravels the Gut–Brain–Skin axis triad and Chap. 3 describes the modulation of gut microbiota by deployment of probiotics for mental health. Next two chapters review the amelioration of autism spectrum disorder (ASD) by probiotic and/or prebiotic therapy and the role of healthy gut microbiota in the manifestation of ASD. Chapter 5 also describes the involvement of Nrf2-Keap1 signalling pathway in amelioration of gut–brain dysbiosis in ASD by probiotics. Next four chapters (Chaps. 6–9) portray the scope of probiotic therapy in prophylaxis and management of various neurological conditions, viz. Alzheimer's disease, major depressive disorders, insomnia and diurnal cycle, and chronic fatigue syndrome. The authors describe the underlying mechanism of action and provide a consolidated overview on the preclinical and clinical evidence supporting probiotic therapy in the management and control of these indications. Next in line is the last chapter of the volume, describing the advantages and challenges of current animal models used in the evaluation of probiotics for neurodegenerative disorders. Emphasis is laid on describing the future directions in designing better animal models for these evaluations.

With state-of-the-art commentaries on all aspects of probiotic research for mental health, from contributors across the globe, the eBook provides an authoritative and timely overview of the field. I hope this book will be a useful educational and scientific tool to academicians, health professionals, students and pharma/biotech businessmen worldwide. As editors of the book, we express our sincere thanks to all the authors for their excellent contributions to the book.

Chandigarh, Punjab, India   Indu Pal Kaur
Ludhiana, Punjab, India   Parneet Kaur Deol
Brampton, ON, Canada   Simarjot Kaur Sandhu

Contents

About the Editors

Indu Pal Kaur is a professor of Pharmaceutics and currently the Chairperson at the University Institute of Pharmaceutical Sciences (UIPS), Panjab University, Chandigarh, India. Her research forte involves enhancing the performance of drugs, including small and large biomolecules, viz. probiotics using active tailored delivery systems. The emphasis of her work lies in industrial and clinical translation, as evidenced by four technologies transferred by her group to the industry. She has 145 high-impact publications, 21 book chapters, four books, and four journal special issues to her credit. She is on the Editorial and Review Board of more than 50 journals.

Parneet Kaur Deol is presently working as Assistant Professor at G.H.G. Khalsa College of Pharmacy, Gurusar Sadhar, Ludhiana, Punjab, India. She has more than 10 years of experience in probiotic research and has published her work in highly reputed peer-reviewed international journals. She has co-edited a particular issue for *Current Pharmaceutical Design* with Prof. Indu Pal Kaur in 2019. Dr. Deol has presented her work at various national and international platforms. She has contributed 21 international publications, with a cumulative impact of more than 50.

Simarjot Kaur Sandhu is presently working as an analytical chemist at Taro Pharmaceutical Industries Ltd., Brampton, Ontario, Canada. She has done her doctorate from the University Institute of Pharmaceutical Sciences, Panjab University, Chandigarh, India. Her area of research lies in improving the profile of phytopharmaceuticals and probiotics by encapsulating them within nanoparticles. She has published six international publications and has filed two India Patent Applications and one European and one US patent application out of her Ph.D. work.

Gut–Brain Axis: Role of Gut Microbiota in Neurodegenerative Disease

Aarti Narang Husarik and Rajat Sandhir

Abstract

The role of gut bacteria in neurodegenerative disease has long been speculated; however, the extent of influence and the exact composition of microflora that mechanistically alter outcomes are less understood.

While aging was thought to be a major contributor to neurodegenerative disease, the role of the immune system started to become more appreciated bringing the hypothesis of "inflammaging" to the forefront. Gut bacteria serve to prime our immune system and therefore play a role in shaping our immune response to infection and disease. The differences in gut flora between healthy individuals and ones suffering from Alzheimer's or Parkinson's disease have been widely documented; however, it is not understood if they are the cause or the effect of the disease. The second hypothesis in the field is the antimicrobial response hypothesis or infection hypothesis, which proposes that the neurodegenerative disease is an undesired outcome of the brain's immune response against pathogens. In this context, it is important to understand whether it is the presence of microbes themselves in the brain or just the microbes in the gut that prime the immune system and cause an amplified immune response in the brain cumulatively leading to neurodegenerative disease. It is also important to understand the concept of pathogen-associated molecular patterns (PAMPs) that serve to trigger innate immunity by engaging toll-like receptors (TLRs) and that these PAMPs or molecular patterns may be present and trigger inflammation without the presence of actual pathogen.

A. N. Husarik (✉)
BD Biosciences, San Diego, CA, USA

R. Sandhir
Panjab University, Chandigarh, India
e-mail: Sandhir@pu.ac.in

© Springer Nature Singapore Pte Ltd. 2022
P. K. Deol, S. K. Sandhu (eds.), *Probiotic Research in Therapeutics*,
https://doi.org/10.1007/978-981-16-6760-2_1

The ultimate goal of delineating these mechanisms is to then use this knowledge to develop treatments. Some approaches that have been tested in preclinical and clinical studies including fecal transplants have been summarized here as well.

Keywords

Alzheimer's disease · Microbiome in neurodegenerative disease · Brain–gut–microbiome axis · Parkinson's disease · Complement · Inflammaging · Inflammasome · Pattern recognition

1.1   Introduction

There has been a paradigm shift in our understanding of the pathophysiology of chronic neurological diseases. Traditionally microorganisms were not considered important players in the cause or effect of neurological disease but more recently a bidirectional signaling between the brain and the gut microbiome has become evident. This communication occurs both ways: the neural network controls the gastrointestinal function via the enteric nervous system (ENS) and the gut communicated with the central nervous system (CNS) via sympathetic (prevertebral ganglia) and parasympathetic (the vagus nerve) signaling. Considering how critical gut bacteria are to early brain development and adult neurogenesis (Dinan and Cryan 2017; Quigley 2017), it is of interest to investigate how this communication between ENS and CNS may be impacted by the nature of the gut microbiome. The human gut is home to highest bacterial density recorded in any microbial ecosystem at about 10^{11} to 10^{12} per milliliter. The bacterial composition has a key impact on human biochemistry and disease pathologies (Aziz et al. 2013). Initial reports pointing towards a gut–brain axis brought attention to the impact that might have on the CNS (Cryan and O'Mahony 2011; Bravo et al. 2012; Cryan and Dinan 2012). Studies in germ-free animals substantiated the fact that gut bacteria or the lack thereof could impact animal behavior and neural biochemistry (Heijtz et al. 2011).

1.2   The Gut–Brain Axis

The effects of microbiota can be summarized as direct and indirect. The direct effects include the secretion of factors by the microbiome that directly impact brain function, whereas the indirect effects include more systemic effects of altered microbiota that indirectly influence brain function and propensity towards disease.

A substantial amount of preclinical studies have shown the indirect association between gut microbiota and brain function. The effect of depletion of gut microbiota using antibiotic Ampicillin for 1 month was tested by Wang et al. (2015) using Sprague-Dawley (SD) rats. They observed that serum corticosterone anxiety was increased and spatial memory decreased and these symptoms were reversed by

Lactobacillus fermentum NS9. This quintessential gain/loss of function study provides evidence for the existence of a gut–brain axis. In a human study of healthy women volunteers, consumption of fermented milk product with probiotics changed activity of resting brain, especially central processing of emotion and sensation (Tillisch et al. 2013; Hu et al. 2016). The evidence of more direct effects of the gut bacteria are through communication with the brain via gut peptides and SCFAs. Gut peptides modulate enteroendocrine cells that further affect hormonal signaling pathways (Borre et al. 2014). Studies in germ-free (GF) mice suggest that the gut microbiota release gut peptides and short chain fatty acids (SCFAs) (Koh et al. 2016; Van der Hee and Wells 2021). SCFAs may influence microglial maturation directly or leak through the intestinal mucosa impacting immune regulation and CNS function. These SCFAs can also influence tight junctions between brain endothelial cells leading to altered blood–brain barrier BBB permeability (Braniste et al. 2014). Another way that microbiota communicate with the brain is by eliciting signals via the vagal nerve causing behavioral alterations (Perez-Burgos et al. 2013). These are just some of the various ways, bacterial metabolites alter pathology and treatment response. Figure 1.1 briefs the mechanism involved in gut–brain axis.

1.3   Alzheimer's Disease (AD)

AD is a neurodegenerative disorder characterized by formation of amyloid-beta (Aβ) plaques and neurofibrillary tangles and while these are considered the hallmarks of the disease new and more challenging perspectives of underlying disease pathology have come to light. As discussed briefly above, alterations in the composition of gut microbiome can have impact on intestinal permeability and immune activation and hence systemic inflammation. Inflammatory state perpetuated by imbalances in the microflora can promote neuroinflammation, neural injury, and ultimately neurodegeneration. Along these lines, Aβ has been speculated to be an antimicrobial peptide participating in the innate immune response with undesirable effects that manifest as disease (Soscia et al. 2010; Pastore et al. 2020). Also, bacterial amyloids may elicit cross-seeding and prime microglia through molecular mimicry. The potential mechanisms of amyloid spreading include neuron-to-neuron spreading, using other cells such as astrocytes, fibroblasts, microglia to cross the blood–brain barrier (Kowalski and Mulak 2019).

1.3.1   Cause and Effect Conundrum

Neurodegenerative disease specifically Alzheimer's disease has historically been associated with aging but as is now understood age is not as critical as was once thought. The pathophysiology has been hypothesized to be influenced by inflammation perhaps due to aging and other factors. The following section describes the major hypotheses that have been put forth to explain the pathophysiology of Alzheimer's disease.

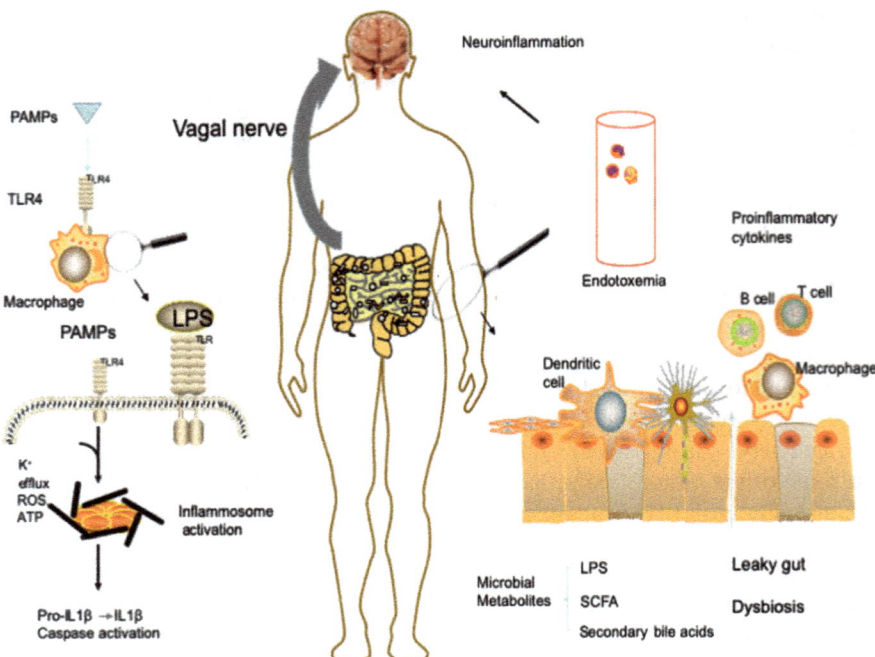

Fig. 1.1 The gut–brain axis is the cross-talk between the intestinal microbiota, immune modulators, and neural circuitry. Various dietary, environmental and immune factors can lead to Leaky gut syndrome which facilitates the translocation of components of gut bacteria, such as LPS to cross the barrier and cause peripheral inflammatory activation. PRRs (Pattern recognition receptors) such as TLRs (Toll like Receptors) sense PAMPs (Pathogen associated molecular patterns). TLR4 present on microglia recognizes PAMPs such as LPS or misfolded proteins and triggers inflammatory pathways ultimately leading to neuroinflammation. The gut microbiome shapes the immune response in various ways by priming the innate immune system as well as B and T cells through antigen presenting cells such as dendritic cells. The primed immune system responds to the triggers such as PAMPs and misfolded proteins leading to gliosis and neuroinflammation which is underlying common factor in neurodegenerative disease

1.3.2   Amyloid Hypothesis

This hypothesis was first put forth by Alois Alzheimer for a 56-years-old woman who had early onset AD (Zilka and Novak 2006). Alois Alzheimer's observations of amyloid plaques and neurofibrillary tangles, from the cortex of his patient set precedence for hallmarks of the disease to be present pathologically before a diagnosis for AD can be made.

Amyloid originates from aberrant processing of amyloid precursor protein (APP) (Perneczky et al. 2013). The amyloidogenic/aberrant APP metabolism leads to the formation of beta-amyloid peptides or fibrils 1–42 and 1–40. The Aβ1–40 is benign, while Aβ1–42 is pathogenic and it is these extracellular Aβ fibrils that lead to formation of the amyloid plaques (Galante et al. 2012). The amyloid hypothesis states that the deposition of fibrils formed from this aberrant proteolysis of APP leads to the formation of classical senile plaques (McGeer and McGeer 2013; Bolós et al. 2017). However, senile plaques may or may not manifest as cognitive pathology perhaps due to varied levels of microglial clearance of Aβ among patients. Pathogen or pattern recognition receptors (PRRs) such as the TLRs and CD36 recognize pathogen-associated molecular patterns (PAMPs) such as Aβ and induce a strong inflammatory response (Venegas and Heneka 2017; Le Page et al. 2018). This inflammatory response causes the hyperphosphorylation of the Tau protein leading to the formation of neurofibrillary tangles (Serrano-Pozo et al. 2011; Holtzman et al. 2016; Leyns and Holtzman 2017).

There are some limitations of this hypothesis that came to light after the years of research in the field. If this hypothesis was true, it would imply that Aβ levels in the blood would correlate with the extent of progression of the disease and that Aβ levels in the blood can be used as a biomarker for various stages of the disease. However, that was not found to be the case and the levels corresponded more with the early stages. Interestingly, more Aβ was found in blood and CSF during the mild cognitive impairment (MCI) stage of the disease than in clinically diagnosed AD (Camponova et al. 2017). Many clinical trials, with the end goal of reducing Aβ levels in the hope to prevent the progression to full blown AD did not show promise (Cummings et al. 2018; Molinuevo et al. 2018).

A deeper understanding of the underlying mechanism suggested that α-secretase may be the driver for production of soluble APP fragments (Chow et al. 2010), while β-secretase (BACE) and γ-secretase act sequentially in amyloidogenic Aβ production (Rivest 2009; Siegel et al. 2017). Several inhibitors of these secretases have failed in clinical trials, further negating the amyloid hypothesis. However, there are other hypotheses such as the infection hypothesis that remain to be fully tested (Bourgade et al. 2016, 2017). Some other pieces of evidence that support amyloid hypothesis include lithium study as it was shown to have a beneficial effect on pathogenic changes of AD mostly by inhibiting glycogen synthase kinase-3 (GSK-3) (Mendes et al. 2009). This enzyme plays a key role in APP metabolism and Tau phosphorylation. Lithium treatment decreased the GSK3 mRNA, thus validating APP metabolism and Tau as critical players in the pathology (Mendes et al. 2009). In another preclinical study, lithium was shown to delay the progress of neurofibrillary tangles in a tau transgenic line (Tg30tau) (Leroy et al. 2010). In another study in Drosophila model of AD, lithium was shown to alleviate the pathology of the amyloid-beta by GSK-3 inhibition (Sofola et al. 2010). So this data suggests that while there is evidence showing some correlation between amyloid and disease pathology, there is also evidence showing that preventing it is not always enough.

1.3.3 The Hygiene Hypothesis

This hypothesis states that the improved sanitation or lower exposure to microorganisms in early life is associated with increased propensity towards autoimmune diseases (Strachan 1989). The use of sanitizing agents such as bleach in cleaning agents, frequent use of antibiotics, and sanitary food preparation practices all result in decreased exposure to microorganisms including those that are beneficial. The lack of this exposure is detrimental to the immune system due to decreased stimulation of immune response. The commensals cause mild stimulation of the immune system that is actually beneficial and promotes regulatory T cells (Tregs), rather than to type 1 T helper cells Th1 and type 2 T helper cells Th2 cells (Rook and Lowry 2008). T regulatory cells, as the name suggests, regulate T cell function while Th1 cells mount immune responses against intracellular pathogens, and Th2 cells against extracellular pathogens (Schwartz et al. 2001). Excessive or improperly directed Th1 or Th2 response leads to autoimmune disease. Lack of commensal bacterial reduces the tolerance inducing Treg cells, thereby predisposing the body to overactive immune system which ultimately manifests as autoimmune disease or chronic inflammatory disease (Romagnani 2004; Rook and Lowry 2008). AD falls into the category of systemic inflammatory disease and shows classic features of Th1-mediated immune response (Pellicanò et al. 2012). Some commensal bacteria that are known to induce the helpful Treg differentiation and function are *Bacteroides fragilis* (Round and Mazmanian 2010) and *Clostridium* clusters XIVa, IV, and XVIII (Atarashi et al. 2011b, 2013). Conversely, the capsule of microbes like *B. fragilis* can polarize towards Th1-type responses which are not so helpful. Thus, the composition of gut microbiome is a key factor in predisposition towards inflammatory disorders.

1.3.4 Infection Hypothesis (Antimicrobial Response Hypothesis Also Known as Antimicrobial Protection Hypothesis)

In the past few years, it has been speculated that β-amyloid deposition occurs in response to genuine or perceived immune challenge. Pathogens or pathogen like molecules known as pathogen-associated molecular patterns (PAMPs) are entrapped by Aβ in β-amyloid. This Aβ fibrillization in turn elicits neuroinflammatory response that helps clear β-amyloid/pathogen deposits. However, prolonged or misdirected activation of this inflammation leads to progressive neurodegenerative disorder. There is increasing evidence that microbe levels in the brain correlate with AD. Thus, the antimicrobial protection hypothesis states that microbial burden in the brain exacerbates inflammation related to β-amyloid deposition, neurodegeneration, and AD progression. In the antimicrobial protection model, Aβ's pathophysiology is shifted towards dysregulated innate immune response as plausible explanation for disease progression; however, β-amyloid deposition still remains at the heart of pathology of neurodegeneration. Thus, this new model only extended the amyloid cascade hypothesis.

1.3.5 A Case for Viruses

There is a lot of correlative evidence between viral infection and Alzheimer's disease, particularly HSV1. The links have been broadly discussed by Itzhaki et al. including the discovery that the viral DNA was associated with AD plaques and the accumulation of Aβ in HSV1-infected cell cultures and infected mice (Wozniak et al. 2007, 2009; Santana et al. 2012; Itzhaki 2018). Further evidence showed the accumulation of Aβ and P-tau HSV1-infected cell cultures (Zambrano et al. 2008; Álvarez et al. 2012; Eimer et al. 2018). Rudy Tanzi's group showed key evidence that Aβ is essentially an immune response, an antimicrobial peptide that is triggered as a result of a perceived infection and the seeding is amplified by an inflammatory response. Tanzi and others suggested that even a small amount of insoluble, aggregation-prone Aβ42 can get seeded and act as a template for otherwise soluble and abundant Aβ40 oligomerization and spreading (Sowade and Jahn 2017). Furthermore, in vitro studies showed that treatment with Acyclovir and other HSV1 replication-inhibitors reduced levels of Aβ and P-tau suggesting the critical role of the viral replication in the pathology. Helicase primase inhibitor was also shown to have promising results in cell culture (Wozniak et al. 2011).

The confounding evidence exists that supports and argues against this connection. Some investigations show that AD patients have anti-HSV1 IgG and IgM antibodies in their serum, suggesting a link between the infection and cognitive decline (Agostini et al. 2016). The argument against it was that the percentage of AD patients and elderly people whose brains had HSV1 was low (Olsson et al. 2016; Pisa et al. 2017). It has been speculated that the age of the samples, the conditions in which they were stored and the sensitivity of the PCR detection may have confounded the results.

Another possible way the virus infection could alter Aβ accumulation may be by impairing the lysosomal function leading to inefficient debris removal (Kristen et al. 2018). Further, improved methods such as transcriptomic analysis of AD patients and control brains showed higher levels of herpes viruses 6A, 7 (HHV6 and 7), and HSV1 were present in elderly and AD brains, than in the controls. These findings corroborated with what had been reported previously (Jamieson et al. 1991; Lin et al. 1997).

The association of viral loads with clinical dementia rating and density of neurofibrillary tangles and plaque pathology (Readhead et al. 2018) suggested that it was not a mere opportunistic infection. Several studies show that β amyloid acts protectively against viruses and is substantiated by the fact that HSV1 DNA is exclusively located within amyloid plaques (Bourgade et al. 2014).

1.3.6 Inflammaging Hypothesis

Inflammaging is the low-grade chronic inflammation characterized by circulating levels of TNF-α, IL-6, and C-reactive protein (CRP). The new knowledge surrounding the networks of lymphatic vessels in meningeal spaces suggests that there is an inflammation continuum between the systemic and CNS immune responses. Bacterial proteins in the gut serve to prime the immune cells and in turn enhance neuroinflammation by a phenomenon called molecular mimicry also referred to as cross-seeding. What this means is that these molecules mimic pathogen molecules and trigger the immune system such that it causes deleterious effects. Another interesting way the microbiome talks to the systemic immune system is through bacterial metabolites which may in turn influence the outcome of treatments. Studies show higher TNF-α production in B cells from elderly individuals than young subjects, and which correlated with serum TNF-α levels in both humans and mice (Frasca and Blomberg 2011). Effect of microbiota on the peripheral immune cell activation and cytokine profile in neurodegenerative disease poses a third dimension to inflammaging hypothesis.

If we take a look at what has been reported about the inflammatory microenvironment of germ-free mice, it becomes clear there is impairment of healthy immune responses in part due to developmental defects in microglia leading increased tendency towards aggregation of Aβ resulting in neurological diseases including AD (Erny et al. 2015). A plausible explanation could be decreased clearance of amyloid. Several studies had previously shown that omega-3 fatty acid consumption and other dietary factors influence the progression of disease. The influence of combination of age and diet on gut bacteria thus altering the absorption of omega-3 also falls under inflammaging and is important to consider in the context of AD. The increased amount or propensity of Aβ to aggregate in the chronic inflammatory state is mediated by complement (Crehan et al. 2012). In support of the key role of complement, genome wide association studies identified single nucleotide polymorphisms (SNPs) in genes encoding complement proteins Clusterin (CLU) and CR1 (CR1) to be associated with risk of late-onset AD. A knockout of complement protein C3 C3KO showed improved outcomes in TauP301S mice without a significant reduction in gliosis and tau pathology (Wu et al. 2019). Increased levels and activation of C3 were found in PSD (post-synaptic densities) fractions and CSF from AD patients in the same study suggesting that C3 plays a critical role in the pathology of AD. Complement (c3d) has also been shown to exacerbate CNS injury by binding to neoepitopes exposed as a result of injury and targeted complement inhibitors such as B4-Crry have resulted in improved outcomes (Narang et. al. 2017). However, how the levels of complement vary with age and stage of disease progression could not be concluded from these studies. Further studies are needed to explore the influence of changing gut microbiome on complement activation in the context of AD to establish how these changes may aggravate or ameliorate disease progression.

1.4   Parkinson's Disease

The connection between gut microbiome and Parkinson's disease (PD) is not new either. Recent preclinical studies have redrawn attention to the subject and with the availability of advanced tools more objective and less correlative evidence is beginning to surface. Braak and colleagues hypothesized both intrinsic and extrinsic pathologies of the enteric neurons in the GI tract, the dorsal motor nucleus of the vagus nerve occurred even before the effect on Substantia Niagra (Braak et al. 2003). Lewy bodies were considered the hallmark of PD and the data pointed towards the presence of a-syn pathology in the gut up to 20 years prior to the onset of the disease. Further, Sampson et al. should the key role of gut microbiota in PD progression in alpha-synuclein-overexpressing (ASO) mice, an animal model for PD (Sampson et al. 2016). In this study, they showed a comparison between germ-free ASO mice and specific pathogen free (SPF) ASO mice. They found that germ-free ASO mice had lesser motor deficit accompanied by decreased alpha-synucleinopathy and microglia activation compared to specific pathogen free ASO mice. However, colonization with microbiota from PD patients led to an increase of symptoms in comparison to healthy individual's microbiota. This study can be considered key evidence for the causal role of gut bacteria in PD.

1.4.1 Dual Hit Hypothesis

Braak et al. proposed that α-syn misfolding and aggregation occurs in peripheral nerve terminals which later spreads centripetally through the vagus nerve based on the early pathology (Scheperjans et al. 2018). There is confounding evidence that supports and argues against the hypothesis as summarized here.

Two epidemiological studies showed that full truncal vagotomy decreases the risk of PD while selective vagotomy did not (Svensson et al. 2015; Liu et al. 2017). Also another study showed α-syn pathology in patient gut tissues up to 20 years prior to PD diagnosis (Hilton et al. 2014; Stokholm et al. 2016). The idea of a prion like cell-to-cell transfer between cells has been shown in animal models (Steiner et al. 2018) and centripetal spreading of α-syn (Breid et al. 2016; Ayers et al. 2017). Another way of transfer of α-syn to neuronal cells from the gut is via epithelial microfold cells (M cells) and dendritic cells found in the Peyer's patches (Friedland 2015). These studies support the gut-to-brain transmission of α-syn pathology. There are arguments that support and refute the gut as the site of origin of PD (Borghammer 2018; Lionnet et al. 2018). The argument refuting Braak's hypothesis came from Adler and Beach et al. who reported the absence of α-syn pathology (Lewy type α-synucleinopathy (LTS)) exclusively in the ENS (but not CNS) in 466 autopsies investigated (Adler and Beach 2016). This, however, does not exclude the possibility that initial α-syn pathology started in the gut, but this was difficult to prove using standard immunohistochemistry as some of these aggregates are sensitive to the proteases (Borghammer 2018; Scheperjans et al. 2018).

1.4.1.1 Evidence Unseen
α-Syn inclusions are detected in ENS of the vast majority of PD patients mostly by immunohistochemistry-based studies. It is speculated that the evidence is confounded by variability in the IHC methods and that the lack of effective methods to analyze GI biopsies is another factor considered (Corbillé et al. 2016). One

argument is that α-syn immunoreactivity is also seen in formalin fixed paraffin embedded healthy samples (Shannon et al. 2012; Aldecoa et al. 2015; Fasano et al. 2015; Sánchez-Ferro et al. 2015; Chung et al. 2016; Shin et al. 2017; Ruffmann et al. 2018; Yan et al. 2018). However similar flaws were also reported with one- and two-dimensional electrophoresis of gastrointestinal biopsies (Corbillé et al. 2017). Another confounding factor is the sheer low amounts of α-syn aggregates in gastrointestinal biopsies (Beach et al. 2016). Therefore, we have to take the IHC based detection of pathology with a grain of salt.

Another factor is that the scoring of gastrointestinal dysfunction in patients is mostly subjective (Knudsen et al. 2017). The use of radio-opaque markers to assess gastrointestinal transit times could help objectively evaluate the two hit hypothesis. Recently, PET scans using ^{11}C-donepezil have been used to assess both premanifest and manifest PD patients and additional methods are warranted to get a clearer picture (Fedorova et al. 2017).

1.4.2   NLRP3 Inflammasome Activation Mechanism

Inflammasomes are protein complexes that recognize stimuli such as pathogen-associated molecular patterns (PAMPs) and danger-associated molecular patterns (DAMPs) (Schroder and Tschopp 2010). Although the activation of the inflammasome during infection is protective and serves to clear the pathogen, unregulated or overactive NLRP3 inflammasome activation can lead to unintended pathological damage. Inflammasomes are more well studied in peripheral immune cells than in the brain; however, they have been now implicated as primary mechanism of α-Syn mediated neuroinflammation in PD (Codolo et al. 2013). One of the most studied inflammasomes is NLRP3 because of the variety of signals that can activate it (Mariathasan et al. 2006; Martinon et al. 2006; Hafner-Bratkovič et al. 2012). It has been hypothesized that fibrillar α-Syn causes microglial TLR activation leading to microglial NLRP3 inflammasome activation (Song et al. 2017; Gordon et al. 2018). This also goes along with the idea that toll-like receptors (TLRs) mediate the gut–brain axis because they serve as sensors for these DAMPS and PAMPs (Caputi and Giron 2018). In response to activation of TLRs, the NLRP3 inflammasome assembles and produces proinflammatory cytokines such as IL-β as well as IL-1α, IL-18, and IL-33 (He et al. 2016).

Several models have been proposed to explain how both direct and indirect signal recognition occur and is mediated by additional accessory proteins. It is important to critically examine the mechanisms to ascertain the role in the context of the gut–brain axis. One idea is that the leaking of lipopolysachharide, LPS, a bacterial protein, causes the activation of NLRP3 inflammasome which causes mitochondrial dysfunction and in turn affects neuronal function (Gordon et al. 2018). Another idea is that there is a "two signal" model of NLRP3 activation that proposes TLR signal is the first signal leading to NF-kB-mediated signaling of pro-IL-1β and pro-IL-18. The second signal is ATP, calcium or potassium flux, or mitochondrial reactive oxygen species (ROS) all of which can be induced by a variety of triggers including gut microbiota (Youm et al. 2013; Elliott and Sutterwala 2015; Giacoppo et al. 2017; Sarkar et al. 2017; Gong et al. 2018; Lee et al. 2019). These two hits of signals cause release of proinflammatory cytokines and apoptotic factors leading to the activation if caspases. The cytokines such as IL-1β and ROS release create sustained oxidative stress which then feeds into neuroinflammation causing more aggregated α-Syn leading to symptoms of PD (Elliott and Sutterwala 2015). Codolo et al. showed that misfolded α-Syn (aggregated α-Syn, α-Syn F) engages TLR2 on the surface of monocytes causing inflammasome activation which is mediated by fibril phagocytosis, cathepsin B release into the cytosol, followed by release of ROS which leads to production of mature IL-1β (Codolo et al. 2013).

Aside from the already discussed pattern recognition by TLRs leading to activation of NLRP3 inflammasomes, the microbiota can indirectly interact with NLRP3 inflammasome by producing secondary bile acids. The term secondary bile acids just imply that they are the by-product of body's bile acid by the bacteria. This indirect interaction is a sort of negative regulation mediated by the TGR5 receptor such that the bacterial metabolism of bile keeps the NLRP3 activation in check. However as a result of Western diet induced dysbiosis, the secondary bile acids are not reduced as much leaving the NLRP3 activation unchecked (Guo et al. 2016; Jena et al. 2018).

There is both clinical and preclinical evidence pointing to a key role of NLRP3 inflammasomes. Postmortem studies found NLRP3 inflammasome in the microglia in substantia nigra of PD patients (Gordon et al. 2018). Further, NLRP3 inhibition was protective in all rodent models of PD tested (Sarkar et al. 2017; Anderson et al. 2018). Similarly, knocking out NLRP3 in an AD animal model was also found to be protective showing a direct link between NLRP3 inflammasomes and neurodegenerative disease (Heneka 2017). Since it is known that NLRP3 inflammasomes are impacted by factors such as diet and gut dysbiosis (Pavillard et al. 2017), it puts into perspective how critical gut microbiome is in neuroinflammation and degenerative disease. Among NLRP3 activation induced cytokines, IL-1β has been speculated as the most relevant in the context of PD. Among other effects, IL-1β levels have been correlated with insulin resistance (Youm et al. 2013).

1.4.3 Insulin Resistance Angle

The link between insulin resistance and neurological disease has been investigated before; however, the involvement of gut bacteria is less understood. We just described how NLRP3 inflammasome activation may be a consequence of dysbiosis. Here we discuss the preclinical studies where knocking out NLRP3 protected against insulin resistance. The NLRP3 inflammasome has also been linked to insulin resistance and obesity (Stienstra et al. 2011; Vandanmagsar et al. 2011). Since IL-1β blocks insulin signaling (Ehses et al. 2009) and impacts insulin resistance as discussed above, it is no surprise that more and more evidence now links insulin resistance disorders, inflammasomes due to gut dysbiosis, and PD and AD brain (Hu et al. 2007; Wahlqvist et al. 2012; Wang et al. 2012; Aviles-Olmos et al. 2013; Karlsson et al. 2013; Santiago and Potashkin 2013; Athauda and Foltynie 2016;

Bloom et al. 2018; De Pablo-Fernandez et al. 2018). Inhibition of NLRP3 has been shown to be protective against insulin resistance and T2DM as well as PD so really looking into gut microbes that cause NLRP3 activation can be a key step in our understanding of this pathology and its treatment (Coll et al. 2015; Abderrazak et al. 2016; Bian et al. 2017; Chen et al. 2019). To sum up the above findings, LPS-TLR interaction (LPS from gut bacteria) leads to NLRP3 inflammasome which causes IL-1β expression that leads to insulin resistance, cognitive decline due to neuroinflammation (Jackson et al. 2019).

1.4.4 Curli Hypothesis

This hypothesis states that Curli fibers (by-products of bacterial biofilms) are composed of unfolded amyloids that modulate the inflammatory responses and increase α-Syn production. These Curli amyloid proteins are produced by Enterobacteriaceae family of gut bacteria (Tursi and Tükel 2018). Curli fibers are formed by bacteria as a defense mechanism against bacteriophages. They secrete CsgA (curli polymer) that mediates adhesion to form biofilms (Vidakovic et al. 2017). Studies show that amyloid protein α-Syn in aged rats and nematodes increased with the administration of these bacteria (Gallo et al. 2015; Chen et al. 2016). Biochemical studies demonstrate that curli fibers accelerate α-Syn aggregation (Chorell et al. 2015; Evans et al. 2015). Sampson et al. showed that curli-producing *Escherichia coli* can exacerbate motor impairment and GI dysfunction in ASO mice due to α-Syn aggregation. However a similar effect was not seen in control, non-susceptible mice. This damage caused by Curli-producing bacteria in ASO mice was shown to be ameliorated by inhibition of amyloid formation (Sampson et al. 2020). This idea although new, still has amyloid at the center of neuropathology, but offers an interesting take on a less known and more direct role gut bacteria might play in neurodegenerative disease.

1.4.5 Leaky Gut or the Intestinal Barrier Mechanism

The intestinal barrier is unique in the sense that it lines one of our largest organs and serves so many important functions. Gut-associated lymphoid tissue (GALT), the largest component of the immune system in the body, interacts with the gut microbiota that shape immune responses both locally and systemically. In the context of gut–brain axis, intestinal lining serves to shield from bacterial LPS from the systemic circulation and increasing inflammation. The role of gut microbiome in maintaining the integrity of intestinal barrier is very underappreciated, and a clear correlation between leaky gut (measured by LPS staining in mucosa) and α-Syn aggregation has been shown (Forsyth et al. 2011; Buford 2017; Obrenovich 2018). These findings are also corroborated by animal studies using both genetic and toxin induced models (Pfeiffer 2012). Thus gut origin of neuropathology in neurodegenerative disease has been shown in more ways than one (Borghammer 2018).

It has been widely documented that PD subjects show signs of leaky gut and related endotoxemia (i.e., LPS in the blood) and LPS binding protein LBP in comparison to age-matched controls (Forsyth et al. 2011; Clairembault et al. 2015; Buford 2017; Schwiertz et al. 2018). However the correlation between disease severity and fecal biomarkers of inflammation is not that straightforward. Only two of the several markers tested were higher in PD patients (alpha-1-antitrypsin and zonulin) compared to age-matched controls. In addition, abnormal intestinal tight junction proteins were found in PD patients compared to age-matched controls (Clairembault et al. 2015; Schwiertz et al. 2018). Parkinson disease patients often co-present with GI related disorders such as with inflammatory bowel disease (IBD), Crohn's disease. There are several studies showing that IBD and Crohn's patients are predisposed to PD; however, this connection is correlative not causative and has been disproven by other studies (Camacho-Soto et al. 2018; Villumsen et al. 2019; Weimers et al. 2019; Zhu et al. 2019).

Studies have attempted to dissect specific microbial metabolites that improve barrier integrity and SCFA appear to be the main ones (Koh et al. 2016). SCFAs such as acetate, propionate, and butyrate, lactate, and succinate interact with GPCRs via specific receptors. For example, GPR41 binds propionate/butyrate, GPR43 acetate/propionate, and GPR109a for acetate, propionate, butyrate, GPR81 binds lactate and GPR91, succinate (Ganapathy et al. 2013). SCFA interacts with GPCRs to induce inflammatory response, for example, GLPR43 signals the recruitment of T effector cells (Ganapathy et al. 2013). GPR109a, however, interacts with butyrate to act as histone deacetylase inhibitor (epigenetic effects). The effect of butyrate on HDAC1 dependent Fas regulation of T cells has been reported as well (Zimmerman et al. 2012). Diet and age-associated dysbiosis are often correlated with reduced SCFA as well. Another group of important bacterial metabolites are quorum sensing molecules (QSMs), which include N-acyl homoserine lactones (AHLs) in Gram-negative bacteria and autoinducing peptides (AIPs) in Gram-positive bacteria. AHLs like 3-oxododeconyl-L-homoserine lactone (C12) produced by *Pseudomonas aeruginosa* modulate the activity of macrophages, fibroblasts, epithelial cells, mast cells, T lymphocytes, B-lymphocytes, and neutrophils (Glucksam-Galnoy et al. 2013).

In the context of AD, similar to PD, there is evidence of dysbiosis, intestinal barrier dysfunction, and endotoxemia. AD patients postmortem brains show 21-fold greater LPS staining than in control brain tissue (Forsyth et al. 2011; Zhao et al. 2017a, b). As discussed previously, LPS has the ability to activate TLR4, especially on brain microglia (Caputi and Giron 2018). This is further recapitulated by the fact that TLR4−/− mice are protected from the pathogenic effects of rotenone and MPTP both of which are otherwise used as PD triggers in animal models (Noelker et al. 2013). This barrier dysfunction may not be the cause or initiator of PD but definitely serves to feed forward the neuroinflammation loop through the LPS-TLR4 axis.

1.5   Fecal Matter Transplantation (FMT)

Upon examination of relative abundance of gut bacterial species in patients with neuroinflammatory disease vs. controls, *Escherichia/Shigella* were higher and *Eubacterium rectale* were lower (Cattaneo et al. 2017). In general, people with AD had lower diversity in their fecal microbiota. *Firmicutes* and *Bifidobacterium* were lower and *Bacteroidetes were higher* (Kowalski and Mulak 2019). An interesting correlation was also found between bacterial abundance and the cerebrospinal fluid markers of AD pathology. FMT in a preclinical model of AD showed promising results including decreased p-Tau, Aβ40 and Aβ42, and improvements in synaptic plasticity. Not much is available in terms of clinical evidence for FMT in AD patients. Preclinical evidence shows that FMT helps with symptoms in PD. As discussed briefly before, Sampson et al. showed motor function improvements in ASO mice when they received healthy human feces compared to PD patients (Sampson et al. 2016). Another study also showed improved motor function accompanied by increased neurotransmitter levels and decreased inflammation after FMT from healthy mice in a model of PD (Sun et al. 2018). More direct evidence was seen in worsened motor performance upon receiving feces from PD mice along with reduction in neurotransmitter levels compared to controls. Clinical evidence, although very limited, showed that the motor improvements such as reduced leg tremors were temporary symptoms returned within 1–2 months post-FMT (Huang et al. 2019).

1.6   Conclusion

Based on the evidence and mechanisms, the way forward is to enhance intestinal epithelial integrity, protect against dysbiosis, reduce proinflammatory response, and inhibit propagation of neuroinflammation and neurodegeneration (Frasca and Blomberg 2016; Plaza-Díaz et al. 2017; Kowalski and Mulak 2019). Probiotic bacteria such as *Enterococcus faecium* and *Lactobacillus* reduced oxidative stress markers in the brain (Divyashri et al. 2015) and Lactobacilli and Bifidobacteria reduced neuroinflammation (Akbari et al. 2016; Kobayashi et al. 2017; Musa et al. 2017; Nimgampalle and Yellamma 2017; Azm et al. 2018). For example, probiotic supplementation has been shown to improve objective measures of AD in humans as measured by Mini-Mental State Examination scores (Akbari et al. 2016). Fecal microbiota transplantation (FMT) approaches from young and healthy donors could have beneficial effects, in the elderly however the longevity of these benefits is not yet known. Other than that, there is accumulating evidence about the therapeutic potential of complement inhibitors in reducing neuroinflammation induced by the body innate immune response to dysbiosis (Carpanini et al. 2019). In light of these upcoming perspectives on AD pathology, perhaps we need to re-evaluate the way we think about biomarkers and treatment strategies for neurodegenerative disease and consider the gut–brain axis as central to many pathways of neuroinflammation and neurodegeneration (Pistollato et al. 2016).

Understanding how and why bacterial species influence immune cells will provide greater ability to model such responses. For example, some bacteria favor a Treg response as previously discussed (*Bacteroides fragilis* and *Clostridium* clusters XIVa, IV, and XVIII) (Round and Mazmanian 2010; Atarashi et al. 2011a), while others favor Th1-type responses (*B. fragilis*) (Mazmanian et al. 2005). While genetic factors are largely responsible for early onset AD, there is evidence that suggests that amyloid deposition starts several years before dementia. Recent studies have shown that amyloid protein may just be antimicrobial peptides that have a physiological function; however, its accumulation further feeds into neuroinflammation by triggering complement and microgliosis. Therefore, if the microbiome can modulate the immune system in a way that produces desirable neurological outcomes, then how can we mimic that using small molecules or biologics. Can we modulate the inflammatory cytokine signaling pathway, the inflammasome activation pathway, or modulate the TLR activation or resultant complement activation and downstream effects to modulate the inflammatory microenvironment. Finally, the gut–brain axis is complex and there is enough evidence to support how critical it is to address the role of microbiome when considering treatments for affected patients. Also, the effect of confounding factors such as diet, concomitant diseases, and drugs requires careful attention when treating patients and in the analyses of the data from existing studies as well. An interdisciplinary approach to utilizing the knowledge available on gut–brain axis in neurological disease is needed for a strategic breakthrough in the treatment and early diagnosis of neurodegenerative disease. Clinincal trials for designer probiotics and fecal microbiota transplants to increase responsiveness are in their infancy. However, profiling and mimicking the beneficial immune response alterations caused by microbiome that lead to favorable outcomes and modeling them using advanced tools in the patients could be a potential alternative.

References

Abderrazak A et al (2016) Inhibition of the inflammasome NLRP3 by arglabin attenuates inflammation, protects pancreatic β-cells from apoptosis, and prevents type 2 diabetes mellitus development in ApoE2Ki mice on a chronic high-fat diet. J Pharmacol Exp Ther 357:487. https://doi.org/10.1124/jpet.116.232934

Adler CH, Beach TG (2016) Neuropathological basis of nonmotor manifestations of Parkinson's disease. Mov Disord 31:1114. https://doi.org/10.1002/mds.26605

Agostini S et al (2016) High avidity HSV-1 antibodies correlate with absence of amnestic Mild cognitive impairment conversion to Alzheimer's disease. Brain Behav Immun 58:254. https://doi.org/10.1016/j.bbi.2016.07.153

Akbari E et al (2016) Effect of probiotic supplementation on cognitive function and metabolic status in Alzheimer's disease: a randomized, double-blind and controlled trial. Front Aging Neurosci 8:256. https://doi.org/10.3389/fnagi.2016.00256

Aldecoa I et al (2015) Alpha-synuclein immunoreactivity patterns in the enteric nervous system. Neurosci Lett 602:145. https://doi.org/10.1016/j.neulet.2015.07.005

Álvarez G et al (2012) Herpes simplex virus type 1 induces nuclear accumulation of hyperphosphorylated tau in neuronal cells. J Neurosci Res 90:1020. https://doi.org/10.1002/jnr.23003

Anderson FL et al (2018) Inflammasomes: an emerging mechanism translating environmental toxicant exposure into neuroinflammation in Parkinson's disease. Toxicol Sci 166:3. https://doi.org/10.1093/toxsci/kfy219

Atarashi K et al (2011a) Induction of colonic regulatory T cells by indigenous Clostridium species. Science 331:337. https://doi.org/10.1126/science.1198469

Atarashi K, Umesaki Y, Honda K (2011b) Microbiotal influence on T cell subset development. Semin Immunol 23:146. https://doi.org/10.1016/j.smim.2011.01.010

Atarashi K et al (2013) Treg induction by a rationally selected mixture of Clostridia strains from the human microbiota. Nature 500:232. https://doi.org/10.1038/nature12331

Athauda D, Foltynie T (2016) Insulin resistance and Parkinson's disease: a new target for disease modification? Prog Neurobiol 145–146:98. https://doi.org/10.1016/j.pneurobio.2016.10.001

Aviles-Olmos I et al (2013) Parkinson's disease, insulin resistance and novel agents of neuroprotection. Brain 136:374. https://doi.org/10.1093/brain/aws009

Ayers JI et al (2017) Robust central nervous system pathology in transgenic mice following peripheral injection of α-synuclein fibrils. J Virol 91:e02095. https://doi.org/10.1128/jvi.02095-16

Aziz Q, Doré J, Emmanuel A, Guarner F, Quigley EMM (2013, January) Gut microbiota and gastrointestinal health: current concepts and future directions. Neurogastroenterol Motil. https://doi.org/10.1111/nmo.12046

Azm SAN et al (2018) Lactobacilli and bifidobacteria ameliorate memory and learning deficits and oxidative stress in β-amyloid (1–42) injected rats. Appl Physiol Nutr Metab 43:718. https://doi.org/10.1139/apnm-2017-0648

Beach TG et al (2016) Multicenter assessment of immunohistochemical methods for pathological alpha-synuclein in sigmoid colon of autopsied Parkinson's disease and control subjects. J Parkinsons Dis 6:761. https://doi.org/10.3233/JPD-160888

Bian F et al (2017) Inhibition of NLRP3 inflammasome pathway by butyrate improves corneal wound healing in corneal alkali burn. Int J Mol Sci 18:562. https://doi.org/10.3390/ijms18030562

Bloom GS, Lazo JS, Norambuena A (2018) Reduced brain insulin signaling: a seminal process in Alzheimer's disease pathogenesis. Neuropharmacology 136:192. https://doi.org/10.1016/j.neuropharm.2017.09.016

Bolós M, Perea JR, Avila J (2017) Alzheimer's disease as an inflammatory disease. Biomol Concepts 8:37. https://doi.org/10.1515/bmc-2016-0029

Borghammer P (2018) How does Parkinson's disease begin? Perspectives on neuroanatomical pathways, prions, and histology. Mov Disord 33:48. https://doi.org/10.1002/mds.27138

Borre YE et al (2014) Microbiota and neurodevelopmental windows: implications for brain disorders. Trends Mol Med 20:509. https://doi.org/10.1016/j.molmed.2014.05.002

Bourgade K et al (2014) β-amyloid peptides display protective activity against the human Alzheimer's disease-associated herpes simplex virus-1. Biogerontology 16:85. https://doi.org/10.1007/s10522-014-9538-8

Bourgade K et al (2016) Protective effect of amyloid-β peptides against herpes simplex virus-1 infection in a neuronal cell culture model. J Alzheimers Dis 50:1227. https://doi.org/10.3233/JAD-150652

Bourgade K et al (2017) Anti-viral properties of amyloid-β peptides. In: Handbook of infection and Alzheimer's disease. IOS Press, Amsterdam. https://doi.org/10.3233/978-1-61499-706-221

Braak H et al (2003) Idiopathic Parkinson's disease: possible routes by which vulnerable neuronal types may be subject to neuroinvasion by an unknown pathogen. J Neural Transm 110:517–536. https://doi.org/10.1007/s00702-002-0808-2

Braniste V et al (2014) The gut microbiota influences blood-brain barrier permeability in mice. Sci Transl Med 6:263ra158. https://doi.org/10.1126/scitranslmed.3009759

Bravo JA et al (2012) Communication between gastrointestinal bacteria and the nervous system. Curr Opin Pharmacol 12:667. https://doi.org/10.1016/j.coph.2012.09.010

Breid S et al (2016) Neuroinvasion of α-synuclein prionoids after intraperitoneal and intraglossal inoculation. J Virol 90:9182. https://doi.org/10.1128/jvi.01399-16

Buford TW (2017) (Dis)Trust your gut: the gut microbiome in age-related inflammation, health, and disease. Microbiome 5:80. https://doi.org/10.1186/s40168-017-0296-0

Camacho-Soto A et al (2018) Inflammatory bowel disease and risk of Parkinson's disease in Medicare beneficiaries. Parkinsonism Relat Disord 50:23. https://doi.org/10.1016/j.parkreldis.2018.02.008

Camponova P et al (2017) Alteration of high-density lipoprotein functionality in Alzheimer's disease patients. Can J Physiol Pharmacol 95:894. https://doi.org/10.1139/cjpp-2016-0710

Caputi V, Giron MC (2018) Microbiome-gut-brain axis and toll-like receptors in Parkinson's disease. Int J Mol Sci 19:1689. https://doi.org/10.3390/ijms19061689

Carpanini SM, Torvell M, Morgan BP (2019) Therapeutic inhibition of the complement system in diseases of the central nervous system. Front Immunol 10:362. https://doi.org/10.3389/fimmu.2019.00362

Cattaneo A et al (2017) Association of brain amyloidosis with pro-inflammatory gut bacterial taxa and peripheral inflammation markers in cognitively impaired elderly. Neurobiol Aging 49:60. https://doi.org/10.1016/j.neurobiolaging.2016.08.019

Chen SG et al (2016) Exposure to the functional bacterial amyloid protein curli enhances alpha-synuclein aggregation in aged fischer 344 rats and Caenorhabditis elegans. Sci Rep 6:34477. https://doi.org/10.1038/srep34477

Chen L et al (2019) PPARβ/δ agonist alleviates NLRP3 inflammasome-mediated neuroinflammation in the MPTP mouse model of Parkinson's disease. Behav Brain Res 356:483. https://doi.org/10.1016/j.bbr.2018.06.005

Chorell E et al (2015) Bacterial chaperones CsgE and CsgC differentially modulate human α-synuclein amyloid formation via transient contacts. PLoS One 10:e0140194. https://doi.org/10.1371/journal.pone.0140194

Chow VW et al (2010) An overview of APP processing enzymes and products. NeuroMolecular Med 12:1. https://doi.org/10.1007/s12017-009-8104-z

Chung SJ et al (2016) Alpha-synuclein in gastric and colonic mucosa in Parkinson's disease: limited role as a biomarker. Mov Disord 31:241. https://doi.org/10.1002/mds.26473

Clairembault T et al (2015) Structural alterations of the intestinal epithelial barrier in Parkinson's disease. Acta Neuropathol Commun 3:12. https://doi.org/10.1186/s40478-015-0196-0

Codolo G et al (2013) Triggering of inflammasome by aggregated α-synuclein, an inflammatory response in synucleinopathies. PLoS One 8:e55375. https://doi.org/10.1371/journal.pone.0055375

Coll RC et al (2015) A small-molecule inhibitor of the NLRP3 inflammasome for the treatment of inflammatory diseases. Nat Med 21:248. https://doi.org/10.1038/nm.3806

Corbillé AG et al (2016) What a gastrointestinal biopsy can tell us about Parkinson's disease? Neurogastroenterol Motil 28:966. https://doi.org/10.1111/nmo.12797

Corbillé AG et al (2017) Biochemical analysis of α-synuclein extracted from control and Parkinson's disease colonic biopsies. Neurosci Lett 641:81. https://doi.org/10.1016/j.neulet.2017.01.050

Crehan H, Hardy J, Pocock J (2012) Microglia, Alzheimer's disease, and complement. Int J Alzheimers Dis 2012:983640. https://doi.org/10.1155/2012/983640

Cryan JF, Dinan TG (2012) Mind-altering microorganisms: the impact of the gut microbiota on brain and behaviour. Nat Rev Neurosci 13:701. https://doi.org/10.1038/nrn3346

Cryan JF, O'Mahony SM (2011) The microbiome-gut-brain axis: from bowel to behavior. Neurogastroenterol Motil 23:187. https://doi.org/10.1111/j.1365-2982.2010.01664.x

Cummings J, Ritter A, Zhong K (2018) Clinical trials for disease-modifying therapies in Alzheimer's disease: a primer, lessons learned, and a blueprint for the future. J Alzheimers Dis 64:S3. https://doi.org/10.3233/JAD-179901

De Pablo-Fernandez E et al (2018) Association between diabetes and subsequent Parkinson disease. Neurology 91:e139. https://doi.org/10.1212/wnl.0000000000005771

Dinan TG, Cryan JF (2017) Gut instincts: microbiota as a key regulator of brain development, ageing and neurodegeneration. J Physiol 595:489. https://doi.org/10.1113/JP273106

Divyashri G et al (2015) Probiotic attributes, antioxidant, anti-inflammatory and neuromodulatory effects of Enterococcus faecium CFR 3003: in vitro and in vivo evidence. J Med Microbiol 64: 1527. https://doi.org/10.1099/jmm.0.000184

Ehses JA et al (2009) IL-1 antagonism reduces hyperglycemia and tissue inflammation in the type 2 diabetic GK rat. Proc Natl Acad Sci U S A 106:13998. https://doi.org/10.1073/pnas.0810087106

Eimer WA et al (2018) Alzheimer's disease-associated β-amyloid is rapidly seeded by herpesviridae to protect against brain infection. Neuron 99:56. https://doi.org/10.1016/j.neuron.2018.06.030

Elliott EI, Sutterwala FS (2015) Initiation and perpetuation of NLRP3 inflammasome activation and assembly. Immunol Rev 265:35. https://doi.org/10.1111/imr.12286

Erny D et al (2015) Host microbiota constantly control maturation and function of microglia in the CNS. Nat Neurosci 18:965. https://doi.org/10.1038/nn.4030

Evans ML et al (2015) The bacterial curli system possesses a potent and selective inhibitor of amyloid formation. Mol Cell 57:445. https://doi.org/10.1016/j.molcel.2014.12.025

Fasano A et al (2015) Gastrointestinal dysfunction in Parkinson's disease. Lancet Neurol 14:625. https://doi.org/10.1016/S1474-4422(15)00007-1

Fedorova TD et al (2017) Decreased intestinal acetylcholinesterase in early Parkinson disease. Neurology 88:775. https://doi.org/10.1212/WNL.0000000000003633

Forsyth CB et al (2011) Increased intestinal permeability correlates with sigmoid mucosa alpha-synuclein staining and endotoxin exposure markers in early Parkinson's disease. PLoS One 6: e28032. https://doi.org/10.1371/journal.pone.0028032

Frasca D, Blomberg BB (2011) Aging affects human B cell responses. J Clin Immunol 31:430. https://doi.org/10.1007/s10875-010-9501-7

Frasca D, Blomberg BB (2016) Inflammaging decreases adaptive and innate immune responses in mice and humans. Biogerontology 17:7. https://doi.org/10.1007/s10522-015-9578-8

Friedland RP (2015) Mechanisms of molecular mimicry involving the microbiota in neurodegeneration. J Alzheimers Dis 45:349. https://doi.org/10.3233/JAD-142841

Galante D et al (2012) Differential toxicity, conformation and morphology of typical initial aggregation states of Aβ1-42 and Aβpy3-42 beta-amyloids. Int J Biochem Cell Biol 44:2085. https://doi.org/10.1016/j.biocel.2012.08.010

Gallo PM et al (2015) Amyloid-DNA composites of bacterial biofilms stimulate autoimmunity. Immunity 42:1171. https://doi.org/10.1016/j.immuni.2015.06.002

Ganapathy V et al (2013) Transporters and receptors for short-chain fatty acids as the molecular link between colonic bacteria and the host. Curr Opin Pharmacol 13:869. https://doi.org/10.1016/j.coph.2013.08.006

Giacoppo S, Bramanti P, Mazzon E (2017) Triggering of inflammasome by impaired autophagy in response to acute experimental Parkinson's disease: involvement of the PI3K/Akt/mTOR pathway. NeuroReport 28:996. https://doi.org/10.1097/WNR.0000000000000871

Glucksam-Galnoy Y et al (2013) The bacterial quorum-sensing signal molecule n-3-oxo-dodecanoyl-l-homoserine lactone reciprocally modulates pro- and anti-inflammatory cytokines in activated macrophages. J Immunol 191:337. https://doi.org/10.4049/jimmunol.1300368

Gong Z et al (2018) Mitochondrial dysfunction induces NLRP3 inflammasome activation during cerebral ischemia/reperfusion injury. J Neuroinflammation 15:242. https://doi.org/10.1186/s12974-018-1282-6

Gordon R et al (2018) Inflammasome inhibition prevents - synuclein pathology and dopaminergic neurodegeneration in mice. Sci Transl Med 10:eaah4066. https://doi.org/10.1126/scitranslmed.aah4066

Guo C et al (2016) Bile acids control inflammation and metabolic disorder through inhibition of NLRP3 inflammasome. Immunity 45:802. https://doi.org/10.1016/j.immuni.2016.09.008

Hafner-Bratkovič I et al (2012) NLRP3 inflammasome activation in macrophage cell lines by prion protein fibrils as the source of IL-1β and neuronal toxicity. Cell Mol Life Sci 69:4215. https://doi.org/10.1007/s00018-012-1140-0

He Y, Hara H, Núñez G (2016) Mechanism and regulation of NLRP3 inflammasome activation. Trends Biochem Sci 41:1012. https://doi.org/10.1016/j.tibs.2016.09.002

Heijtz RD et al (2011) Normal gut microbiota modulates brain development and behavior. Proc Natl Acad Sci U S A 108:3047. https://doi.org/10.1073/pnas.1010529108

Heneka MT (2017) Inflammasome activation and innate immunity in Alzheimer's disease. Brain Pathol 27:220. https://doi.org/10.1111/bpa.12483

Hilton D et al (2014) Accumulation of α-synuclein in the bowel of patients in the pre-clinical phase of Parkinson's disease. Acta Neuropathol 127:235. https://doi.org/10.1007/s00401-013-1214-6

Holtzman DM et al (2016) Tau: from research to clinical development. Alzheimer Dement 12:1033. https://doi.org/10.1016/j.jalz.2016.03.018

Hu G et al (2007) Type 2 diabetes and the risk of Parkinson's disease. Diabetes Care 30:842. https://doi.org/10.2337/dc06-2011

Hu X, Wang T, Jin F (2016) Alzheimer's disease and gut microbiota. Sci China Life Sci 59:1006. https://doi.org/10.1007/s11427-016-5083-9

Huang H et al (2019) Fecal microbiota transplantation to treat Parkinson's disease with constipation: a case report. Medicine 98:e16163. https://doi.org/10.1097/MD.0000000000016163

Itzhaki RF (2018) Corroboration of a major role for herpes simplex virus type 1 in Alzheimer's disease. Front Aging Neurosci 10:324. https://doi.org/10.3389/fnagi.2018.00324

Jackson A et al (2019) Diet in Parkinson's disease: critical role for the microbiome. Front Neurol 10:1245. https://doi.org/10.3389/fneur.2019.01245

Jamieson GA et al (1991) Detection of herpes simplex virus type 1 DNA sequences in normal and Alzheimer's disease brain using polymerase chain reaction. Biochem Soc Trans 19: 122S. https://doi.org/10.1042/bst019122s

Jena PK et al (2018) Dysregulated bile acid synthesis and dysbiosis are implicated in Western diet-induced systemic inflammation, microglial activation, and reduced neuroplasticity. FASEB J 32:2866. https://doi.org/10.1096/fj.201700984RR

Karlsson FH et al (2013) Gut metagenome in European women with normal, impaired and diabetic glucose control. Nature 498:99. https://doi.org/10.1038/nature12198

Knudsen K et al (2017) Objective colonic dysfunction is far more prevalent than subjective constipation in Parkinson's disease: a colon transit and volume study. J Parkinsons Dis 7:359. https://doi.org/10.3233/JPD-161050

Kobayashi Y et al (2017) Therapeutic potential of Bifidobacterium breve strain A1 for preventing cognitive impairment in Alzheimer's disease. Sci Rep 7:13510. https://doi.org/10.1038/s41598-017-13368-2

Koh A et al (2016) From dietary fiber to host physiology: short-chain fatty acids as key bacterial metabolites. Cell 165:1332. https://doi.org/10.1016/j.cell.2016.05.041

Kowalski K, Mulak A (2019) Brain-gut-microbiota axis in Alzheimer's disease. J Neurogastroenterol Motil 25:48. https://doi.org/10.5056/jnm18087

Kristen H et al (2018) The lysosome system is severely impaired in a cellular model of neurodegeneration induced by HSV-1 and oxidative stress. Neurobiol Aging 68:5. https://doi.org/10.1016/j.neurobiolaging.2018.03.025

Le Page A et al (2018) Role of the peripheral innate immune system in the development of Alzheimer's disease. Exp Gerontol 107:59. https://doi.org/10.1016/j.exger.2017.12.019

Lee E et al (2019) MPTP-driven NLRP3 inflammasome activation in microglia plays a central role in dopaminergic neurodegeneration. Cell Death Differ 26:213. https://doi.org/10.1038/s41418-018-0124-5

Leroy K et al (2010) Lithium treatment arrests the development of neurofibrillary tangles in mutant tau transgenic mice with advanced neurofibrillary pathology. J Alzheimers Dis 19:705. https://doi.org/10.3233/JAD-2010-1276

Leyns CEG, Holtzman DM (2017) Glial contributions to neurodegeneration in tauopathies. Mol Neurodegener 12:50. https://doi.org/10.1186/s13024-017-0192-x

Lin WR et al (1997) Neurotropic viruses and Alzheimer's disease: a search for varicella zoster virus DNA by the polymerase chain reaction. J Neurol Neurosurg Psychiatry 62:586. https://doi.org/10.1136/jnnp.62.6.586

Lionnet A et al (2018) Does Parkinson's disease start in the gut? Acta Neuropathol 135:1. https://doi.org/10.1007/s00401-017-1777-8

Liu B et al (2017) Vagotomy and Parkinson disease (A Swedish register-based matched-cohort study). Neurology 88:1996. https://doi.org/10.1212/WNL.0000000000003961

Mariathasan S et al (2006) Cryopyrin activates the inflammasome in response to toxins and ATP. Nature 440:228. https://doi.org/10.1038/nature04515

Martinon F et al (2006) Gout-associated uric acid crystals activate the NALP3 inflammasome. Nature 440:237. https://doi.org/10.1038/nature04516

Mazmanian SK et al (2005) An immunomodulatory molecule of symbiotic bacteria directs maturation of the host immune system. Cell 122:107. https://doi.org/10.1016/j.cell.2005.05.007

McGeer PL, McGeer EG (2013) The amyloid cascade-inflammatory hypothesis of Alzheimer disease: implications for therapy. Acta Neuropathol 126:479. https://doi.org/10.1007/s00401-013-1177-7

Mendes CT et al (2009) Lithium reduces Gsk3b mRNA levels: implications for Alzheimer disease. Eur Arch Psychiatry Clin Neurosci 259:16. https://doi.org/10.1007/s00406-008-0828-5

Molinuevo JL et al (2018) The rationale behind the new Alzheimer's disease conceptualization: lessons learned during the last decades. J Alzheimers Dis 62:1067. https://doi.org/10.3233/JAD-170698

Musa NH et al (2017) Lactobacilli-fermented cow's milk attenuated lipopolysaccharide-induced neuroinflammation and memory impairment in vitro and in vivo. J Dairy Res 84:488. https://doi.org/10.1017/S0022029917000620

Narang A, Qiao F, Atkinson C et al (2017) Natural IgM antibodies that bind neoepitopes exposed as a result of spinal cord injury, drive secondary injury by activating complement. J Neuroinflammation 14:120. https://doi.org/10.1186/s12974-017-0894-6

Nimgampalle M, Yellamma K (2017) Anti-Alzheimer properties of probiotic, Lactobacillus plantarum MTCC 1325 in Alzheimer's disease induced albino rats. J Clin Diagn Res 11:KC01. https://doi.org/10.7860/JCDR/2017/26106.10428

Noelker C et al (2013) Toll like receptor 4 mediates cell death in a mouse MPTP model of Parkinson disease. Sci Rep 3:1393. https://doi.org/10.1038/srep01393

Obrenovich M (2018) Leaky gut, leaky brain? Microorganisms 6:107. https://doi.org/10.3390/microorganisms6040107

Olsson J et al (2016) HSV presence in brains of individuals without dementia: the TASTY brain series. DMM Dis Models Mech 9:1349. https://doi.org/10.1242/dmm.026674

Pastore A et al (2020) Why does the Aβ peptide of Alzheimer share structural similarity with antimicrobial peptides? Commun Biol 3:135. https://doi.org/10.1038/s42003-020-0865-9

Pavillard LE et al (2017) NLRP3-inflammasome inhibition prevents high fat and high sugar diets-induced heart damage through autophagy induction. Oncotarget 8:99740. https://doi.org/10.18632/oncotarget.20763

Pellicanò M et al (2012) Immune profiling of Alzheimer patients. J Neuroimmunol 242:52. https://doi.org/10.1016/j.jneuroim.2011.11.005

Perez-Burgos A et al (2013) Psychoactive bacteria Lactobacillus rhamnosus (JB-1) elicits rapid frequency facilitation in vagal afferents. Am J Physiol Gastrointest Liver Physiol 304:G211. https://doi.org/10.1152/ajpgi.00128.2012

Perneczky R et al (2013) Soluble amyloid precursor protein β as blood-based biomarker of Alzheimer's disease. Transl Psychiatry 3:e227. https://doi.org/10.1038/tp.2013.11

Pfeiffer RF (2012) Gastrointestinal dysfunction in Parkinson's disease. In: Parkinson's disease, 2nd edn. Wiley-Blackwell, London. https://doi.org/10.1201/b12948

Pisa D et al (2017) Polymicrobial infections in brain tissue from Alzheimer's disease patients. Sci Rep 7:5559. https://doi.org/10.1038/s41598-017-05903-y

Pistollato F et al (2016) Alzheimer disease research in the 21st century: past and current failures, new perspectives and funding priorities. Oncotarget 7:38999. https://doi.org/10.18632/oncotarget.9175

Plaza-Díaz J et al (2017) Evidence of the anti-inflammatory effects of probiotics and synbiotics in intestinal chronic diseases. Nutrients 9:555. https://doi.org/10.3390/nu9060555

Quigley EMM (2017) Microbiota-brain-gut axis and neurodegenerative diseases. Curr Neurol Neurosci Rep 17:94. https://doi.org/10.1007/s11910-017-0802-6

Readhead B et al (2018) Multiscale analysis of independent Alzheimer's cohorts finds disruption of molecular, genetic, and clinical networks by human herpesvirus. Neuron 99:64.e7. https://doi.org/10.1016/j.neuron.2018.05.023

Rivest S (2009) Regulation of innate immune responses in the brain. Nat Rev Immunol 9:429. https://doi.org/10.1038/nri2565

Romagnani S (2004) The increased prevalence of allergy and the hygiene hypothesis: missing immune deviation, reduced immune suppression, or both? Immunology 112:352. https://doi.org/10.1111/j.1365-2567.2004.01925.x

Rook GAW, Lowry CA (2008) The hygiene hypothesis and psychiatric disorders. Trends Immunol 29:150. https://doi.org/10.1016/j.it.2008.01.002

Round JL, Mazmanian SK (2010) Inducible Foxp3+ regulatory T-cell development by a commensal bacterium of the intestinal microbiota. Proc Natl Acad Sci U S A 107:12204. https://doi.org/10.1073/pnas.0909122107

Ruffmann C et al (2018) Detection of alpha-synuclein conformational variants from gastrointestinal biopsy tissue as a potential biomarker for Parkinson's disease. Neuropathol Appl Neurobiol 44:722. https://doi.org/10.1111/nan.12486

Sampson TR et al (2016) Gut microbiota regulate motor deficits and neuroinflammation in a model of Parkinson's disease. Cell 167:1469. https://doi.org/10.1016/j.cell.2016.11.018

Sampson TR et al (2020) A gut bacterial amyloid promotes α-synuclein aggregation and motor impairment in mice. elife 9:e53111. https://doi.org/10.7554/eLife.53111

Sánchez-Ferro Á et al (2015) In vivo gastric detection of α-synuclein inclusions in Parkinson's disease. Mov Disord 30:517. https://doi.org/10.1002/mds.25988

Santana S et al (2012) Herpes simplex virus type I induces an incomplete autophagic response in human neuroblastoma cells. J Alzheimers Dis 30:815. https://doi.org/10.3233/JAD-2012-112000

Santiago JA, Potashkin JA (2013) Integrative network analysis unveils convergent molecular pathways in Parkinson's disease and diabetes. PLoS One 8:e83940. https://doi.org/10.1371/journal.pone.0083940

Sarkar S et al (2017) Mitochondrial impairment in microglia amplifies NLRP3 inflammasome proinflammatory signaling in cell culture and animal models of Parkinson's disease. NPJ Parkinson Dis 3:30. https://doi.org/10.1038/s41531-017-0032-2

Scheperjans F, Derkinderen P, Borghammer P (2018) The gut and Parkinson's disease: hype or hope? J Parkinsons Dis 8:S31. https://doi.org/10.3233/JPD-181477

Schroder K, Tschopp J (2010) The inflammasomes. Cell 140:821. https://doi.org/10.1016/j.cell.2010.01.040

Schwartz JC, Zhang X, Fedorov AA, Nathenson SG, Almo SC (2001) Structural basis for co-stimulation by the human CTLA-4/B7-2 complex. Nature 410(6828):604–608

Schwiertz A et al (2018) Fecal markers of intestinal inflammation and intestinal permeability are elevated in Parkinson's disease. Parkinsonism Relat Disord 50:104. https://doi.org/10.1016/j.parkreldis.2018.02.022

Serrano-Pozo A et al (2011) Reactive glia not only associates with plaques but also parallels tangles in Alzheimer's disease. Am J Pathol 179:1373. https://doi.org/10.1016/j.ajpath.2011.05.047

Shannon KM et al (2012) Alpha-synuclein in colonic submucosa in early untreated Parkinson's disease. Mov Disord 27:709. https://doi.org/10.1002/mds.23838

Shin C et al (2017) Fundamental limit of alpha-synuclein pathology in gastrointestinal biopsy as a pathologic biomarker of Parkinson's disease: comparison with surgical specimens. Parkinsonism Relat Disord 44:73. https://doi.org/10.1016/j.parkreldis.2017.09.001

Siegel G et al (2017) The Alzheimer's disease γ-secretase generates higher 42:40 ratios for β-amyloid than for p3 peptides. Cell Rep 19:1967. https://doi.org/10.1016/j.celrep.2017.05.034

Sofola O et al (2010) Inhibition of GSK-3 ameliorates Aβ pathology in an adult-onset Drosophila model of Alzheimer's disease. PLoS Genet 6:e1001087. https://doi.org/10.1371/journal.pgen.1001087

Song L et al (2017) NLRP3 inflammasome in neurological diseases, from functions to therapies. Front Cell Neurosci 11:63. https://doi.org/10.3389/fncel.2017.00063

Soscia SJ et al (2010) The Alzheimer's disease-associated amyloid β-protein is an antimicrobial peptide. PLoS One 5:e9505. https://doi.org/10.1371/journal.pone.0009505

Sowade RF, Jahn TR (2017) Seed-induced acceleration of amyloid-β mediated neurotoxicity in vivo. Nat Commun 8:512. https://doi.org/10.1038/s41467-017-00579-4

Steiner JA, Quansah E, Brundin P (2018) The concept of alpha-synuclein as a prion-like protein: ten years after. Cell Tissue Res 373:161. https://doi.org/10.1007/s00441-018-2814-1

Stienstra R et al (2011) Inflammasome is a central player in the induction of obesity and insulin resistance. Proc Natl Acad Sci U S A 108:15324. https://doi.org/10.1073/pnas.1100255108

Stokholm MG et al (2016) Pathological α-synuclein in gastrointestinal tissues from prodromal Parkinson disease patients. Ann Neurol 79:940. https://doi.org/10.1002/ana.24648

Strachan DP (1989) Hay fever, hygiene, and household size. Br Med J 299:1259. https://doi.org/10.1136/bmj.299.6710.1259

Sun MF et al (2018) Neuroprotective effects of fecal microbiota transplantation on MPTP-induced Parkinson's disease mice: gut microbiota, glial reaction and TLR4/TNF-α signaling pathway. Brain Behav Immun 70:48. https://doi.org/10.1016/j.bbi.2018.02.005

Svensson E et al (2015) Vagotomy and subsequent risk of Parkinson's disease. Ann Neurol 78:522. https://doi.org/10.1002/ana.24448

Tillisch K et al (2013) Consumption of fermented milk product with probiotic modulates brain activity. Gastroenterology 144:1394. https://doi.org/10.1053/j.gastro.2013.02.043

Tursi SA, Tükel Ç (2018) Curli-containing enteric biofilms inside and out: matrix composition, immune recognition, and disease implications. Microbiol Mol Biol Rev 82:e00028. https://doi.org/10.1128/mmbr.00028-18

Van der Hee B, Wells JM (2021) Microbial regulation of host physiology by short-chain fatty acids. Trends Microbiol 29(8):700–712. https://doi.org/10.1016/J.TIM.2021.02.001

Vandanmagsar B et al (2011) The NLRP3 inflammasome instigates obesity-induced inflammation and insulin resistance. Nat Med 17:179. https://doi.org/10.1038/nm.2279

Venegas C, Heneka MT (2017) Danger-associated molecular patterns in Alzheimer's disease. J Leukoc Biol 101:87. https://doi.org/10.1189/jlb.3mr0416-204r

Vidakovic L et al (2017) Dynamic biofilm architecture confers individual and collective mechanisms of viral protection. Nat Microbiol 3:26. https://doi.org/10.1038/s41564-017-0050-1

Villumsen M et al (2019) Inflammatory bowel disease increases the risk of Parkinson's disease: a Danish nationwide cohort study 1977-2014. Gut 68:18. https://doi.org/10.1136/gutjnl-2017-315666

Wahlqvist ML et al (2012) Metformin-inclusive sulfonylurea therapy reduces the risk of Parkinson's disease occurring with Type 2 diabetes in a Taiwanese population cohort. Parkinsonism Relat Disord 18:753. https://doi.org/10.1016/j.parkreldis.2012.03.010

Wang J et al (2012) A metagenome-wide association study of gut microbiota in type 2 diabetes. Nature 490:55. https://doi.org/10.1038/nature11450

Wang T et al (2015) Lactobacillus fermentum NS9 restores the antibiotic induced physiological and psychological abnormalities in rats. Benefic Microbes 6:707. https://doi.org/10.3920/BM2014.0177

Weimers P et al (2019) Inflammatory bowel disease and Parkinson's disease: a nationwide Swedish cohort study. Inflamm Bowel Dis 25:111. https://doi.org/10.1093/ibd/izy190

Wozniak MA et al (2007) Herpes simplex virus infection causes cellular β-amyloid accumulation and secretase upregulation. Neurosci Lett 429:95. https://doi.org/10.1016/j.neulet.2007.09.077

Wozniak MA, Frost AL, Itzhaki RF (2009) Alzheimer's disease-specific tau phosphorylation is induced by herpes simplex virus type 1. J Alzheimers Dis 16:341. https://doi.org/10.3233/JAD-2009-0963

Wozniak MA et al (2011) Antivirals reduce the formation of key Alzheimer's disease molecules in cell cultures acutely infected with herpes simplex virus type 1. PLoS One 6:e25152. https://doi.org/10.1371/journal.pone.0025152

Wu T et al (2019) Complement C3 is activated in human AD brain and is required for neurodegeneration in mouse models of amyloidosis and tauopathy. Cell Rep 28:2111. https://doi.org/10.1016/j.celrep.2019.07.060

Yan F et al (2018) Gastrointestinal nervous system a-synuclein as a potential biomarker of Parkinson disease. Medicine 97:e11337. https://doi.org/10.1097/MD.0000000000011337

Youm YH et al (2013) Canonical Nlrp3 inflammasome links systemic low-grade inflammation to functional decline in aging. Cell Metab 18:519. https://doi.org/10.1016/j.cmet.2013.09.010

Zambrano Á et al (2008) Neuronal cytoskeletal dynamic modification and neurodegeneration induced by infection with herpes simplex virus type 1. J Alzheimers Dis 14:259. https://doi.org/10.3233/JAD-2008-14301

Zhao Y et al (2017a) Microbiome-derived lipopolysaccharide enriched in the perinuclear region of Alzheimer's disease brain. Front Immunol 8:1064. https://doi.org/10.3389/fimmu.2017.01064

Zhao Y, Jaber V, Lukiw WJ (2017b) Secretory products of the human GI tract microbiome and their potential impact on Alzheimer's disease (AD): detection of lipopolysaccharide (LPS) in AD hippocampus. Front Cell Infect Microbiol 7:318. https://doi.org/10.3389/fcimb.2017.00318

Zhu F et al (2019) The risk of Parkinson's disease in inflammatory bowel disease: a systematic review and meta-analysis. Dig Liver Dis 51:38. https://doi.org/10.1016/j.dld.2018.09.017

Zilka N, Novak M (2006) The tangled story of Alois Alzheimer. Bratisl Lek Listy 107:343

Zimmerman MA et al (2012) Butyrate suppresses colonic inflammation through HDAC1-dependent fas upregulation and fas-mediated apoptosis of T cells. Am J Physiol Gastrointest Liver Physiol 302:G1405. https://doi.org/10.1152/ajpgi.00543.2011

Critical Inspection of the Gut–Brain–Skin Triangle and Its Modulation Through Probiotics

Parul Chugh, Shivani Sood, and Mahesh S. Dhar

Abstract

Homeostasis is the dynamic process of maintaining self-balance or to put alternatively the body's natural tendency to poise equilibrium in response to external stimuli and reorient itself toward undulating environmental circumstances. Various physiochemical and biological processes work in tandem to maintain homeostasis. The microbiota allied to host is an important biological intermediary that plays a pivotal role in maintaining this homeostasis. It consists of a myriad of commensal and symbiotic microbes which are an important dynamic determinant of human health. The host-health promoting microbes, known as probiotics, are the center of intense research which has been extensively reviewed in the previous chapters. It has become patently evident in the recent years that the gut microbiome and the brain communicate in a bidirectional fashion with a great possibility of mutual implication pertaining to each other's functions. Modulation of intestinal microbiota with probiotics is used as a therapeutic modality in prevention and treatment of a multitude of ailments. Emerging evidence from interdisciplinary research corroborates the incident of communication axis between various organs, for instance, the brain–gut or brain–skin axis. In this chapter with reliance placed on prior published compelling evidence from extant literature, we aim to endeavor and comprehensively comprehend the underlying mechanism which further helps to unravel the interrelationship between gut–

Parul Chugh and Shivani Sood contributed equally with all other contributors.

P. Chugh
Amity Institute of Biotechnology, Amity University, Noida, India

S. Sood
National Centre for Disease Control, Delhi, India

M. S. Dhar (✉)
Microbiology Division, National Centre for Disease Control, Delhi, India

© Springer Nature Singapore Pte Ltd. 2022
P. K. Deol, S. K. Sandhu (eds.), *Probiotic Research in Therapeutics*,
https://doi.org/10.1007/978-981-16-6760-2_2

brain–skin axis and the modulation of their microbiota by deployment of designer probiotics. The chapter also presents fresh impetus on the evidence demonstrating how the gut microbiome may possibly affect brain function in adults, thereby having an extensive impact on neurological disorders.

Keywords

Designer probiotics · Gut–brain–skin axis · Microbiota · Homeostasis

2.1 Probiotics: In and Out

Ecological communities of commensal, pathogenic and symbiotic microorganisms, living on a multicellular organism, represent the microbiota. It plays a crucial role in maintaining the metabolic, hormonal, and immunologic homeostasis of their host (Salvucci 2016; Hooper et al. 2012). The human gut microbiome is diverse and dynamic in nature comprising of large number of bacteria, archaea, fungi, protozoa, and viruses. Microbiologists estimate that 10^{14} bacteria exist in and on an average adult human which is almost 10 times more than the total number of human cells (Thursby and Juge 2017). There exists an intricate relationship among the host cells and the microbiota. The disruption of this interaction can perturb the homeostasis and can eventually contribute toward the development of severe disorders such as diarrhea, PSC (primary sclerosing cholangitis), stomach cancer, etc. (Lv et al. 2016; Barrett et al. 2013; Herrera and Parsonnet 2009).

The host immune system is significantly influenced by the gut microbiome which provides protection against exogenous pathogens and also formulates immunoprotective responses (Kosiewicz et al. 2011). Thus, a modified gut microbiome can significantly contribute toward the development of autoimmune and inflammatory disease, even in organs like brain and skin which are isolated from the gut (Bowe and Logan 2011). The mechanisms by which intestinal microbiota exerts its impact on skin health, though not yet fully understood, are hypothesized to have a propensity to be linked to the modulatory effect of gut enteric bacteria on systemic immunity (O'Neill et al. 2016). Circulation of cytokines and primed immune cells from the Peyer's patches to the skin implicates a possible link in gut–skin communication (Spahn and Kucharzik 2004). These immune-modulators are known to alter the immune status and perk up the defense mechanism. Recent research indicates a correlation between microbial imbalance condition, i.e. "intestinal dysbiosis" and widespread skin disorders involving the "gut–skin axis" (Hadian et al. 2020). Usually, dysbiosis is a result of unbalance of the microflora in the gastrointestinal (GI) tract including the stomach and intestines. Dysbiosis in GI tract is usually the product of dietary change, high levels of stress, poor dental hygiene, alcohol abuse, or unprotected sex (Tomasello et al. 2016). Furthermore, a hypothesis states that factors such as alteration in diet, lifestyle, and use of antibiotics may alter the normal gut microflora, which in turn may lead to alterations in bacterial metabolism and overgrowth of potential pathogenic microorganisms. Growth of pathogenic microbes in intestine may result in release of toxic products leading to disease condition (Myers and Hawrelak 2004). For instance, segmented filamentous bacteria living in the gut has been associated with a variety of Th17-mediated diseases like psoriasis along with inflammatory bowel disease, asthma, and rheumatoid arthritis (Lee et al. 2018; O'Neill et al. 2016; Korn et al. 2009). Furthermore, recent evidences suggest direct effect of intestinal microbiota and their metabolites on the skin physiology, pathology, and immune response (O'Neill et al. 2016; Samuelson et al. 2015). Short chain fatty acids (SCFAs) such as propionate, acetate, and butyrate, produced during fiber fermentation in the intestine, play an integral role in determining the prevalence of certain skin microbiome profiles that subsequently influence skin immune defense mechanisms. For instance, Propionibacterium is a genus which can produce SCFAs, primarily acetate and propionic acid. The propionic acid shows antimicrobial effect against methicillin-resistant *Staphylococcus aureus* USA300 (Schwarz et al. 2017; Samuelson et al. 2015; Shu et al. 2013). Butyrate subdues immune responses by suppressing the proliferation, migration, adhesion, and production of cytokines in inflammatory cells. SCFAs produced in the gut control both the activation and apoptosis of immune cells by inhibition of histone deacetylase and inactivation of NF-κB signaling pathways. The histone deacetylase inhibition facilitates the proliferation of regulatory cells that perform numerous physiological functions including wound healing (Samuelson et al. 2015; Loser and Beissert 2012; Meijer et al. 2010). These studies are supporting evidence for an existing interactive functional mechanism between the gut and the skin. Similarly, compelling evidence lies to put forth the concept of an intricate communication network between the gut, skin, and brain. Originally the theory on "gut–brain–skin axis" was proposed by John H. Stokes and Donald M. Pillsbury in 1930, conceptualizing an interrelationship between the emotional state, intestinal flora, and skin inflammation. During fetal development, neural crest cells almost simultaneously differentiate into the CNS (central nervous system) and ENS (enteric nervous system). The gut being the biggest digestive organ, immune organ, and endocrine organ of the human body also possesses ENS, which is relatively independent of the brain (Bercik et al. 2011a, b; Cryan and O'Mahony 2011; Rhee et al. 2009). Soon after birth the gut and gut microbiota work together to perform the tasks of digestion, immune and endocrine functions, and neurotransmission. It is now known that the benefits of human microbe symbiosis can be extended to human mental health, and in recent years evidences have shown that the gut–brain axis or the bidirectional communication between the resident microbes of the GI tract and the brain plays a key role in maintaining brain health (Mayer et al. 2014). The GI microbiota influences human behavior and may affect the pathophysiology of mental illnesses (Foster and Neufeld 2013). The knowledge gained in recent years about the function and importance of the microbiome has broadened the concept of the gut–brain axis to the "microbiota-gut–brain axis," emphasizing the prominence of the microbiome in the regulation of gut–brain communication (Bercik et al. 2011a, b; Cryan and O'Mahony 2011; Rhee et al. 2009). NIH-JCVI's Human Microbiome Project was a prominent effort toward the better understanding of this bidirectional relationship. The project was initiated with an intention to understand the effect of microbiota on human health by exploring and mining the associated microbiome from 18 body-sites of more than 250 individuals. The study isolated, sequenced, and enumerated novel microbial strains, namely *Fusobacterium nucleatum* CT-1, *Prevotella copri* DSM 18205, *Roseburia* M72/1, *Eubacterium plautii*, *Bilophila* sp., etc. from the gut and *Acinetobacter* species, *Anaerococcus vaginalis*, *Erysipelothrix rhusiopathiae*, *Rhodococcus equi* ATCC33707, *Staphylococcus epidermis*, etc. from the skin were observed prominently (Peterson et al. 2009). Though the role of the said strains in human health is yet to be explored, still these natural residents can condense our enslavement on chemical based interventions for daily upkeep of overall health and disease prevention.

Presently, the use of oral probiotics as a treatment or alternate strategy to treat various skin and gut related disorders are not very uncommon. Suggested mechanisms for the action of probiotics are described in different ways as modification or improvement of the immune system, interaction with other intestinal flora, and development of some organic acids such as lactic acid, selective substitution or alteration of existing microbiome, improvement of the mucosal barrier, and production of valuable biomolecules such as bacteriocins (Hemarajata and Versalovic 2013). Extended studies in this direction will assist in providing a new puddle of acquaintance to the dynamic subject of what is referred to as "healthy gut microbiome." In this chapter we will dissect gut–brain–skin axis and will attempt to examine the mechanisms of action of gut–brain–skin axis communication and how the microbiome's influence can be harnessed for therapeutic purpose via probiotic supplementation.

2.2 Skin to Brain via Gut

The study of unified relationship between the skin and rest of the body is a continuous subject of research for clinicians and scientists around the world. The gut–brain–skin axis theory highlights the interconnection between the GI system, skin, and mental health. The gut–brain–skin axis theory originated in 1930 as a result of clinical studies and observations by John H. Stokes and Donald M. Pillsbury. Their study had linked emotional states with GI disorders through different mechanisms like diet and neuronal responses. They reported cases of patients suffering from colitis along with urticarial and dermatographism. It was proposed that modifications in gut microflora increase the gut permeability and eventually lead to systemic inflammation eventually ending in altered cutaneous physiology. This significant relationship was sustained by the observation that hypochlorhydria is associated with multiple skin conditions like acne, psoriasis, eczema, etc. (Stokes and Pillsbury 1930). It was also revealed that modification of gastric acids and fluctuation of the gut microflora could have further supplementary effects beyond

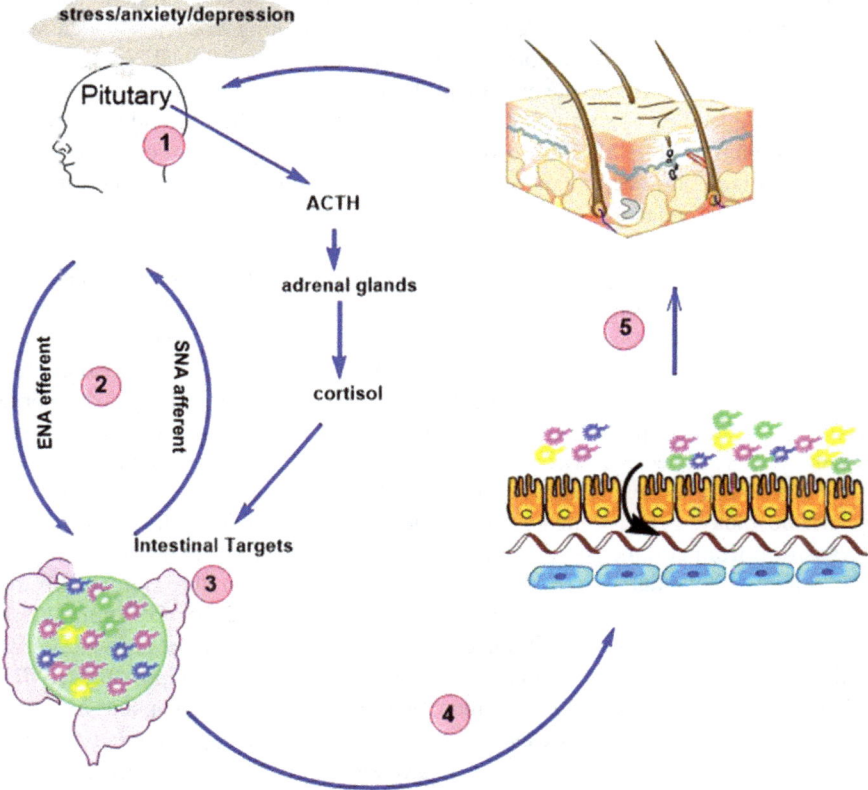

Fig. 2.1 Proposed mechanism of gut–brain–skin triangle

the GI system. Thus, use of cod-liver oil and bacillus was proposed in restoring gut homeostasis, which is parallel to present day probiotics usage (Lallès 2020).

One of the possible mechanisms of gut–brain–skin axis involves a cascade of events (Fig. 2.1) and focuses on CNS axis getting stimulated in response to environmental factors like stress, anxiety, or depression. It is a highly regulated process and involves a dynamic interaction between amygdala, hippocampus, and hypothalamus. The interaction results in the release of hypothalamus corticotrophin-releasing factor which further stimulates ACTH (adrenocorticotropic hormone) secretion from the pituitary gland which in due course leads to the release of cortisol from the adrenal glands. Cortisol further stimulates intestinal targets which ultimately lead to the loss of normal microbial biofilm. The events further accelerate the risk of excess sebum production and increased psychological distress and skin related disorders.

The GI tract contains a large number of commensal bacteria and is the primary focus of the gut–brain–skin axis. SIBO (small intestinal bacterial overgrowth) and hypochlorhydria are two conditions that represent a connection of cutaneous pathology and mental health. Modifications in gastric acid secretion, such as hypochlorhydria increase the risk of SIBO which has long been associated with psychological disorders like anxiety, depression, etc. (Barrett et al. 2008). In an Australian study on SIBO, the use of probiotic *Lactobacillus casei* successfully reduced SIBO leading to significant improvement in GI, cutaneous and psychological symptoms (Parodi et al. 2008; Addolorato et al. 2008; Pimentel et al. 2000). In recent years there has been an increase in our understanding and perception that, in addition to local intestinal effects, there is a constant interaction between the intestine, its microbiota, and the brain. For instance, food intake signals communicate with this axis at the hypothalamus level, while gut peptides, in turn, are engaged in controlling central processes like satiety and feeding behavior (Martin et al. 2018).

Furthermore, a recent study showed that commensal bacteria can detect a myriad of neurohormones, as may communicate with skin microflora via sweat and epidermis. Substance P (SP) is a neuropeptide excessively produced by skin nerve terminals. In a research pertaining to the mechanisms of the response of SP among different bacterial species it was seen that the response was variable, but they all lead to an increase in adhesion and/or virulence (N'Diaye et al. 2017). In addition, bacteria nowadays perceive the hormones and neurotransmitters in the gut as stimuli to transform from commensal to pathogenic behavior (Sandrini et al. 2015). Hence SP exhibits substantial variations in local concentration in the skin under the influence stimuli such as pain, stress, or infection (Nakano 2004; O'Connor et al. 2004; Harrison and Geppetti 2001) and even in case of nervous breakdown (Cizza et al. 2008). Variations in resident skin SP concentration, and possibly CGRP (calcitonin gene-related peptide), can then lead to an increase in cutaneous bacterial virulence. These studies provide evidence of existence of a connection between the central nervous system, the cutaneous microbiota, and skin homeostasis. Several other neuropeptides are found in the skin, and their possible effects on bacteria in the skin are yet to be explored.

Numerous other processes, including blood sugar control, insulin secretion, and insulin sensitivity, are also regulated by gut peptides. Moreover, gut peptides influence distant tissue functions, like bone metabolism. Metabolic disorders and gastrointestinal infection affect this axis as observed in patients with type 2 diabetes or chronic *Helicobacter pylori* infection in mice (Yang and Sheu 2016). We assume that several unexplored pathways and factors that affect the gut–brain–skin axis persist but are yet to be unraveled.

1. CNS axis (HPA (hypothalamic pituitary adrenal)) perhaps gets stimulated in response to environmental factors such as anxiety, depression, or stress.

2. HPA initializes cortisol production and is regulated by a dynamic interaction between amygdala, hippocampus, and hypothalamus. Release of hypothalamus corticotrophin-releasing factor secretion further stimulates ACTH (adrenocorticotropic hormone) secretion from the pituitary gland which in due course leads to the release of cortisol from the adrenal glands.

3. CNS interacts with various intestinal targets such as ENS, muscle layers, and gut mucosa in both afferent and efferent autonomic pathways, altering surface attributes of exchange, motility, immunity, permeability, and mucous secretion. The enteric microbiome exhibits bidirectional

(continued)

communication with the intestinal targets, curbing gastrointestinal functions, and being modulated by interactions between the brain and the gut itself.

4. Loss of normal microbial biofilm unties the barrier for control of endotoxins and they gain systemic access.

5. The events further increase the risk of excess sebum production and increased psychological distress and skin related disorders such as acne vulgaris, pimples, eczema, psoriasis, etc.

2.3   The Relationship Between Skin and Its Microbiome: Beyond Skin Deep

Composition and role of skin microflora have emerged as one of the dermal biology's most fascinating and growing fields (Cogen et al. 2008). Skin microflora plays a significant role in the selective exclusion of harmful microbes that cause infection of the skin (Cinque et al. 2011). Natural skin microbiota is likely to contribute to the competitive elimination of pathogens, a role that could probably be improved by the use of probiotics which contribute to the modulation of cutaneous microflora, lipid barrier, and skin immune system, resulting in skin homeostasis preservation (Cinque et al. 2011; Ouwehand et al. 2003). Probiotics have been shown to improve several skin disorders like atopic eczema (Rusu et al. 2019), wound and scar healing (Sinha et al. 2019), and promote skin rejuvenation (Roudsari et al. 2015). Majamaa and Isolauri (1997) conducted the first double-placebo experiment on dermal pathologies to improve atopic dermatitis/eczema using probiotics. They examined the immunological and clinical implications of exclusion of cow's milk in children with or without *L. rhamnosus* GG strain in a hydrolyzed formula that was given to mothers of ten breast-fed babies with atopic eczema and cow's milk allergy. Significant improvement in atopic dermatitis within 1 month of treatment was found children receiving *L. rhamnosus* GG. They concluded that probiotic strains enhanced endogenous barrier processes in patients with atopic dermatitis and food allergy. Effects of probiotics in treating skin diseases are extensively studied via both oral administration and topical application. SCORing Atopic Dermatitis (SCORAD) index was reduced in Atopic dermatitis patients by probiotics treatment consisting of *L. casei*, *L. salivarius*, *L. acidophilus*, and *Bifidobacterium bifidum* strains (Yeşilova et al. 2012). Similarly, ECN (*E. coli* Nissle 1917) was revealed to be beneficial in treating various chronic inflammatory disorders. The oral administration of ECN induced the immune regulatory mechanisms in allergen-induced dermatitis mouse model by stimulating cytokine production. In addition, decline in epidermal thickness and proliferation of immune cells was observed along with an increase in forkhead box P3 (Foxp3) (+) cells and a trend of increased expression of IFNÿ, IL-10, and TGFβ in eczematous skin (Rather et al. 2016). Furthermore, research findings demonstrate that certain immune responses could be evoked and skin barrier functions can be improved by bacterial components. In comparison to viable cells, the bacterial components are stable at room temperature, thus making them more suitable for topical applications

(Guéniche et al. 2010a, b). Bacterial compounds such as lipoteichoic acid, peptidoglycan, hyaluronic acid, and sphingomyelinase are known to exert beneficial dermal effects. *S. thermophilus* has been shown to increase the production of ceramide (ceramide sphingolipids) when applied to skin for 7 days. These molecules wield anti-inflammatory activity and have antimicrobial activity against *P. acnes* (Di Marzio et al. 1999, 2003, 2008). Similarly, in a study by Iordache et al. (2008), the expression of opportunistic bacterial virulence factors could be suppressed by the presence of soluble molecules produced by lactic acid-producing bacteria (Lew and Liong 2013).

Probiotic microbes use various mechanisms, such as lowering pH, to maintain skin health and prevent pathogen multiplication (Arck et al. 2010). Bacterial probiotic extracts have been shown to display anti-adhesion and antimicrobial properties on cutaneous and mucous surfaces. Hansen and Jespersen (2010) developed a tissue dressing consisting of lactic acid-producing bacteria (*L. sporogenes*, *L. acidophilus*, *L. plantarum*, etc.) for use in wound healing. Results from previous studies show promising outcomes in the treatment or management of skin health; although these effects were dependent on strain, dosage, and application. Various skin health-related claims of individual or combined probiotic microorganisms based on animal models, in vivo studies, and topical applications on human subjects are enlisted in Tables 2.1, 2.2, and 2.3, respectively. The skin health claims are very diverse and range from managing acne to improving AD scores. Probiotics reduce erythema, modulate wrinkle development in the presence of UV, and help in recovering skin immune homeostasis. Although further validation is needed for evaluating the particular interactions and underlying mechanism which may occur. Further study should be aimed at investigating the potential to meet demand for probiotic formulations as dermal products.

2.4   Understanding the Gut–Brain Axis

The gut–brain axis is the bidirectional communication between the GI tract and the brain (Grenham et al. 2011). This axis is controlled at the hormonal, neural, and immunological levels for keeping homeostasis and any flaw in the axis can lead to pathophysiological outcomes. The regular co-event of stress-related mental disorders, for example, gastrointestinal disorders and nervousness/anxiety has likewise additionally underlined the significance of the gut–brain axis (Matsumoto et al. 2013; Cryan and Dinan 2012). The platform of the gut–brain axis comprises the central nervous system (CNS), the enteric nervous system (ENS), the sympathetic and parasympathetic arms of the autonomic nervous system (ANS), the neuroendocrine and neuroimmune systems, and additionally the gut microbiota (Grenham et al. 2011). An intricate reflex system is shaped to encourage signaling along the axis, with afferent fiber projections to integrative CNS structures and efferent fiber projections that are undertaken to the smooth muscles in the intestinal wall (Cryan and Dinan 2012). Through this bidirectional communication, brain signals can influence the sensory, motor, and secretory functions of the GI tract and

Table 2.1 Probiotics for treating skin related disorders in animals

Probiotic strain	Dosage	Animal model	Clinical response	Proposed mechanism	References
L. pentosus GMNL-77	5×10^7 cfu/0.2 mL/day	BALB/c imiquimod-induced psoriasis-like mice	Decrease in erythema and scaling	Decrease in expression of pro-inflammatory cytokines mediated by suppression of antigen presenting cells	Chen et al. (2017)
L. rhamnosus IDCC 3201	1×10^{10} cells/day $\times$ 8 weeks	NC/Nga mice	Decrease in epidermal thickness and frequency of scratching	Suppression of inflammation mediated by mast cells	Lee et al. (2016)
L. plantarum HY7714	100 μL PBS/day with 1×10^9 cfu, 1 h prior to UVB irradiation	Hairless mice	Decrease in development of wrinkles following UVB radiation	Inhibition of matrix metalloproteinase -13 expression	Kim et al. (2014)
L. reuteri ATCC 6475	3.5×10^5 organisms/day $\times$ 20–24 weeks	C57BL/6 wild type and IL-10-deficient mice	Increase in dermal thickness, thicker, and shinier fur	IL-10 dependent anti-inflammatory pathway	Rattanaprasert et al. (2019)
L. rhamnosus	1×10^9 cfu/day	SKH-1 hairless mice	Decrease in transepidermal water loss	Decrease in IL-4 and Thymic Stromal Lymphopoietin via mechanism involving increase in CD4+CD25+Foxp3+ regulatory T cells	Kim et al. (2012)

Table 2.2 Oral Probiotics for treating skin related disorders in Humans

Probiotic strain	Dosage	Study group	Clinical response	Proposed mechanism	References
L. johnsonii	1×10^{10} cfu/day $\times$ 6 weeks	54 healthy subjects	Increase in recovery of skin immune homeostasis following UV-induced immunosuppression	Normalization of epidermal expression of CD1a	Peguet-Navarro et al. (2008)
B. infantis 35624	1×10^{10} cfu/day $\times$ 8 weeks	26 subjects with plaque psoriasis	Decrease in systemic inflammation	Decrease in C-reactive proteins and Tumor necrosis factor-alpha	Groeger et al. (2013)
L. acidophilus, LB-51, B. bifidum	5×10^9, 5×10^9, 20×10^9 cfu $2\times$/day $\times$ 12 weeks	45 females with acne	Significant decrease in number of acne lesions using probiotic together with antibodies	Synergistic anti-inflammatory effect	Jung et al. (2013)
L. plantarum HY7714	1×10^{10} cfu/day $\times$ 12 weeks	129 females with dry skin and wrinkles	Increase in skin hydration and decrease in wrinkle depth	Molecular control of signaling pathways and gene expression in skin cells	Lee et al. (2015)
L. rhamnosus SP1	3×10^9 cfu/day (75 mg/day) $\times$ 12 weeks	20 adults with acne	Improved appearance of adult acne	Decrease in Insulin like growth factor-1 expression and increase in Forkhead box protein O1	Fabbrocini et al. (2016)

Table 2.3 Topical Probiotics for treating skin related disorders in Humans

Probiotic used	Dose	Skin condition/ disease	Findings	References
S. thermophiles	1.7 g/5 mL in 20 mL lotion, twice a day.	Atopic dermatitis	Increased skin ceramides and improvement in all aspects of atopic dermatitis	Di Marzio et al. (2003)
E. faecalis cell-free supernatant	6400 AU twice a day for 8 weeks	Acne	Reduced number of acne lesions	Kang et al. (2008), Sharma et al. (2020)
B. longum	10% extract twice a day for 2 months	Reactive skin	Decreased skin sensitivity, improved resistance to physical aggression, decrease in dryness but took 57 days	Guéniche et al. (2010a, b)
L. plantarum	1% and 5% twice a day for 2 months	Acne	Reduced erythema, repaired the skin barrier and reduced skin microflora to reduce acne at 5%	Muizzuddin et al. (2012)
L. johnsonii	0.3% twice a day for 21 days	Atopic dermatitis	Improved atopic dermatitis scores	Blanchet-Réthoré et al. (2007)

conflictingly, the GI tract signals can influence normal brain functioning (Grenham et al. 2011). Numerous studies suggest that any modifications in the gut microbiota can extraordinarily impact the association between the gut and the brain, influence brain functioning, and alter host behavior. Many investigations have utilized germ-free animals to examine the gut–brain axis. Neufeld et al. (2011) observed a higher plasma corticosterone level in the GF (germ-free) mice which showed higher stress reaction compared with the SPF (specific pathogen-free) mice. The expression level of BDNF (brain-derived neurotrophic factor), glutamate, and serotonin receptors suggesting anxiety was additionally seen in the GF mice. The study also highlighted the effect of intestinal microbiota on behavior and neurochemical vicissitudes in the brain. Such studies may help us in devising potential restorative solutions for treating disorders like depression and anxiety because of the emanate concern on gut–brain communication and its capacity to influence mental disorders. Similar studies likewise implicated probiotics in modulation and improvement of mood, stress response, etc. (Lakhan and Kirchgessner 2010). An in vivo investigation on the impact of psychotropic-like properties of probiotic in rodent and human subjects was reported by Messaoudi et al. (2011). Daily usage of the probiotics blends of *L. helveticus* R0052 and *B. longum* R0175 (10^9 cfu) reduced anxiety-like behavior in rats and relieved stress in human subjects ($p < 0.05$). Researchers proposed that the probiotic blend may exhibit a remedial impact on depressive behavior through decrease of pro-inflammatory cytokines, which subsequently leads to induction of depression and restores intestinal integrity by apoptotic inhibition (Arseneault-Bréard et al. 2012). Another study indicated that comparable probiotics composition can lead to decrease in post-myocardial infarction depressive behavior and an improvement in intestinal penetrability in rats (Trudeau et al. 2019). It was also revealed that probiotics are suitable to regulate gut microbiota as well as engage with stress, anxiety, and depression management which can be utilized as a novel tool in treatment of mental disorders. Depression-induced Sprague-Dawley rats showed increase in the serotonergic precursor (tryptophan) and a decline in pro-inflammatory immune responses, upon administration of *B. infantis* for 14 days (Desbonnet et al. 2009). Similarly, *L. rhamnosus* (JB-1) showed decrease in stress-induced corticosterone, anxiety levels, depressive behavior, and induced region-dependent alterations in gamma-amino butyric acid receptors (GABA-A and GABA-B) mRNA expressions via the vagus nerve in mice suggesting antidepressant effect (Bravo et al. 2011). Furthermore, copious evidences on the brain-skin axis emerged from recent study where stress perceived by mice led to major neurogenic skin inflammation and altered the epidermal barrier function like sound stress causing nerve growth factor and mast cell dependent neurogenic inflammation in skin (Choi and Di Nardo 2018). It also influences the neuropeptide production by dorsal root ganglia sensory neurons (DRG) present in the dorsal root of a spinal nerve (Purves et al. 2001) and is responsible for performing essential functions like nociception and presynaptic control (Lorenzo et al. 2014; Huang et al. 2007). Interestingly, formation of functional associations between nerve fibers and mast cells appears to be similar in skin and gut. For instance, perceived stress in the part of skin via neurogenic inflammation can consequently inhibit hair growth by premature hair follicle regression (Pavlovic et al. 2008).

CFS (chronic fatigue syndrome) is an intricate and incapacitating disorder involving intense fatigue that might be accelerated by physical or mental exercises. Around 97% of CFS patients asserted neuropsychological complications like migraine. In a pilot study, CFS patients consuming *L. casei* strain *Shirota* (LcS) (24×10^9 cfu) every day for 2 months indicated a significant ($p < 0.01$) decline in anxiety levels (Rao et al. 2009). This examination offered further support in the presence of the gut–brain interconnection which can be mediated by the gut microbiota. In another study, human subjects were required to take either a cultured drink containing *L. casei Shirota* (108 cfu/mL) or a placebo control every day for 3 weeks. Human subjects with a poor state of mind toward the start of the test showed a critical ($p < 0.05$) improvement after the probiotic treatment (Benton et al. 2007). Furthermore, Tillisch et al. (2013) assessed the impact of consuming fermented milk containing a blend of probiotics (*B. animalis* subsp *Lactis*, *S. thermophiles*, *L. bulgaricus*, and *Lactococcus lactis* subsp. *Lactis*) on gut–brain communication in humans. Results uncovered that brain activity, which plays a role in controlling feelings and sensation in women was affected after the consumption of fermented milk. This study again shows the relationship of probiotics on the alteration of brain activity and furthermore gives proof to the modulatory impact of probiotics in gut–brain communications. In a study by Bercik et al. (2011a, b), the administration of *B. longum* NCC3001 was determined to normalize the anxiety-like behavior of the dextran sodium sulfate-induced colitis in mice. Accumulating confirmations demonstrate the presence of gut–brain communication and its significance in altering brain activity and behavior. Abilities of specific probiotics to regulate different aspects of the gut–brain axis provide potential advantages in controlling stress, anxiety, and depression. Various oral probiotic strains for treatment of various psychosomatic conditions ranging from stress to memory impairment in animals and humans have been summarized in Tables 2.4 and 2.5, respectively. Multiple studies on specific gut microbes are still in the primary stages and further investigations are justified to inspect the specific interactions that may take place.

2.5   That Gut-Feeling: In Skin and Brain

Owing to its dynamic nature, the human microbiome digresses with surrounding environmental stress and retorts to feedback from other systems (Volkova et al. 2001). Even slight changes in the microbiome can influence the levels of systemic inflammation. Though long-term dietary habits are established to influence bacterial composition, similarly short-term and drastic dietary modification can swiftly amend gut microbiome. This opens the doors for actively amending the existing microbiome for therapeutic purposes, considering the consequence of the gut microbiome on inflammatory ailments (Huang et al. 2017).

Curative exploration of probiotics via administration of live enteric bacteria has a potential in preventing and controlling various skin and brain conditions (Grant and Baker 2017; Sánchez et al. 2017; Sarao and Arora 2017; Farris 2016; Hill et al. 2014; Krutmann 2009). As discussed earlier, any modulation of gut microbiome with probiotics leads to altering levels of inflammatory cytokines in the blood as they interact with the GI mucosal immune system. GI bacteria, including probiotic strains have been shown to bind to the MHC complexes and modify their expressions. Livingston et al. (2010) demonstrated that *Lactobacillus paracasei* NCC2461 administered in mice induced T regulatory cells and inhibited CD_{4+} T cell proliferation while increasing the secretions of anti-inflammatory cytokines. These observations clearly indicate promising role of *L. paracasei* in reducing disorders like allergic contact dermatitis.

As mentioned, there had been older commentaries and clinical anecdotes suggesting that orally consumed lactic acid bacteria might be of benefit in alleviating depressive symptoms. It was also reported that patients with mental health disorders appeared to have very low levels of *L. acidophilus*. In a series of case reports in 1924, one Illinois physician reported value of oral *L. acidophilus* for the treatment of both acne and mental health disorders; in addition to *L. acidophilus* improving the complexion, it was stated that 'in certain patients it even seemingly contributes to mental improvement'. It was also reported that the yeast Saccharomyces cerevisiae could improve both acne vulgaris and constipation.

Multiple previous studies and clinical reports revealed that lactic acid bacteria ingested orally may be of use in mitigating depressive symptoms (Nguyen et al. 2019; George et al. 2018). In a set of clinical studies in 1924, an Illinois physician mentioned benefits of oral *L. acidophilus* in the treatment of acne as well as mental health disorders. In addition to improving the complexion, it was claimed that in

Table 2.4 Probiotics for treatment of various psychosomatic conditions in animals

Probiotic strain	Psychosomatic conditions	Animal model	Clinical response	Proposed mechanism	References
L. rhamnosus	Anxiety	Germ-free mice	Reduced anxiety- and depression-related behavior	Higher expression of serotonin receptor 1a	Stilling et al. (2015)
L. rhamnosus R0011 and L. helveticus R0052	Memory deficit, anxiety, dysbiosis	B- And T-cell-deficient Rag1$^{-/-}$ Mice	Improved baseline impairment	Modulation of intestinal microbiota	Smith et al. (2014)
L. acidophilus, B. lactis, L. fermentum	Memory impairment related to diabetes mellitus	Diabetic rats	Improved the impaired spatial memory in diabetic animals	Stimulation of schaffer collaterals in hippocampus	Davari et al. (2013)
L. helveticus ROO52	Altered anxiety-like behavior	Wt and Il-10-deficient 129/Svev mice	Probiotics alone decreased anxiety-like behavior in Wt mice on a chow diet	Inflammatory pathways	Ohland et al. (2013)
L. rhamnosus R0011 and L. helveticus R0052 (5%)	Stress due to maternal separation leading to dysbiosis	Rat pups	Change in gut flora composition observed	Normalization of Hpa axis activity	Gareau et al. (2007)

Table 2.5 Probiotics for treating various psychosomatic conditions in human beings

Probiotic strain supplementation	Study group	Psychosomatic conditions	Clinical response	References
L. acidophilus, L. casei, B. bifidum	Patients with major depressive disorder	Depression	Improvement in clinical signs	Akkasheh et al. (2016)
L. helveticus IDCC3801	Healthy elderly individuals	Cognition	Improvement in cognitive functioning during cognitive fatigue tests	Chung et al. (2014)
Fermented milk product with probiotic containing B. animalis subsp lactis, S. thermophilus, L. bulgaricus, L. lactis subsp. lactis	Healthy women	Emotion, attention	Altered activity of brain regions that control central processing of emotions	Tillisch et al. (2013)
Probiotic yogurt or a multispecies probiotic capsule containing L. acidophilus La5 and B. lactis Bb12	Normal population	Depression, anxiety, stress	Improvement in participants supplemented with probiotic yogurt or probiotic capsule	Mohammadi et al. (2016)
L. casei Shirota	Normal elderly individuals with decreased mood	Mood status/ swings	Improvement in participants in bottom third of the depressed/elated dimension at baseline	Benton et al. (2007)

some patients it also appears to help in mental improvement (Saunders 1924). The possible physiological mechanisms by which mental well-being could be affected by deliberate manipulation of the intestinal flora were finally examined in the last decade, following the publication of two important hypotheses. According to the first hypothesis, stress acts as a critical factor in major depressive disorder and is believed to modify the GI microflora by reducing *Lactobacilli* and *Bifidobacterium* levels (Logan and Katzman 2005). The second hypothesis states that lactic acid bacteria may have a therapeutic role in CFS (chronic fatigue syndrome) management (Logan et al. 2003). In the frontal cortex and limbic system, oral administration of probiotics has been shown to improve peripheral tryptophan levels, as well as modify serotonin and dopamine turnover (Hawk et al. 1917). The existence of non-pathogenic bacteria like bifidobacteria in the intestinal tract tends to modulate an elevated stress response, maintaining levels of BDNF, a neuropeptide reported to be reduced in depression (Sudo et al. 2004).

The effect of probiotics as a means of modulating the release of neurogenic substance P, both in the intestinal tract and in the skin (Guéniche et al. 2010a, b; Verdú et al. 2006), cannot be underestimated as an important pathway that links the nervous system to the gut and skin. Experimental changes in the normal gut flora can increase substance P release into the nervous system and promote behaviors reflective of anxiety (Collins et al. 2009). Moreover, even minute increase in substance P can cause anxiety and depression (Herpfer et al. 2007). Additionally, those who respond to antidepressant pharmacotherapy are reported to have serum substance P reductions in accordance with improved mood conditions (Lieb et al. 2004). Furthermore, Philippe et al. (2011) concluded that oral administration of a specific strain of *L. paracasei* ST11 can lead to a decrease in secretions of substance P, which enhanced the skin barrier function and decreased local skin inflammation. Similarly, stress due to sound causes substantial substance P, nerve growth factor, and mast cell dependent neurogenic inflammation in mice skin and affected neuropeptide production by dorsal root ganglia sensory neurons. Interestingly, formation of associations between nerve fibers and mast cells appears to be very similar in skin and gut (Pavlovic et al. 2008). It has been known for quite a while that biologically active peptides like substance P not only interact within the gut, brain, and skin but also have a specific embryonic origin (Teitelman et al. 1981). However, the underlying mechanism still needs to be unraveled to understand the comprehensive interaction between the two.

An alternate pathway by which probiotics can influence both brain and skin disorders is through glycemic control regulation. It has become quite apparent in recent years that there may indeed be a link between dietary components, most notably low-fiber carbohydrates, and the risk of acne (Bowe et al. 2010). Regional diets, for example, low in refined foods and sugars are linked with a reduced risk of acne. This is important because recent research indicates that the gut microbiota contributes tolerance to glucose (Kleerebezem and Vaughan 2009) and that *Bifidobacterium lactis* administered orally can increase fasting insulin levels and glucose turnover rates, in the midst of a high-fat diet (Burcelin 2010). The mechanisms tend to include the ability of bifidobacteria to prevent the efflux of lipopolysaccharide (LPS) endotoxins into systemic circulation, though further research is needed in this regard. Loss of *Bifidobacteria* through poor dietary choices/patterns—high fat, sugar—contributes to increased intestinal permeability, invasion of LPS endotoxins via the intestinal barrier, which in addition leads to low-grade inflammation, oxidative stress, and insulin resistance (Cani and Delzenne 2009). Probiotic treatment in humans can decrease systemic access to gut-derived LPS endotoxins and decrease reactivity to these endotoxins (Schiffrin et al. 2009). Recent studies indicate that acne is linked to increased intake of highly appetizing, sweet, processed, carb-rich foods with low nutrient density and that a period of insulin resistance occurs during puberty is well known, one coinciding with the development of acne, depression, and/or anxiety (Ghodsi et al. 2009). The strong research evidences support the role of gut microbiota in managing the gut–brain–skin axis. Also probiotic based modulation of gut may play a significant role in management of the associated skin and brain disorders.

2.6   Probiotics 2.0

So far there lies a strong case in favor of the microbial modulation of brain functioning and skin health. Natural microbiota of human body is an imperative determinant of multiple critical processes such as hematopoiesis, aging, infectious disease immunity, and behavior. Hence, elucidating the taxonomic and functional composition of a healthy gut microbiome is a pre-requisite for deciphering the relationship between natural microbiome and existing health condition. While the typical microbiota sometimes may not defend the host against pathogens, several disturbed/altered conditions such as inflammation, obesity, and neuropsychiatric disorders are correlated with disrupted gut microbiota. Moreover in the twenty-first century, it is becoming progressively ostensive that en route toward alternative approaches is quite obligatory to control infectious diseases in humans and animals, over conventional antibiotic therapy. Furthermore, new technologies are needed to develop novel probiotic products that contain strains of human origin. A list of 12 pathogenic bacteria families, resistant to antibiotics, named as "priority pathogens," poses a serious threat (WHO 2017). It has led to increased demand of alternative therapeutic approaches. This is where "designer probiotics" or "probiotics 2.0" comes to rescue by bridging the gap between demand and dearth of novel and effective new-fangled antibiotics (Maxmen 2017; Kumar et al. 2016; Paton et al. 2012; Braat et al. 2006). In this context different microbes like *Bifidobacteria, E. coli* Nissle 1917, and *Saccharomyces cerevisiae, S. boulardii, Kluyveromyces lactis,* and *Pichia pastoris* have been exploited as future prospective probiotics expressing heterologous genes encoding antimicrobial and anti-inflammatory biomolecules (Rodríguez-Nogales et al. 2018). Exploiting the designer probiotics based approach offers an added advantage over conventional antibiotics, as it is independent from selective pressure for progression/development of microbial resistance.

Braat et al. (2006) exploited the potential of a live and genetically modified bacterium for topical delivery of recombinant proteins in the human gut, where the concept of safety and biological containment was also taken into account. Thus in this post-genomic era, the study of probiotics also holds promise for emerging research fields such as functional nutraceuticals, oncotherapy, and psychoneuroendocrinology. In this light, the probiotics are emerging as powerful tool for drug delivery thereby circumventing the after effects associated with the systemic administration of the drugs. For instance, the treatment of mice with recombinant *L. gasseri* NM713 expressing streptococcal M6 protein (CRR6) prevented them from streptococcus group A infections (Mansour and Abdelaziz 2016) and recombinant *Lactococcus lactis* (LL-12) expressing human interleukin-10 (IL-10) relieved Crohn's disease (Braat et al. 2006). Similarly, in human dendritic cells, *L. lactis* strains developing native immunomodulatory SpaCBA (surface piliation appendages) were observed to stimulate toll-like receptor-2-dependent signaling in cell lines and to modulate anti-inflammatory cytokine synthesis (TNF-a, IL-6, IL-10, and IL-12) (Von Ossowski et al. 2013). Recombinant probiotics that could produce fibrin for fast wound healing or probiotics with collagen activating factors can be a possibility in the near future.

Widespread chronic disorders such as Alzheimer's disease, autism, hyperactivity disorder with attention deficits, stroke, atopic dermatitis, eczema, psoriasis, and vitiligo are growing increasingly in human populations. Exploiting the gut ecosystem via recombinant probiotics containing therapeutic biomolecules or phage therapy presents potential solutions to existing lifestyle diseases and chronic disorders (Divya Ganeshan and Hosseinidoust 2019). Designer probiotics either used as dietary supplements or administered topically may significantly promote normal physiology as well as immunity to protect the host against pathogens, oxidative stress, inflammatory diseases, and autoimmune responses. Formulations/preparations comprehending engineered probiotics that generate human pro-insulin and anti-inflammatory cytokines will provide a means to overcome the diabetes threat. In non-obese diabetic mice, *L. lactis* recombinant protein lHSP65-6IA2P2 has been found to prevent hyperglycemia, decrease insulitis, and enhance glucose tolerance and regulatory immune reactions in type 1 diabetes mellitus (Liu et al. 2016). Likewise, oral administration of *E. coli* Nissle 1917 expressing *N*-acylphosphatidylethanolamines inhibited mice obesity by regulating the consumption of food. Because of resemblances in mice and humans "feeding behavior," this approach can be simulated/counterfeit in humans too (Chen et al. 2014). In addition, some different strains with proven health benefits may also be considered as options for next-generation probiotics and other drugs based on microbiota. However, more research is needed to better understand the interactions between these strains, with the goal of developing an effective therapeutic solution.

2.7   Future Aspects and Undercarpet Challenges

This chapter sought to provide a cross-section of the gut–brain–skin axis and also gives an overview of the promising role of probiotics in disease prevention and treatment. Though a lot has been done, however, significant hurdles still remain to be overcome. The gut microbiota is the most imperative part of the commensal microbiota and works together with the gut in an unabridged manner to respond to endogenous as well as exogenous signals. Multiple investigations and clinical studies have corroborated the gut microbiome's influence on host homeostasis, allostasis, and the pathogenesis of disease. Through complex immune mechanisms, the influence of the gut microbiome extends to organs like skin and brain. With determined and focused modulation of the microbiome, probiotics have upheld their benefits in the prevention and/or treatment of inflammatory skin diseases including acne vulgaris, AD, psoriasis, and neurological disorders including depression, cognition, anxiety, and stress. Novel probiotic entities, specifically targeting gut–brain–skin axis, constitute a new class of products that are a great leap beyond the conventional over the counter solutions. Identification of strains from the gut–brain–skin axis followed by rigorous pre-clinical and clinical trials can help generate unique and anticipated probiotics with desired confounding effects. Various human microbiome projects held across countries have helped identify numerous novel microbial strains that are consistently present in and on the general population. Similarly, mining of the individual components of "healthy" gut microbiota could succor forthcoming data to actually predict the role of these microbial strains in human health and disease and help in developing interventions for daily upkeep of overall health and disease prevention.

In parallel, genetically modified microbes may open up a whole new horizon to the concept of probiotics in the upcoming near future. New DNA based technologies offer exciting opportunities to explicate mechanisms by which probiotics mediate their unconventional/atypical effects (including identification of microbial signals that fortify mucosal barrier and regulate mucosal immune function) and develop genetically modified designer probiotics with precise immunomodulatory properties or as biological delivery vehicles.

The regulation of probiotics in human diet varies by topography as well as governing regulatory authorities; still the outpouring of the functional food is likely to boom the existing probiotic market. Since its use, the concept of the therapeutic micro-organism is gaining popularity owing to the increasing links between health, diet, and nutrition. Thus, the practice of bioengineered microbes may circumvent side effects of antibiotics and could allow long-term embankments against various chronic diseases. Still challenges like regulating their proliferation and spread in environment particularly from feces, strain characterization, dose optimization, and lateral gene transfer to normal microflora, etc. are still to be addressed.

Nevertheless, consumer protection and the requirement for health prerogatives, with scientific evidence need to be established. Considering that the worldwide definition and classification of probiotics by regulatory authorities is different, the status of probiotic products is still unclear. So there may be reservations among regulatory bodies, manufacturers, and consumers about claims regarding probiotic products. Since the probiotic concept is invading the globe, more research is needed on probiotic traits. In addition, most probiotics contain only LAB which has minimal phylogenetic diversity and functionality. Therefore, crucial optimization of the experiments that regulatory agents require is the need of the hour. The application of synthetic biology techniques, namely introduction of synthetic genes that allow design and construction of reliable genetic circuits, precise fine tuning of transgene expression may offer new frontiers toward evolving the expansion of designer probiotics. This can further stretch new paradigms to the existing platform of disease prevention and treatment options or therapy, to shift the balance toward the beneficial microbiota with confounding attributes.

2.8 Conclusion

With recent advances in probiotic research, the development of designer probiotics does not seem so distant. Since little is known regarding the probable mechanisms of probiotics modulation of various physiological functions and factors (optimum dose, frequency, and duration of treatment for different probiotic strains), impending research in this upcoming field will ameliorate better understanding of the complex mechanisms underlying the gut–brain–skin axis and investigate the therapeutic potential of long-term modulation of the gut microbiome. Furthermore, it will also potentially expand therapeutic manipulation to include commensal gut fungi and viruses in order to fully harness the gut microbiome's influence in disease and health. The use of probiotics in the field of mental health seems an exciting proposition and may be a reality in the near future. The challenges associated with looming evolution of high throughput designer probiotics pose a yardstick to restrict their overuse and in turn thereby controlling their widespread misuse.

References

Addolorato G, Mirijello A, D'Angelo C, Leggio L, Ferrulli A, Abenavoli L et al (2008) State and trait anxiety and depression in patients affected by gastrointestinal diseases: psychometric evaluation of 1641 patients referred to an internal medicine outpatient setting. Int J Clin Pract 62(7):1063–1069. https://doi.org/10.1111/j.1742-1241.2008.01763.x

Akkasheh G, Kashani-Poor Z, Tajabadi-Ebrahimi M, Jafari P, Akbari H, Taghizadeh M et al (2016) Clinical and metabolic response to probiotic administration in patients with major depressive disorder: a randomized, double-blind, placebo-controlled trial. Nutrition 32(3):315–320. https://doi.org/10.1016/j.nut.2015.09.003

Arck P, Handjiski B, Hagen E, Pincus M, Bruenahl C, Bienenstock J, Paus R (2010) Is there a 'gut-brain-skin axis'? Exp Dermatol 19(5):401–405. https://doi.org/10.1111/j.1600-0625.2009.01060.x

Arseneault-Bréard J, Rondeau I, Gilbert K, Girard S-A, Tompkins TA, Godbout R et al (2012) Combination of *Lactobacillus helveticus* R0052 and *Bifidobacterium longum* R0175 reduces post-myocardial infarction depression symptoms and restores intestinal permeability in a rat model. Br J Nutr 107(12):1793–1799. https://doi.org/10.1017/s0007114511005137

Barrett JS, Canale KE, Gearry RB, Irving PM, Gibson PR (2008) Probiotic effects on intestinal fermentation patterns in patients with irritable bowel syndrome. World J Gastroenterol 14(32): 5020–5024. https://doi.org/10.3748/wjg.14.5020

Barrett K, Ghishan FK, Merchant J, Said H, Wood J (2013) Physiology of the gastrointestinal tract, vol 1–2. Elsevier, Cambridge, MA

Benton D, Williams C, Brown A (2007) Impact of consuming milk drink containing a probiotic on mood and cognition. Eur J Clin Nutr 61(3):355–361. https://doi.org/10.1038/sj.ejcn.1602546

Bercik P, Denou E, Collins J, Jackson W, Lu J, Jury J et al (2011a) The intestinal microbiota affects central levels of brain-derived neurotropic factor and behavior in mice. Gastroenterology 141(2):599–609. https://doi.org/10.1053/j.gastro.2011.04.052

Bercik P, Park AJ, Sinclair D, Khoshdel A, Lu J, Huang X et al (2011b) The anxiolytic effect of *Bifidobacterium longum* NCC3001 involves vagal pathways for gut-brain communication. J Neurogastroenterol Motil 23(12):1132–1139. https://doi.org/10.1111/j.1365-2982.2011.01796.x

Blanchet-Réthoré S, Bourdès V, Mercenier A, Haddar CH, Verhoeven PO, Andres P (2007) Effect of a lotion containing the heat-treated probiotic strain *Lactobacillus johnsonii* NCC 533 on *Staphylococcus aureus* colonization in atopic dermatitis. Clin Cosmet Investig Dermatol 10: 249–257. https://doi.org/10.2147/ccid.s135529

Bowe WP, Logan AC (2011) Acne vulgaris, probiotics and the gut-brain-skin axis-back to the future? Gut Pathog 3:1. https://doi.org/10.1186/1757-4749-3-1

Bowe WP, Joshi SS, Shalita AR (2010) Diet and acne. J Am Acad Dermatol 63(1):124–141. https://doi.org/10.1016/j.jaad.2009.07.043

Braat H, Rottiers P, Hommes DW, Huyghebaert N, Remaut E, Remon JP et al (2006) A phase I trial with transgenic bacteria expressing interleukin-10 in Crohn's disease. Clin Gastroenteral Hepatol 4(6):754–759. https://doi.org/10.1016/j.cgh.2006.03.028

Bravo JA, Forsythe P, Chew MV, Escaravage E, Savignac HM, Dinan TG et al (2011) Ingestion of lactobacillus strain regulates emotional behavior and central GABA receptor expression in a mouse via the vagus nerve. Proc Natl Acad Sci 108(38):16050–16055. https://doi.org/10.1073/pnas.1102999108

Burcelin R (2010) Intestinal microflora, inflammation, and metabolic diseases. Abstract 019. In: Keystone Symposia-Diabetes Whistler, British Columbia, Canada

Cani PD, Delzenne NM (2009) Interplay between obesity and associated metabolic disorders: new insights into the gut microbiota. Curr Opin Pharmacol 9(6):737–743. https://doi.org/10.1016/j.coph.2009.06.016

Chen Z, Guo L, Zhang Y, Walzem RL, Pendergast JS, Printz RL et al (2014) Incorporation of therapeutically modified bacteria into gut microbiota inhibits obesity. J Clin Invest 124(8):3391–3406. https://doi.org/10.1172/JCI72517

Chen YH, Wu CS, Chao YH, Lin CC, Tsai HY, Li YR et al (2017) *Lactobacillus pertosus* GMNL-77 inhibits skin lesions in imiquimod-induced psoriasis-like mice. J Food Drug Anal 25(3):559–566. https://doi.org/10.1016/j.jfda.2016.06.003

Choi JE, Di Nardo A (2018) Skin neurogenic inflammation. Semin Immunopathol 40(3):249–259. https://doi.org/10.1007/s00281-018-0675-z

Chung YC, Jin HM, Cui Y, Jung JM, Park JI, Jung ES et al (2014) Fermented milk of *Lactobacillus helveticus* IDCC3801 improves cognitive functioning during cognitive fatigue tests in healthy older adults. J Funct Foods 10:465–474. https://doi.org/10.1016/j.jff.2014.07.007

Cinque B, La Torre C, Melchiorre E, Marchesani G, Zoccali G, Palumbo P et al (2011) Use of probiotics for dermal applications. In: Liong MT (ed) Probiotics: microbiology monographs. Springer, Berlin, pp 221–241. https://doi.org/10.1007/978-3-642-20838-6_9

Cizza G, Marques AH, Eskandari F, Christie IC, Torvik S, Silverman MN et al (2008) Elevated neuro immune biomarkers in sweat patches and plasma of premenopausal women with major depressive disorder in remission: the POWER study. Biol Psychiatry 64(10):907–911. https://doi.org/10.1016/j.biopsych.2008.05.035

Cogen AL, Nizet V, Gallo RL (2008) Skin microbiota: a source of disease or defense? Br J Dermatol 158(3):442–455. https://doi.org/10.1111/j.1365-2133.2008.08437.x

Collins S, Verdu E, Denou E, Bercik P (2009) The role of pathogenic microbes and commensal bacteria in irritable bowel syndrome. Dig Dis 27(Suppl 1):85–89. https://doi.org/10.1159/000268126

Cryan JF, Dinan TG (2012) Mind-altering microorganisms: the impact of the gut microbiota on brain and behavior. Nat Rev Neurosci 13(10):701–712. https://doi.org/10.1038/nrn3346

Cryan JF, O'Mahony SM (2011) The microbiome-gut-brain axis: from bowel to behavior. Neurogastroenterol Motil 23(3):187–192. https://doi.org/10.1111/j.1365-2982.2010.01664.x

Davari SA, Talaei SA, Alaei HO (2013) Probiotics treatment improves diabetes induced impairment of synaptic activity and cognitive function: behavioral and electrophysiological proofs for microbiome–gut–brain axis. Neuroscience 240:287–296. https://doi.org/10.1016/j.neuroscience.2013.02.055

Desbonnet L, Garrett L, Clarke G, Bienenstock J, Dinan TG (2009) The probiotic *Bifidobacteria infantis*: an assessment of potential antidepressant properties in the rat. J Psychiatr Res 43(2):164–174. https://doi.org/10.1016/j.jpsychires.2008.03.009

Di Marzio L, Cinque B, De Simone C, Cifone MG (1999) Effect of the lactic acid bacterium *Streptococcus thermophilus* on ceramide levels inhuman keratinocytes in vitro and stratum corneum in vivo. J Invest Dermatol 113(1):98–106. https://doi.org/10.1046/j.1523-1747.1999.00633.x

Di Marzio L, Centi C, Cinque B, Masci S, Giuliani M, Arcieri A et al (2003) Effect of the lactic acid bacterium *Streptococcus thermophilus* on stratum corneum ceramide levels and signs and symptoms of atopic dermatitis patients. Exp Dermatol 12:615–620. https://doi.org/10.1034/j. 1600-0625.2003.00051.x

Di Marzio L, Cinque B, Cupelli F, De Simone C, Cifone MG, Giuliani M (2008) Increase of skin-ceramide levels in aged subjects following a short-term topical application of bacterial sphingomyelinase from *Streptococcus thermophilus*. Int J Immunopathol Pharmacol 21:137–143. https://doi.org/10.1177/039463200802100115

Divya Ganeshan S, Hosseinidoust Z (2019) Phage therapy with a focus on the human microbiota. Antibiotics 8(3):131

Fabbrocini G, Bertona M, Picazo Ó, Pareja-Galeano H, Monfrecola G, Emanuele E (2016) Supplementation with *Lactobacillus rhamnosus* SP1 normalises skin expression of genes implicated in insulin signalling and improves adult acne. Benefic Microbes 7:625–630. https://doi.org/10.3920/BM2016.0089

Farris PK (2016) Are skincare products with probiotics worth the hype? Dermatol Times

Foster JA, Neufeld KM (2013) Gut-brain axis: how the microbiome influences anxiety and depression. Trends Neurosci 36:305–312. https://doi.org/10.1016/j.tins.2013.01.005

Gareau MG, Jury J, MacQueen G, Sherman PM, Perdue MH (2007) Probiotic treatment of rat pups normalizes corticosterone release and ameliorates colonic dysfunction induced by maternal separation. Gut 56(11):1522–1528. https://doi.org/10.1136/gut.2006.117176

George F, Daniel C, Thomas M, Singer E, Guilbaud A, Tessier FJ et al (2018 Nov) Occurrence and dynamism of lactic acid bacteria in distinct ecological niches: a multifaceted functional health perspective. Front Microbiol 27(9):2899

Ghodsi SZ, Orawa H, Zouboulis CC (2009) Prevalence, severity, and severity risk factors of acne in high school pupils: a community-based study. J Invest Dermatol 129(9):2136–2141. https://doi. org/10.1038/jid.2009.47

Grant MC, Baker JS (2017) An overview of the effect of probiotics and exercise on mood and associated health conditions. Crit Rev Food Sci Nutr 57(18):3887–3893

Grenham S, Clarke G, Cryan JF, Dinan TG (2011) Brain-gut-microbe communication in health and disease. Front Physiol 2(94):1–15. https://doi.org/10.3389/fphys.2011.00094

Groeger D, O'Mahony L, Murphy EF, Bourke JF, Dinan TG, Kiely B et al (2013) *Bifidobacterium infantis* 35624 modulates host inflammatory processes beyond the gut. Gut Microbes 4:325–339. https://doi.org/10.4161/gmic.25487

Guéniche A, Bastien P, Ovigne JM, Kermici M, Courchay G, Chevalier V et al (2010a) *Bifidobacterium longum* lysate, a new ingredient for reactive skin. Exp Dermatol 19(8):e1–e8. https://doi.org/10.1111/j.1600-0625.2009.00932.x

Guéniche A, Benyacoub J, Philippe D, Bastien P, Kusy N, Breton L et al (2010b) Lactobacillus paracasei CNCM I-2116 (ST11) inhibits substance P-induced skin inflammation and accelerates skin barrier function recovery in vitro. Eur J Dermatol 20:731–737. https://doi.org/10.1684/ejd. 2010.1108

Hadian Y, Fregoso D, Nguyen C, Bagood MD, Dahle SE, Gareau MG et al (2020) Microbiome-skin-brain axis: a novel paradigm for cutaneous wounds. Wound Repair Regen 28(3):282–292. https://doi.org/10.1111/wrr.12800

Hansen JE, Jespersen LK (2010) Wound or tissue dressing comprising lactic acid bacteria. 10 June United States patent application US 12/519,548. 2010 Jun 10

Harrison S, Geppetti P (2001) Substance P. Int J Biochem Cell Biol 33(6):555–576. https://doi.org/ 10.1016/S1357-2725(01)00031-0

Hawk PB, Knowles FC, Rehfuss ME, Clarke JA, Bergeim O, Fishback HR et al (1917) The use of baker's yeast in diseases of the skin and of the gastrointestinal tract. JAMA 69(15):1243–1247. https://doi.org/10.1001/jama.1917.02590420035009

Hemarajata P, Versalovic J (2013) Effects of probiotics on gut microbiota: mechanisms of intestinal immunomodulation and neuro modulation. Ther Adv Gastroenterol 6(1):39–51. https://doi.org/ 10.1177/1756283X12459294

Herpfer I, Katzev M, Feige B, Fiebich BL, Voderholzer U, Lieb K (2007) Effects of substance P on memory and mood in healthy male subjects. Hum Psychopharmacol 22:567–573. https://doi.org/10.1002/hup.876

Herrera V, Parsonnet J (2009) Helicobacter pylori and gastric adenocarcinoma. Clin Microbiol Infect 15(11):971–976. https://doi.org/10.1111/j.1469-0691.2009.03031.x

Hill C, Guarner F, Reid G, Gibson GR, Merenstein DJ, Pot B et al (2014) Expert consensus document: the International Scientific Association for Probiotics and Prebiotics consensus statement on the scope and appropriate use of the term probiotic. Nat Rev Gastroenterol Hepatol 11(8):506–514. https://doi.org/10.1038/nrgastro.2014.66

Hooper LV, Littman DR, Macpherson AJ (2012) Interactions between the microbiota and the immune system. Science 336(6086):1268–1273. https://doi.org/10.1126/science.1223490

Huang CW, Tzeng JN, Chen YJ, Tsai WF, Chen CC, Sun WH (2007) Nociceptors of dorsal root ganglion express proton-sensing G-protein-coupled receptors. Mol Cell Neurosci 36(2): 195–210. https://doi.org/10.1016/j.mcn.2007.06.010. PMID 17720533

Huang YJ, Marsland BJ, Bunyavanich S, O'Mahoney L, Leung DYM, Muraro A et al (2017) The microbiome in allergic disease: current understanding and future opportunities – 2017 PRACTALL document of the American Academy of Allergy, Asthma & Immunology and the European Academy of Allergy and Clinical Immunology. J Allergy Clin Immunol 139(4): 1099–1110. https://doi.org/10.1016/j.jaci.2017.02.007

Iordache F, Iordache C, Chifiriuc MC, Bleotu C, Pavel M, Smarandache D (2008) Antimicrobial and immunomodulatory activity of some probiotic fractions with potential clinical application. Archiva Zootechnica 11(3):41–51

Jung GW, Tse JE, Guiha I, Rao J (2013) Prospective, randomized, open-label trial comparing the safety, efficacy, and tolerability of an acne treatment regimen with and without a probiotic supplement and minocycline in subjects with mild to moderate acne. J Cutan Med Surg 17(2): 114–122. https://doi.org/10.2310/7750.2012.12026

Kang CM. Nyayapathy S, Lee JY, Suh JW, Husson RN (2008) Wag31, a homologue of the cell division protein DivIVA, regulates growth, morphology and polar cell wall synthesis in mycobacteria. Microbiology 154(3):725–735. https://doi.org/10.1099/mic.0.2007/014076-0

Kim HJ, Kim YJ, Kang MJ, Seo JH, Kim HY, Jeong SK et al (2012) A novel mouse model of atopic dermatitis with epicutaneous allergen sensitization and the effect of Lactobacillus rhamnosus. Exp Dermatol 21(9):672–675. https://doi.org/10.1111/j.1600-0625.2012.01539.x

Kim HM, Lee DE, Park SD, Kim Y, Kim YJ, Jeong JW et al (2014) Oral administration of Lactobacillus plantarum HY7714 protects hairless mouse against ultraviolet B-induced photo-aging. J Microbiol Biotechnol 24(11):1583–1591. https://doi.org/10.4014/jmb.1406.06038

Kleerebezem M, Vaughan EE (2009) Probiotic and gut lactobacilli and bifidobacteria: molecular approaches to study diversity and activity. Annu Rev Microbiol 63:269–290

Korn T, Bettelli E, Oukka M, Kuchroo VK (2009) IL-17 and Th17 cells. Annu Rev Immunol 27: 485–517

Kosiewicz MM, Zirnheld AL, Alard P (2011) Gut microbiota, immunity, and disease: a complex relationship. Front Microbiol 2:180. https://doi.org/10.3389/fmicb.2011.00180

Krutmann J (2009) Pre- and probiotics for human skin. J Dermatol Sci 54(1):1–5. https://doi.org/10.1016/j.jdermsci.2009.01.002

Kumar M, Yadav AK, Verma V, Singh B, Mal G, Nagpal R, Hemalatha R (2016) Bioengineered probiotics as a new hope for health and diseases: potential and prospects: an overview. Future Microbiol 11(4):585–600. https://doi.org/10.2217/fmb.16.4

Lakhan SE, Kirchgessner A (2010) Gut inflammation in chronic fatigue syndrome. Nutr Metab 7(79):1–10. https://doi.org/10.1186/1743-7075-7-79

Lallès JP (2020) Intestinal alkaline phosphatase in the gastrointestinal tract of fish: biology, ontogeny, and environmental and nutritional modulation. Rev Aquac 12:555. https://doi.org/10.1111/raq.12340

Lee DE, Huh C, Ra J, Choi I, Jeong J, Kim S (2015) Clinical evidence of effects of *Lactobacillus plantarum* HY7714 on skin aging: a randomized, double blind, placebo-controlled study. J Microbiol Biotechnol 25:2160–2168. https://doi.org/10.4014/jmb.1509.09021

Lee S, Yoon J, Kim Y, Jeong D, Park S, Kang D (2016) Therapeutic effect of tyndallized *Lactobacillus rhamnosus* IDCC 3201 on atopic dermatitis mediated by down-regulation of immunoglobulin E in NC/Nga mice. Microbiol Immunol 60:468–476. https://doi.org/10.1111/1348-0421.12390

Lee SY, Lee E, Park YM, Hong SJ (2018) Microbiome in the gut-skin axis in atopic dermatitis. Allergy, Asthma Immunol Res 10(4):354–362. https://doi.org/10.4168/aair.2018.10.4.354

Lew LC, Liong MT (2013) Bioactives from probiotics for dermal health: functions and benefits. J Appl Microbiol 114(5):1241–1253. https://doi.org/10.1186/1757-4749-3-1

Lieb K, Walden J, Grunze H, Fiebich BL, Berger M, Normann C (2004) Serum levels of substance P and response to antidepressant pharmacotherapy. Pharmacopsychiatry 37:238–239

Liu KF, Liu XR, Li GL, Lu SP, Jin L, Wu J (2016) Oral administration of Lactococcus lactis-expressing heat shock protein 65 and tandemly repeated IA2P2 prevents type 1 diabetes in NOD mice. Immunol Lett 1(174):28–36

Livingston M, Loach D, Wilson M, Tannock GW, Baird M (2010) Gut commensal *lactobacillus reuteri* 100-23 stimulates an immune regulatory response. Immunol Cell Biol 88(1):99–102. https://doi.org/10.1016/j.mehy.2004.08.019

Logan AC, Katzman M (2005) Major depressive disorder: probiotics may be an adjuvant therapy. Med Hypotheses 64:533–538. https://doi.org/10.1016/j.mehy.2004.08.019

Logan AC, Venket RA, Irani D (2003) Chronic fatigue syndrome: lactic acid bacteria may be of therapeutic value. Med Hypotheses 60:915–923. https://doi.org/10.1016/s0306-9877(03)00096-3

Lorenzo LE, Godin AG, Wang F, St-Louis M, Carbonetto S, Wiseman PW et al (2014) Gephyrin clusters are absent from small diameter primary afferent terminals despite the presence of GABAA receptors. J Neurosci 34(24):8300–8317. https://doi.org/10.1523/jneurosci.0159-14.2014

Loser K, Beissert S (2012) Regulatory T cells: banned cells for decades. J Invest Dermatol 132:864–871. https://doi.org/10.1038/jid.2011.375

Lv LX, Fang DQ, Shi D, Chen DY (2016) Alterations and correlations of the gut microbiome, metabolism and immunity in patients with primary biliary cirrhosis. Environ Microbiol 18:2272–2286. https://doi.org/10.1111/1462-2920.13401

Majamaa H, Isolauri E (1997) Probiotics: a novel approach in the management of food allergy. J Allergy Clin Immunol 99:179–185

Mansour NM, Abdelaziz SA (2016) Oral immunization of mice with engineered *Lactobacillus gasseri* NM713 strain expressing *Streptococcus pyogenes* M6 antigen. Microbiol Immunol 60:527–532

Martin CR, Osadchiy V, Kalani A, Mayer EA (2018) The brain-gut-microbiome axis. Cell Mol Gastroenterol Hepatol 6(2):133–148. https://doi.org/10.1016/j.jcmgh.2018.04.003

Matsumoto M, Kibe R, Ooga T, Aiba Y, Sawaki E, Koga Y et al (2013) Cerebral low-molecular metabolites influenced by intestinal microbiota: a pilot study. Front Syst Neurosci 7(9):1–19. https://doi.org/10.3389/fnsys.2013.00009

Maxmen A (2017) Living therapeutics: scientists genetically modify bacteria to deliver drugs. Nat Med 23:5–7

Mayer EA, Knight R, Mazmanian SK (2014) Gut microbes and the brain: paradigm shift in neuroscience. J Neurosci 34:15490–15496

Meijer K, de Vos P, Priebe MG (2010) Butyrate and other short-chain fatty acids as modulators of immunity: what relevance for health? Curr Opin Clin Nutr Metab Care 13:715–721. https://doi.org/10.1097/MCO.0b013e32833eebe5

Messaoudi M, Lalonde R, Violle N, Javelot H, Desor D, Nejdi A et al (2011) Assessment of psychotropic-like properties of a probiotic formulation (*Lactobacillus helveticus* R0052 and

Bifidobacterium longum R0175) in rats and human subjects. Br J Nutr 105(5):775–764. https://doi.org/10.1017/S0007114510004319

Mohammadi AA, Jazayeri S, Khosravi-Darani K (2016) The effects of probiotics on mental health and hypothalamic–pituitary–adrenal axis: a randomized, double-blind, placebo-controlled trial in petrochemical workers. Nutr Neurosci 19:387–395

Muizzuddin N, Maher W, Sullivan M, Schnittger S, Mammone T (2012) Physiological effect of a probiotic on skin. J Cosmet Sci 63(6):385–395

Myers SP, Hawrelak JA (2004 Jun) The causes of intestinal dysbiosis: a review. Altern Med Rev 9(2):180–197

N'Diaye A, Gannesen A, Borrel V (2017) Substance P and calcitonin gene-related peptide: key regulators of cutaneous microbiota homeostasis. Front Endocrinol 8:15. https://doi.org/10.3389/fendo.2017.00015

Nakano Y (2004) Stress-induced modulation of skin immune function: two types of antigen-presenting cells in the epidermis are differentially regulated by chronic stress. Br J Dermatol 151(1):50–64. https://doi.org/10.1111/j.1365-2133.2004.05980.x

Neufeld KM, Kang N, Bienenstock J, Foster JA (2011) Reduced anxiety-like behavior and central neurochemical change in germ-free mice. Neurogastroenterol Motil 23(3):255–264. https://doi.org/10.1111/j.1365-2982.2010.01620.x

Nguyen TT, Hathaway H, Kosciolek T, Knight R, Jeste DV (2019) Gut microbiome in serious mental illnesses: a systematic review and critical evaluation. Schizophr Res 234:24. https://doi.org/10.1016/j.schres.2019.08.026

O'Connor TM, O'Connell J, O'Brien DI, Goode T, Bredin CP, Shanahan F (2004) The role of substance P in inflammatory disease. J Cell Physiol 201(2):167–180. https://doi.org/10.1002/jcp.20061

O'Neill CA, Monteleone G, McLaughlin JT, Paus R (2016) The gut-skin axis in health and disease: a paradigm with therapeutic implications. BioEssays 38:1167–1176. https://doi.org/10.1002/bies.201600008

Ohland CL, Kish L, Bell H (2013) Effects of *Lactobacillus helveticus* on murine behavior are dependent on diet and genotype and correlate with alterations in the gut microbiome. Psychoneuroendocrinology 38:1738–1747

Ouwehand AC, Båtsman A, Salminen S (2003) Probiotics for the skin: a new area of potential application? Lett Appl Microbiol 36:327–331

Parodi A, Paolino S, Greco A, Drago F, Mansi C, Rebora A (2008) Small intestinal bacterial overgrowth in rosacea: clinical effectiveness of its eradication. Clin Gastroenterol Hepatol 6(7):759–764

Paton AW, Morana R, Paton JC (2012) Bioengineered microbes in disease therapy. Trends Mol Med 18:417–425

Pavlovic S, Daniltchenko M, Tobin DJ (2008) Further exploring the brain-skin connection: stress worsens dermatitis via substance P-dependent neurogenic inflammation in mice. J Invest Dermatol 128(2):434–446. https://doi.org/10.1038/sj.jid.5701079

Peguet-Navarro J, Dezutter-Dambuyant C, Buetler TM, Leclaire J, Smola H, Blum S (2008) Supplementation with oral probiotic bacteria protects human cutaneous immune homeostasis after UV exposure – double blind, randomized, placebo controlled clinical trial. Eur J Dermatol 18:504–511. https://doi.org/10.1684/ejd.2008.0496

Peterson J, Garges S, Giovanni M, McInnes P, Wang L, Schloss JA et al (2009) The NIH human microbiome project. Genome Res 19(12):2317–2323. https://doi.org/10.1101/gr.096651.109

Philippe D, Blum S, Benyacoub J (2011) Oral *lactobacillus paracasei* improves skin barrier function recovery and reduces local skin inflammation. Eur J Dermatol 21:279–280

Pimentel M, Chow EJ, Lin HC (2000) Eradication of small intestinal bacterial overgrowth reduces symptoms of irritable bowel syndrome. Am J Gastroenterol 95(12):3503–3506. https://doi.org/10.1111/j.1572-0241.2000.03368.x

Purves D, Augustine GJ, Fitzpatrick D et al (eds) (2001) Neuroscience, 2nd edn. Sinauer Associates, Sunderland, MA. https://www.ncbi.nlm.nih.gov/books/NBK10799

Rao AV, Bested AC, Beaulne TM, Katzman MA, Iorio C, Berardi JM et al (2009) A randomized, double-blind, placebo-controlled pilot study of a probiotic in emotional symptoms of chronic fatigue syndrome. Gut Pathog 1(1):1–6. https://doi.org/10.1186/1757-4749-16

Rather IA, Bajpai VK, Kumar S, Lim J, Paek WK, Park YH (2016) Probiotics and atopic dermatitis: an overview. Front Microbiol 7:507. https://doi.org/10.3389/fmicb.2016.00507

Rattanaprasert M, van Pijkeren JP, Ramer-Tait AE, Quintero M, Kok CR, Walter J et al (2019) Genes involved in galactooligosaccharide metabolism in Lactobacillus reuteri and their ecological role in the gastrointestinal tract. Appl Environ Microbiol 85(22):e01788. https://doi.org/10.1128/aem.01788-19

Rhee SH, Pothoulakis C, Mayer EA (2009) Principles and clinical implications of the brain–gut–enteric microbiota axis. Nat Rev Gastroenterol Hepatol 6:306–314. https://doi.org/10.1038/nrgastro.2009.35

Rodríguez-Nogales A, Algieri F, Garrido-Mesa J, Vezza T, Utrilla MP, Chueca N et al (2018) The administration of Escherichia coli Nissle 1917 ameliorates development of DSS-induced colitis in mice. Front Pharmacol 9:468. https://doi.org/10.3389/fphar.2018.00468

Roudsari MR, Karimi R, Sohrabvandi S, Mortazavian AM (2015) Health effects of probiotics on the skin. Crit Rev Food Sci Nutr 55(9):1219–1240. https://doi.org/10.1080/10408398.2012.680078

Rusu E, Enache G, Cursaru R, Alexescu A, Radu R, Onila O et al (2019) Prebiotics and probiotics in atopic dermatitis. Exp Ther Med 18(2):926–931. https://doi.org/10.3892/etm.2019.7678

Salvucci E (2016) Microbiome, holobiont and the net of life. Crit Rev Microbiol 42:485–494. https://doi.org/10.3109/1040841X.2014.962478

Samuelson DR, Welsh DA, Shellito JE (2015) Regulation of lung immunity and host defense by the intestinal microbiota. Front Microbiol 6:1085. https://doi.org/10.3389/fmicb.2015.01085

Sánchez B, Delgado S, Blanco-Míguez A, Lourenço A, Gueimonde M, Margolles A (2017) Probiotics, gut microbiota, and their influence on host health and disease. Mol Nutr Food Res 61:1600240. https://doi.org/10.1002/mnfr.201600240

Sandrini S, Aldriwesh M, Alruways M, Freestone P (2015) Microbial endocrinology: host-bacteria communication within the gut microbiome. J Endocrinol 225(2):R21–R34. https://doi.org/10.1530/joe-14-0615

Sarao LK, Arora M (2017) Probiotics, prebiotics, and microencapsulation: a review. Crit Rev Food Sci Nutr 57:344–371. https://doi.org/10.1080/10408398.2014.887055

Saunders AM (1924) The Bacillus acidophilus treatment. Institut Q 15:85–88

Schiffrin EJ, Parlesak A, Bode C, Bode JC, van't Hof MA, Grathwohl D et al (2009) Probiotic yogurt in the elderly with intestinal bacterial overgrowth: endotoxaemia and innate immune functions. Br J Nutr 101:961–966

Schwarz A, Bruhs A, Schwarz T (2017) The short-chain fatty acid sodium butyrate functions as a regulator of the skin immune system. J Investig Dermatol 1:855–864. https://doi.org/10.1016/j.jid.2016.11.014

Sharma K, Sultana T, Liao M, Dahms T, Dillon JAR (2020) EF1025, a hypothetical protein from Enterococcus faecalis, interacts with DivIVA and affects cell length and cell shape. Front Microbiol 11:83. https://doi.org/10.3389/fmicb.2020.00083

Shu M, Wang Y, Yu J, Kuo S, Coda A, Jiang Y (2013) Fermentation of Propionibacterium acnes, a commensal bacterium in the human skin microbiome, as skin probiotics against methicillin-resistant Staphylococcus aureus. PLoS One 8:e55380. https://doi.org/10.1371/journal.pone.0055380

Sinha A, Sagar S, Osborne J (2019) Probiotic bacteria in wound healing; An in-vivo study. Iran J Biotechnol 17(4):e2188. https://doi.org/10.30498/ijb.2019.85188

Smith CJ, Emge JR, Berzins K (2014) Probiotics normalize the gut-brain-microbiota axis in immunodeficient mice. Am J Physiol Gastrointest Liv Physiol 307:G793–G802

Spahn TW, Kucharzik T (2004) Modulating the intestinal immune system: the role of lymphotoxin and GALT organs. Gut 53(3):456–465. https://doi.org/10.1136/gut.2003.023671

Stilling RM, Ryan FJ, Hoban AE (2015) Microbes & neurodevelopment-absence of microbiota during early life increases activity-related transcriptional pathways in the amygdala. Brain Behav Immun 50:209–220. https://doi.org/10.1016/j.bbi.2015.07.009

Stokes JH, Pillsbury DM (1930) The effect on the skin of emotional and nervous states: III. Theoretical and practical consideration of a gastro-intestinal mechanism. Arch Dermatol Syphilol 22:962–993. https://doi.org/10.1001/archderm.1930.01440180008002

Sudo N, Chida Y, Aiba Y, Sonoda J, Oyama N, Yu XN et al (2004) Postnatal microbial colonization programs the hypothalamic-pituitary-adrenal system for stress response in mice. J Physiol 558: 263–275. https://doi.org/10.1113/jphysiol.2004.063388

Teitelman G, Joh TH, Reis DJ (1981) Linkage of the brain-skin-gut axis: islet cells originate from dopaminergic precursors. Peptides 2(Suppl 2):157–168. https://doi.org/10.1016/0196-9781(81) 90026-7

Thursby E, Juge N (2017) Introduction to the human gut microbiota. Biochem J 474(11): 1823–1836. https://doi.org/10.1042/BCJ20160510

Tillisch K, Labus J, Kilpatrick L, Jiang ZG, Stains J, Ebrat B et al (2013) Consumption of fermented milk product with probiotic modulates brain activity. Gastroenterology 144(7):1394–1401. https://doi.org/10.1053/j.gastro.2013.02.043

Tomasello G, Mazzola M, Leone A, Sinagra E, Zummo G, Farina F, Damiani P, Cappello F, Gerges Geagea A, Jurjus A, Bou AT (2016) Nutrition, oxidative stress and intestinal dysbiosis: influence of diet on gut microbiota in inflammatory bowel diseases. Biomed Pap Med Fac Univ Palacky Olomouc Czech Repub 160(4):461–466

Trudeau F, Gilbert K, Tremblay A, Tompkins TA, Godbout R, Rousseau G (2019) *Bifidobacterium longum* R0175 attenuates post-myocardial infarction depressive-like behaviour in rats. PLoS One 14(4):e0215101. https://doi.org/10.1371/journal.pone.0215101

Verdú EF, Bercik P, Verma-Gandhu M, Huang XX, Blennerhassett P, Jackson W et al (2006) Specific probiotic therapy attenuates antibiotic induced visceral hypersensitivity in mice. Gut 55:182–190. https://doi.org/10.1136/gut.2005.066100

Volkova LA, Khalif IL, Kabanova IN (2001) Impact of the impaired intestinal microflora on the course of acne vulgaris. Klinicheskaia Meditsina (Mosk) 79(6):39–41

Von Ossowski I, Pietilä TE, Rintahaka J, Nummenmaa E, Mäkinen VM, Reunanen J et al (2013) Using recombinant *Lactococci* as an approach to dissect the immunomodulating capacity of surface piliation in probiotic *Lactobacillus rhamnosus* GG. PLoS One 8(5):e64416. https://doi.org/10.1371/journal.pone.0064416

WHO (2017). https://www.who.int/news-room/detail/27-02-2017-who-publishes-list-of-bacteria-for-which-new-antibiotics-are-urgently-needed. Accessed 6 Apr 2020

Yang YJ, Sheu BS (2016) Metabolic interaction of Helicobacter pylori infection and gut microbiota. Microorganisms 4(1):15. https://doi.org/10.3390/microorganisms4010015

Yeşilova Y, Çalka Ö, Akdeniz N, Berktaş M (2012) Effect of probiotics on the treatment of children with atopic dermatitis. Ann Dermatol 24(2):189–193. https://doi.org/10.5021/ad.2012.24.2.189

Therapeutic Mechanisms of Gut Microbiota and Probiotics in the Management of Mental Disorders

Monu Yadav, Anil Kumar, and Amarjeet Shandil

Abstract

Probiotics are live microorganisms which work for the human benefit by improving their intestinal microbial balance. Originally they were used to enhance both animals and humans health by modulating intestinal microbiota. There is a biological connection between probiotics and brain as they can communicate via neurotransmitters system, anti-oxidative defence mechanism and neuroinflammatory pathways. Probiotic bacteria are also involved in the production of neuroactive molecules that act on the brain–gut axis. Probiotic treatments that help to improve mood, anxiety and strengthen memory using them in the form of food or supplements to alter the gut microbiota and treat psychiatric conditions are considered as psychobiotics. Dietary ingestion such as prebiotics, probiotics and polyphenol can influence gut microbiota composition. Dysbiosis of the gut microbiota is associated with brain dysfunctions. Regulation of microbiota by probiotics and prebiotics may help to restore gut equilibrium. The impact of nutrients on microbiota composition strengthens the reports that regulating a therapeutic microbiota is essential for a healthy brain. They could be useful as novel therapeutics to protect the brain from neurodegeneration. But

M. Yadav
Department of Pharmacy, School of Medical and Allied Sciences, GD Goenka University, Gurgaon, India

University Institute of Pharmaceutical Sciences, UGC Centre of Advanced Studies (UGC-CAS), Panjab University, Chandigarh, India

A. Kumar (✉)
University Institute of Pharmaceutical Sciences, UGC Centre of Advanced Studies (UGC-CAS), Panjab University, Chandigarh, India

A. Shandil
Department of Pharmaceutical Sciences, Chaudhary Bansi Lal University, Bhiwani, Haryana, India

© Springer Nature Singapore Pte Ltd. 2022
P. K. Deol, S. K. Sandhu (eds.), *Probiotic Research in Therapeutics*,
https://doi.org/10.1007/978-981-16-6760-2_3

research is still needed, mainly clinical and translational studies to determine pharmacokinetics and pharmacodynamics of probiotics.

Keywords

Gut microbiota · Probiotics · Psychobiotic · Neuropsychiatric disorders · Mental health

3.1   Introduction

The human gut comprises 10^{14} microorganisms, which have been observed to be involved in many biological processes such as energy balance, immunomodulation and stimulation of the enteric nervous system (ENS) (Gill et al. 2006; Hooper et al. 2012). Microbiota affects the development of brain health, diabetes, gastrointestinal malignancy, obesity, liver diseases, infectious diseases and allergies (Novik and Savich 2020). Logan et al. proposed several characteristics of probiotics that can positively influence the activity and function of brain (Logan et al. 2003; Logan and Katzman 2005). Dysbiosis of the gut microbiota is associated with neuropsychiatric disorders. Low level of specific microbiota is supposed to be entailed in the development of various mental problems such as anxiety, autism disorder, bipolar disorder, cognitive dysfunction, depression, Parkinson's disorders, schizophrenia (Sarkar et al. 2016; Nguyen et al. 2018). Dinan et al. (2013) modified the terminology of probiotics into psychobiotics that propose their therapeutic role to improve mental health (Fig. 3.1) (Dinan et al. 2013). In preclinical animal studies, many probiotics or psychobiotics has been investigated against anxiety, depression and memory (Misra and Mohanty 2019). Probiotic is reported to control several neurotransmitters such as acetylcholine, dopamine, gamma-aminobutyric acid,

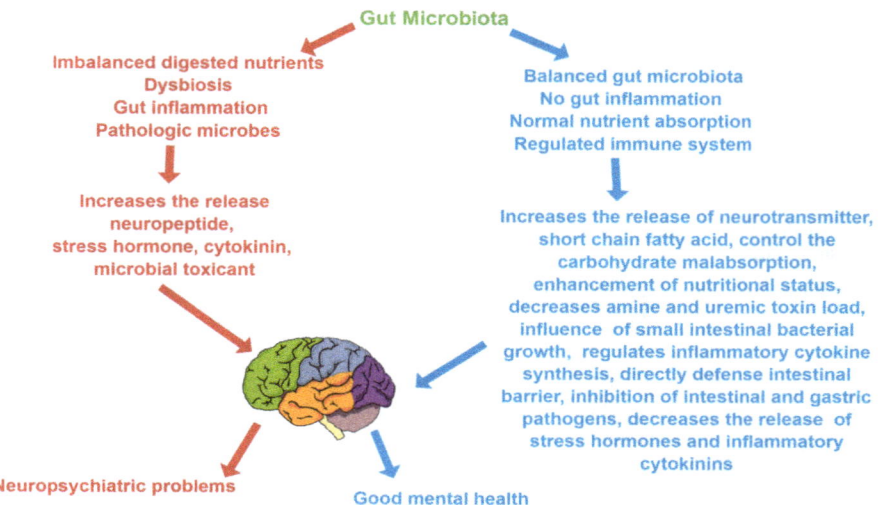

Fig. 3.1 Possible characteristics of gut microbiota

glutamate and serotonin which regulate neural-inhibitory-excitatory mechanisms and cognitive system (Sudo et al. 2004; Lu et al. 2008). Biologically there is a strong connection between the microbiota in hypothalamic-pituitary-adrenal (HPA) axis. It has been studied that acute restraint stress increases the adrenocorticotropic and corticosterone release. Moreover, overstimulated proinflammatory cytokines stimulate HPA, increase the penetrability of the blood–brain barrier (BBB) and reduce serotonin level, which lead to cause depression-like symptoms (Dowlati et al. 2010). Strains such as *Bifidobacterium dentium, Lactobacillus plantarum* and *Lactobacillus brevis* produce GABA and serotonin. Moreover, *Lactobacillus plantarum* and *Lactobacillus odontolyticus* increase the level of acetylcholine (Barrett et al. 2012). It has also been seen that spore-forming bacteria help in the synthesis of gut serotonin from gut enterochromaffin cells. Some intestinal bacterial species are directly linked with the central nervous system (CNS) and can affect neurotransmission by the immune and endocrine system as well as vagal sensory nerve fibres (Borre et al. 2014; Smith and Vale 2006). CNS may influence the gut microbiota by increasing the synthesis of signalling chemical into gut lumen from neurons, enterochromaffin cells and immune cells in the lamina propria which may change motility of the colon and permeability of the intestine (Alonso et al. 2011). Both physical and psychological stress can influence the gastrointestinal tract. Stress mediators can influence any physiological functions of the gastrointestinal tract including mucosal permeability and the functioning of the intestinal barrier, visceral sensitivity and release of neuroendocrine factors which turn to cause the imbalance of gut microbiota and deteriorate the host immunity. This process may lead to stimulating the immune modulators which may worsen the depressive and psychiatric symptoms (Lutgendorff et al. 2008; Codling et al. 2010). Additionally, it has been purposed that long term exposure to stress can produce detrimental changes in HPA and increase the corticotropin-releasing factor (CRF) in cerebrospinal fluid and plasma cortisol of depression vulnerable patients by modulating the physiological function of the immune system (Arborelius et al. 1999). CRF is a neurotransmitter and a peptide hormone that contributes to the response of stress by activating the pathways for the synthesis of adrenocorticotropic hormone (ACTH) (Hauger et al. 2006). Increased CRF level in cerebrospinal fluid and plasma cortisol in the samples of patients with depression has been allocated to the continuous vagal sensory nerve fibre arbitrated information resulting in frequent stimulation of feedback mechanism between forebrain and noradrenergic locus coeruleus neurons that contribute to the response of stress and produce CRF which lead to change neural activity in depression and anxiety (Chyun et al. 1984; Tache and Bonaz 2007). Frequent stimulation of locus coeruleus also increases the concentration of norepinephrine that affects the sleep-wake cycle, body balance, emotions, attention, memory and cognitive system. Additionally, norepinephrine has been found to increase the propagation of pathogens such as *Escherichia coli, Pseudomonas aeruginosa* and *Yersinia enterocolitica* in the gut (Soderholm and Perdue II 2001). Moreover, increased levels of cortisol inhibit the production of T-cells, which may facilitate the growth of pathogens, latent viruses, microbial infections, and the synthesis of proinflammatory cytokines that lead to the activation of the hypothalamic-pituitary-adrenal axis. CRF is also important for the release of colonic mucus and cyclooxygenase-2 by activating the mast cells (Elsenbruch et al. 2010). These reports support the theory that suggests the involvement of stress on gut microbiota in mental disorders. There are many probiotics that have been studied against various neuropsychiatric disorders. Functional flavonoids have been reported to control the metabolism of gut microbiota (Selhub et al. 2014). Dietary polyphenols are considered as xenobiotics for humans but their availability is low as compared to macro- and micro-nutrients. Furthermore, structural complexity also influences their absorption in small intestine. Only 5–10% of the ingested polyphenol gets absorbed in the small intestine and the rest of polyphenols can accumulate in the large intestine and are exposed to the gut microbial enzymatic activities. The small intestine absorbs small percentage of dietary polyphenols after de-conjugation. Polyphenolics having less complexity can pass by biotransformation in the enterocytes and then in the hepatocytes by phase I and phase II reactions. These reactions yield a chain of water-soluble conjugated metabolites which are released in the systemic circulation for consequent distribution to other organs and elimination by the urine. Unabsorbed polyphenols catabolize in the large intestine by the colonic bacterial enzymes and generate metabolites having various physiological effects. Colonic microflora can convert the polyphenols into bioactive compounds, which influence the intestinal biology and human health. Preclinical and clinical evidences suggested that polyphenolics may alter the gut microflora status. Figure 3.2 highlights the possible characteristics of gut microbiota.

3.2   Interaction of Probiotics with Neurotransmitter System

Glutamine is the main precursor of glutamate which is a precursor of GABA. Pharmacologically, low level of these neurotransmitters in the hippocampus and prefrontal cortex (PFC) of the brain has been found to be involved in the pathophysiology of various mental disorders (Rasheed and Alghasham 2012). In an in vitro study, increased level of these neurotransmitters hippocampus and PFC was observed with the ingestion of *L. rhamnosus* (Floyd 1999). Moreover, supplementation of *L. rhamnosus* has also found to decrease anxiety and depression-like symptoms in mice by affecting central mRNA expression of GABA-A and GABA-B receptors. *L. rhamnosus* synthesized glutamate and GABA from microbial glutaminase and glutamate decarboxylase, respectively (Janik et al. 2016). In vitro has also been suggested that gut microbial GABA may cross the intestinal barrier through H^+/GABA symporter, which further interacts with GABA transporters and receptors that are expressed on vagus afferents and enteric neurons. Intriguingly, administration of *L. rhamnosus* reverses the stress-induced imbalance in corticosterone in mice probably due to desensitization of GABAergic synapses by stress (Janik et al. 2016; Bravo et al. 2011). A decrease in GABA signalling causes the release of CRF, resulting in cortisol oversecretion, which may lead to an increase in the activity of the HPA axis. Dysfunctional GABA receptors also retard the neurogenesis in the hippocampus region of the brain and induce cognitive decline, anxiety and depressive behaviours in mice, which may possibly recover with the treatment of

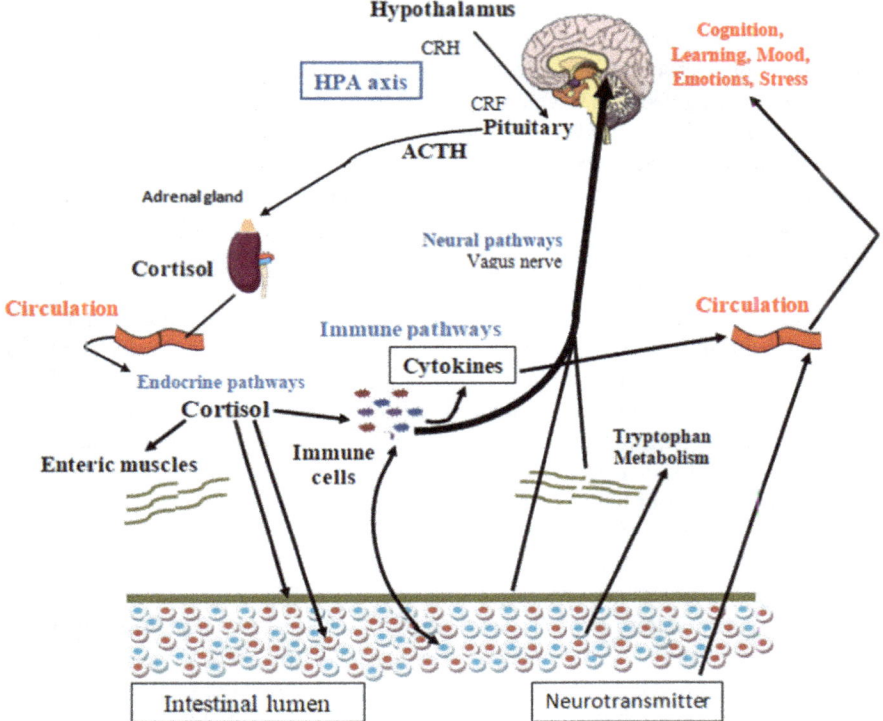

Fig. 3.2 Potential mechanisms of gut microbiota in neuroprotection

L. rhamnosus. Similarly, *L. brevis*-fermented milk also showed to increase GABA concentration and produced synergistic anti-depressive effective when given along with fluoxetine in rats (McVey Neufeld et al. 2018). Interestingly, administration of *L. brevis* improved sleep duration in mice. Moreover, treatment of *L. helveticus* enhanced cognitive function in stressed rats, probably may increase the hippocampal norepinephrine (NE) and plasma IL-10 levels and decrease the adrenocorticotropic hormone (ACTH) and plasma corticosterone levels (Ko et al. 2013; Liang et al. 2015). A preclinical study has reported that consumption of *L. helveticus* increased cognition and reduced gut inflammation in neuroinflammed rats. While another report suggested that cognitive improvement in *L. helveticus* treated mice did not correlate with gut inflammation. Due to the dilemma, it is documented that the HPA axis and hippocampal NE system interact with glucose metabolism in the hippocampus. This mechanism can be influenced by microbial NE as *L. helveticus* released NE that increases in the human bloodstream (Maehata et al. 2019). In an in vivo, it has also been observed that gut-bacteria convert conjugated NE into its neuroactive form. Serotonergic signalling is considered as one of the important factors involved in the pathophysiology of anxiety and mood disorders. Altered levels of 5-HT have been reported in the striatum and hippocampus, which suggests an association between the microbiota and serotonergic signalling. *L. helveticus* has increased hippocampal serotonin levels in depressed rats similar to citalopram (Ko et al. 2013). The gut microbiota generates biologically active serotonin by deconjugating glucuronide-conjugated 5-HT which acts via 5-HT-3A receptors of enteric neurons by activating glial cell-derived neurotrophic factor of intestinal epithelial cells (IECs). Thus, it can be hypothesized that *L. helveticus* affects the central 5-HT circuitry by the neural route. Maehata et al. (2019) also reported that intake of *L. helveticus* raised the expression of 5-HT1A receptors especially in the nucleus accumbens and improve depressive behaviours of mice (Liang et al. 2015). *L. plantarum* administration has shown increased dopamine and its metabolites in the PFC and striatum region of mice brain and alleviated depression and anxiety-like behaviours (Oleskin et al. 2014). It was also suggested that *L. plantarum* increases the dopamine concentration in the PFC to reduce the hyperactivation of the HPA axis (Liu et al. 2015). Dopamine neurons in the ventral tegmental area and PFC regulate reward system behaviours by forming the mesocortical pathway and HPA axis. *L. plantarum* intake is reported to reduce depressive behaviours in mice by reducing glucocorticoid-induced monoamine oxidases (MAOs) activity in the brain tissues (Sullivan and Dufresne 2006; Feenstra et al. 1992; Grunewald et al. 2012). Treatment with *L. plantarum* PS128 (PS128) decreased the anxiety and depression by reducing the corticosterone and inflammation levels (Oleskin et al. 2014). Furthermore, administration of PS128 was also found effective to increase dopamine and serotonin levels in the striatum and PFC region of the brain. *L. helveticus* NS8 has also been investigated to treat anxiety, cognitive dysfunction and depression by increasing the BDNF, norepinephrine and serotonin levels in the hippocampal brain. *B. longum* 1714 and *L. rhamnosus* (JB-1) were found to be potential to treat anxiety and depression (Ko et al. 2013). Especially, JB-1 strain can lead to altering the expression of the GABA receptor in the brain and decreases the level of plasma corticosterone. Similarly, *Bifidobacterium longum* NCC3001 upregulates the expression of hippocampal BDNF and potential against anxiety. In a clinical study, administration of *Bifidobacterium longum* 1714 for 4 weeks showed a significant increase in memory and reduced stress (Butts and Phillips 2013). Probiotic in yoghurt (*Lactobacillus acidophilus* LA5 and *Bifidobacterium lactis* BB12) and probiotic capsules (*Bifidobacterium breve, Bifidobacterium longum, Lactobacillus casei, Lactobacillus acidophilus, Lactobacillus rhamnosus, Lactobacillus bulgaricus* and *Streptococcus thermophiles*) in clinical voluntaries have been observed to improve their mental health by decreasing the anxiety and depression and increasing memory. Furthermore, a combination of *L. helveticus* R0052 along with *B. longum* R0175 was found to decrease anxiety and depression in healthy voluntaries as compared to the control (Allen et al. 2016). Moreover, this combination therapy was the potential to reduce the urinary cortisol level resulting in an improvement of the symptoms of anxiety and depression. There are various probiotics of *Lactobacillus strain* that are found to regulate neuronal functions by maintaining the neurotransmitter system and may be beneficial to manage mental health (Table 3.1).

Table 3.1 The neurotransmitters produced by *Lactobacillus* strain probiotics and their regulatory functions

Probiotics	Neurotransmitter	Regulatory function
L. brevis	Gamma-aminobutyric acid	Hippocampal neurogenesis
L. bulgaricus		HPA axis regulation
L. casei		Sleep
L. helveticus		
L. paracasei		
L. plantarum		
L. reuteri		
L. rhamnosus		
L. bulgaricus	Norepinephrine	Aggression
L. casei		Cognitive function
L. helveticus		HPA axis regulation
		Mood
		Sleep
		Sympathetic activity
L. bulgaricus	Dopamine	Mood
L. casei		Motivation and pleasure
L. helveticus		Psychomotor function
L. plantarum		
L. bulgaricus	Glutamate	Gastrointestinal reflexes
L. casei		HPA axis regulation
L. helveticus		Intestinal motility
L. paracasei		Mood
L. plantarum		Neuronal excitement
L. reuteri		
L. rhamnosus		
L. helveticus	Serotonin	Aggression
L. plantarum		Appetite
		HPA axis regulation
		Impulsivity
		Memory
		Mood
		Sleep
L. plantarum	Acetylcholine	HPA axis regulation
		Memory
		Mood
		Sleep
		Synaptic plasticity
L. plantarum	Histamine	Appetite
L. reuteri		Memory
		Mood
		Motivation
		Sleep
		Sympathetic activity

3.3   Interaction of Probiotics with Metabolism and Production of Fatty Acids

Lipids and fatty acids are required in the physiological functioning and development of the brain. Gut microbiota increases the hydrolysis of proteins and lipids, decreases the releases of immunogenic peptides and raises the level of amino acids and fatty acids, thus stimulates the synthesis of various neurotransmitters (Messaoudi et al. 2011; Roy et al. 2006; MacFabe 2012). Strain-specific effect of probiotic bacterial influences the metabolic rates of membrane lipids that leads to change the neuronal sensitivities. *Bifidobacterium* and *Lactobacillus* bacteria produce the short-chain fatty acids such as acetic, butyric and propionic acids and conjugated linolenic, linoleic poly-unsaturated fatty acids which are required for the development of brain and cognition (Marrone and Coccurello 2020; Owada 2008). All these short-chain fatty acids possess excellent anti-inflammatory effects; moreover, acetate is also entailed in the cholesterol synthesis. Cholesterol is the main component for the development of dendrite and synapse formation. Omega-3 fatty acid, a poly-unsaturated fatty acid, is essential for neurogenesis, neurotransmission and cognitive consolidation. Linoleic acid is reported to increase the level of cholecystokinin and plasma glucagon-like peptide-1 and peptide YY (El-Ansary and Al-Ayadhi 2014). These peptides regulate the digestive mechanisms by secreting digestive enzymes and alter the food preference and total intake of calories by the vagal nerve and midbrain hypothalamic neural pathways. On aged rats, a combination of *Bifidobacterium lactis*, *Lactobacillus delbrueckii* subsp. *Rhamnosus*, and inulin was tested to evaluate how it affected intestinal microbes as well as plasma neuropeptide Y and peptide YY levels. The combination decreased neuropeptide Y levels but had no effect on peptide YY levels (Ogawa et al. 2012). This hypothesis clearly demonstrates that microbial released peptides can lead to decrease in the stress-induced depression. Moreover, cells having neuropeptide Y have been observed to be high in elders and patients with constipated as compared to healthy subjects. Peptide YY and neuropeptide Y have shown an inhibitory effect on the release of pancreatic enzyme. Neuropeptide Y has also been found to modify the gastrointestinal smooth muscle contraction and its motility (Bermúdez-Humarán et al. 2019; Hellström 1987). Furthermore, some reports showed the prevalence of anxiety and stress like symptoms in patients with constipation. Therefore, fatty acids synthesis and lipid metabolism by probiotics are found to decrease psychiatric problems (Hansen 2003).

3.4   Interaction of Probiotics with Inflammatory and Antioxidant System

Oxidative species play a pathophysiological role in psychiatric disorders by causing cellular damage and apoptosis (Gayathri and Rashmi 2017). An increased level of protein carbonylation (an oxidative species) has been found in the depression-prone subjects. Endogenous antioxidants are found to be therapeutically effective to reduce oxidative stress by scavenging and eliminating oxidative species from the cells

(Maes et al. 1993; Taverniti and Guglielmetti 2011; Tsilingiri and Rescigno 2013). Now researchers are also giving attention to anti-oxidants due to their potential effect against oxidative species induced ageing-related degenerative disorders. Probiotics have shown anti-oxidant property as they are significant to produce endogenous enzymes, bioactive peptides and vitamins. *Bifidobacterium animalis*, *Lactobacillus acidophilus* and *Lactococcus brevis* are proven to hold anti-oxidant property. Chronic administration of the formulation of these probiotics to rats showed to decrease doxorubicin-induced oxidative stress. *Bifidobacterial* and *Lactobacilli* are observed to produce ascorbic acid, vitamin K and vitamin B. Biotin is also significant for the metabolism of amino acids and production of fatty acids, which are important for the production of neurotransmitters (Kim et al. 2018). *Lactococcus lactis*, *Lactobacillus plantarum*, *Leuconostoc* sp. and *Streptococcus thermophiles* are also known to produce carotenoids, folate and anti-oxidative enzymes such as glutathione reductase and superoxide dismutase; hence, it would be worthwhile to use probiotics as adjuvant therapy to treat stress-induced psychiatric disorders (de LeBlanc et al. 2011; Pompei et al. 2007). Specific hormones, neurotransmitters and inflammatory cytokines like IL-1, IL-6 and TNF-α promote communication between immune cells and brain cells mediate neurogenesis, synaptic modelling, neuronal integrity and memory formation. Stress, infections and injuries increase the prostaglandins, nitric oxide (NO), reactive oxidative stress (ROS) and inflammatory cytokines in the brain by stimulating microglial cells found in the macrophages that express the chemokine receptors which lead to cause the neuron damage. Lipopolysaccharide (LPS) released by gram-negative bacteria including *Haemophilus influenza* and *Neisseria meningitides* triggers various ROS and inflammatory cytokines in the microglial cells and are involved in the neuronal damage leading to psychiatric disorders such as depression. There are various evidences demonstrating the pathophysiological role of oxidative species in patients with depression. Increased oxidative species and decreased endogenous antioxidants such as coenzyme Q10, omega-3 and vitamins A, B, C and E in the serum of depressed patients have also been observed (Saduakhasova et al. 2013). Therefore, it is important to maintain homeostasis of our body, it is essential to strengthen physical and mental health, as the involvement of gut microorganisms and inflammatory cytokines and ROS in depression has been elucidated. The gut peptides participate in gut–brain communication from intestinal microorganisms and enteroendocrine cells. Gut microbiota affects the production of peptides and also involved to control epithelial cells and gastrointestinal endocrine. These microorganisms also secrete specific peptides which structurally and functionally imitate the peptide hormones. Moreover, the ingested nutrients influence the gut microbiota as well as alter the release of gut peptides such as cholecystokinin, galanin, gastrin, ghrelin, leptin, orexin, peptide YY and neuropeptide Y which finally change the hormone response. The reports on administration of probiotic *Pediococcus acidilactici* to piglets reported increased levels of calcitonin and galanin gene-related peptide immune-reactive neurons in the submucosal plexus ganglia of the ileum. Impairment of kynurenine/tryptophan pathway contributes to the pathogenesis of various mental disorders like depression (Gautam et al. 2012). Increased levels of plasma

Table 3.2 Potential role of probiotics in mental health

Probiotics	Physiological changes	Behavioural changes	References
Bifidobacterium infantis	• Decreased stress-induced increase in plasma IL-6 and corticotrophin-releasing factor mRNA expression in the amygdala	Anti-depression	Ganesh et al. (2018), Desbonnet et al. (2010)
	• Increased stress-induced decreased in norepinephrine levels in the brainstem		
	• Increased alpha diversity of gut microbiota		
	• Increased glutamatergic synapse		
Bifidobacterium breve	• Increased the expression of Tph1 mRNA in RIN14B cells	Anti-depression	Tian et al. (2019), Desbonnet et al. (2010)
	• Increased BDNF levels in the coretex	Anti-anxiety	
	• Increased stress-induced decreased in alpha diversity of the gut microbiota		
	• Increased glutamatergic synapse		
	• Increased phenylalanine/tyrosine/TRP biosynthesis		
Clostridium butyricum	• Increased stress-induced decreased in brain 5-HT and BDNF levels	Anti-depression	Savignac et al. (2014), Sun et al. (2018)
	• Increased stress-induced decreased in intestinal GLP-1 secretion and cerebral expression of GLP-1 receptor		
Lactobacillus rhamnosus	• Increased the expression of GABAAa2 and GABAB1b mRNA in the cortical regions	Anti-anxiety	Janik et al. (2016), Bravo et al. (2011)
	• Recovery toward basal corticosterone levels	Anti-depression	
		Memory improvement	
Lactobacillus reuteri	• Decreased stress-induced increase in intestinal IDO1 expression	Anti-anxiety	Miyaoka et al. (2018), Marin et al. (2017)
	• Decreased stress-induced increased in KYN levels	Anti-depression	
	• Decreased stress-induced increased in activated microglia infiltration into the cortex		
Lactobacillus paracasei	• Increased stress-induced decreased in BDNF levels in the cortex	Reduced age-related	Jang et al. (2019), Wei et al. (2019)
	• Decreased in MAOA activity	Anhedonia	
	• Increased 5-HT levels in brain	Anti-anxiety	
	• Increased DA, 5-HT and 5-HIAA levels in the coretex and striatum	Cognitive improvement	
		Anti-depression	

(continued)

Table 3.2 (continued)

Probiotics	Physiological changes	Behavioural changes	References
Lactobacillus kefiranofaciens	Decreased stress-induced increased in serum CORT levels and KYN/TRP ratio	Anti-depression	Savignac et al. (2014)
		Anti-anhedonia	

kynurenine/tryptophan have been seen in depression patients. Supplement treated with *L. reuteri* reduced stress-induced depression behaviour in mice. Administration of kynurenine reduced anti-depressive effect *L. reuteri* by producing H_2O_2 which catalyses peroxidase facilitated reactions that inhibit indoleamine 2,3-dioxygenase (IDO) activity in immune cells leading to decrease in the circulating kynurenine (Réus et al. 2015; Thomas et al. 2012; Gao et al. 2015). Various stains are reported to have protective effective by managing inflammatory system (Table 3.2).

3.5 Interaction of Probiotics with Histamine, Diacylglycerol Kinase and Brain-Derived Neurotrophic Factor (BDNF) Expression

Lactobacillus reuteri has histidine decarboxylase which converts L-histidine to histamine that suppresses the synthesis of TNF-α. The animal study found that microbial histamine reduces the activity of proinflammatory cytokines while increasing the histamine-2 receptor signalling pathway. Moreover, *Lactobacillus reuteri* was also found to produce diacylglycerol kinase that metabolized diacylglycerol into phosphatidic acid and showed anti-inflammatory effect (Gao et al. 2015). Oral administration of *Lactobacillus reuteri* was reported to reduce stress-induced anxiety and depression in mice probably by increasing the expression of the hippocampal BDNF. Ketamine and selected serotonin reuptake inhibitors (SSRIs) also increase expression of BDNF in the hippocampus region and produce anti-depressant action (Corpuz et al. 2018).

Acknowledgement The authors are thankful for the financial support of SERB-DST (New Delhi) Letter number PDF/2018/002605 and University Institute of Pharmaceutical Sciences (UIPS) for the infrastructure facilities to carry out this work.

References

Allen AP, Hutch W, Borre YE, Kennedy PJ, Temko A, Boylan G, Murphy E, Cryan JF, Dinan TG, Clarke G (2016) Bifidobacterium longum 1714 as a translational psychobiotic: modulation of stress, electrophysiology and neurocognition in healthy volunteers. Transl Psychiatry 6(11):1–7
Alonso C, Vicario M, Santos J, Azpiroz F (2011) Stress, intestinal inflammation, and irritable bowel syndrome. Ann Gastroenterol Hepatol:1–10

Arborelius L, Owens MJ, Plotsky PM, Nemeroff CB (1999) The role of corticotropin-releasing factor in depression and anxiety disorders. J Endocrinol 160(1):1–2

Barrett E, Ross RP, O'Toole PW, Fitzgerald GF, Stanton C (2012) Gamma-aminobutyric acid production by culturable bacteria from the human intestine. J Appl Microbiol 113(2):411–417

Bermúdez-Humarán LG, Salinas E, Ortiz GG, Ramirez-Jirano LJ, Morales JA, Bitzer-Quintero OK (2019) From probiotics to psychobiotics: live beneficial bacteria which act on the brain-gut axis. Nutrients 11(4):890

Borre YE, O'Keeffe GW, Clarke G, Stanton C, Dinan TG, Cryan JF (2014) Microbiota and neurodevelopmental windows: implications for brain disorders. Trends Mol Med 20(9): 509–518

Bravo JA, Forsythe P, Chew MV, Escaravage E, Savignac HM, Dinan TG, Bienenstock J, Cryan JF (2011) Ingestion of *Lactobacillus* strain regulates emotional behavior and central GABA receptor expression in a mouse via the vagus nerve. Proc Natl Acad Sci 108(38):16050–16055

Butts KA, Phillips AG (2013) Glucocorticoid receptors in the prefrontal cortex regulate dopamine efflux to stress via descending glutamatergic feedback to the ventral tegmental area. Int J Neuropsychopharmacol 16(8):1799–1807

Chyun YS, Kream BE, Raisz LG (1984) Cortisol decreases bone formation by inhibiting periosteal cell proliferation. Endocrinology 114(2):477–480

Codling C, O'Mahony L, Shanahan F, Quigley EM, Marchesi JR (2010) A molecular analysis of fecal and mucosal bacterial communities in irritable bowel syndrome. Dig Dis Sci 55(2): 392–397

Corpuz HM, Ichikawa S, Arimura M, Mihara T, Kumagai T, Mitani T, Nakamura S, Katayama S (2018) Long-term diet supplementation with *Lactobacillus paracasei* K71 prevents age-related cognitive decline in senescence-accelerated mouse prone 8. Nutrients 10(6):762

Desbonnet L, Garrett L, Clarke G, Kiely B, Cryan JF, Dinan TG (2010) Effects of the probiotic *Bifidobacterium infantis* in the maternal separation model of depression. Neuroscience 170(4): 1179–1188

Dinan TG, Stanton C, Cryan JF (2013) Psychobiotics: a novel class of psychotropic. Biol Psychiatry 74:720–726

Dowlati Y, Herrmann N, Swardfager W, Liu H, Sham L, Reim EK, Lanctot KL (2010) A meta-analysis of cytokines in major depression. Biol Psychiatry 67:446–457

El-Ansary A, Al-Ayadhi L (2014) Relative abundance of short chain and polyunsaturated fatty acids in propionic acid-induced autistic features in rat pups as potential markers in autism. Lipids Health Dis 13(1):140

Elsenbruch S, Rosenberger C, Enck P, Forsting M, Schedlowski M, Gizewski ER (2010) Affective disturbances modulate the neural processing of visceral pain stimuli in irritable bowel syndrome: an fMRI study. Gut 59(4):489–495

Feenstra MG, Kalsbeek A, Van Galen H (1992) Neonatal lesions of the ventral tegmental area affect monoaminergic responses to stress in the medial prefrontal cortex and other dopamine projection areas in adulthood. Brain Res 596(1–2):169–182

Floyd RA (1999) Antioxidants, oxidative stress, and degenerative neurological disorders. Proc Soc Exp Biol Med 222(3):236–245

Ganesh BP, Hall A, Ayyaswamy S, Nelson JW, Fultz R, Major A, Haag A, Esparza M, Lugo M, Venable S, Whary M (2018) Diacylglycerol kinase synthesized by commensal *Lactobacillus reuteri* diminishes protein-kinase C phosphorylation and histamine-mediated signaling in the mammalian intestinal epithelium. Mucosal Immunol 11(2):380–393

Gao C, Major A, Rendon D, Lugo M, Jackson V, Shi Z, Mori-Akiyama Y, Versalovic J (2015) Histamine H2 receptor-mediated suppression of intestinal inflammation by probiotic Lactobacillus reuteri. Microbiology 6(6):e01358–e01315

Gautam M, Agrawal M, Gautam M, Sharma P, Gautam AS, Gautam S (2012) Role of antioxidants in generalised anxiety disorder and depression. Indian J Psychiatry 54(3):244

Gayathri D, Rashmi BS (2017) Mechanism of development of depression and probiotics as adjuvant therapy for its prevention and management. Mental Health Prevent 5:40–51

Gill SR, Pop M, Deboy RT, Eckburg PB, Turnbaugh PJ, Samuel BS, Jeffrey IG, David AR, Claire MF, Karen EN (2006) Metagenomic analysis of the human distal gut microbiome. Science 312: 1355–1359

Grunewald M, Johnson S, Lu D, Wang Z, Lomberk G, Albert PR, Stockmeier CA, Meyer JH, Urrutia R, Miczek KA, Austin MC (2012) Mechanistic role for a novel glucocorticoid-KLF11 (TIEG2) protein pathway in stress-induced monoamine oxidase A expression. J Biol Chem 287(29):24195–24206

Hansen MB (2003) Neurohumoral control of gastrointestinal motility. Physiol Res 52(1):1–30

Hauger RL, Risbrough V, Brauns O, Dautzenberg FM (2006) Corticotropin releasing factor (CRF) receptor signaling in the central nervous system: new molecular targets. CNS Neurologic Disord Drug Targets 5(4):453–479

Hellström PM (1987) Mechanisms involved in colonic vasoconstriction and inhibition of motility induced by neuropeptide Y. Acta Physiol Scand 129(4):549–556

Hooper LV, Littman DR, Macpherson AJ (2012) Interactions between the microbiota and the immune system. Science 336:1268–1273

Jang HM, Lee KE, Kim DH (2019) The preventive and curative effects of Lactobacillus reuteri NK33 and Bifidobacterium adolescentis NK98 on immobilization stress-induced anxiety/ depression and colitis in mice. Nutrients 11(4):819

Janik R, Thomason LA, Stanisz AM, Forsythe P, Bienenstock J, Stanisz GJ (2016) Magnetic resonance spectroscopy reveals oral Lactobacillus promotion of increases in brain GABA, N-acetyl aspartate and glutamate. NeuroImage 125:988–995

Kim GY, Lim JS, Lee SP (2018) Fortification of γ-aminobutyric acid and bioactive compounds in whey by co-fermentation using Bacillus subtilis and Lactobacillus plantarum. Kr J Food Sci Technol 50(6):572–580

Ko CY, Lin HT, Tsai GJ (2013) Gamma-aminobutyric acid production in black soybean milk by Lactobacillus brevis FPA 3709 and the antidepressant effect of the fermented product on a forced swimming rat model. Process Biochem 48(4):559–568

de LeBlanc AM, Del Carmen S, Zurita-Turk M, Santos Rocha C, Van de Guchte M, Azevedo V, Miyoshi A, LeBlanc JG (2011) Importance of IL-10 modulation by probiotic microorganisms in gastrointestinal inflammatory diseases. ISRN Gastroenterol 2011:892971

Liang S, Wang T, Hu X, Luo J, Li W, Wu X, Duan Y, Jin F (2015) Administration of Lactobacillus helveticus NS8 improves behavioral, cognitive, and biochemical aberrations caused by chronic restraint stress. Neuroscience 310:561–577

Liu J, Sun J, Wang F, Yu X, Ling Z, Li H, Zhang H, Jin J, Chen W, Pang M, Yu J (2015) Neuroprotective effects of Clostridium butyricum against vascular dementia in mice via metabolic butyrate. Biomed Res Int 2015:1–12

Logan AC, Katzman M (2005) Major depressive disorder: probiotics may be an adjuvant therapy. Med Hypotheses 64(3):533–538

Logan AC, Rao AV, Irani D (2003) Chronic fatigue syndrome: lactic acid bacteria may be of therapeutic value. Med Hypotheses 60(6):915–923

Lu Y, Christian K, Lu B (2008) BDNF: a key regulator for protein synthesis-dependent LTP and long-term memory? Neurobiol Learn Mem 89:312–323

Lutgendorff F, Akkermans L, Soderholm JD (2008) The role of microbiota and probiotics in stress-induced gastrointestinal damage. Curr Mol Med 8(4):282–298

MacFabe DF (2012) Short-chain fatty acid fermentation products of the gut microbiome: implications in autism spectrum disorders. Microb Ecol Health Dis 23(1):19260

Maehata H, Kobayashi Y, Mitsuyama E, Kawase T, Kuhara T, Xiao JZ, Tsukahara T, Toyoda A (2019) Heat-killed Lactobacillus helveticus strain MCC1848 confers resilience to anxiety or depression-like symptoms caused by subchronic social defeat stress in mice. Biosci Biotechnol Biochem 83(7):1239–1247

Maes M, Meltzer HY, Scharpè S, Bosmans E, Suy E, De Meester I, Calabrese J, Cosyns P (1993) Relationships between lower plasma L-tryptophan levels and immune-inflammatory variables in depression. Psychiatry Res 49(2):151–165

Marin IA, Goertz JE, Ren T, Rich SS, Onengut-Gumuscu S, Farber E, Wu M, Overall CC, Kipnis J, Gaultier A (2017) Microbiota alteration is associated with the development of stress-induced despair behavior. Sci Rep 7:43859

Marrone MC, Coccurello R (2020) Dietary fatty acids and microbiota-brain communication in neuropsychiatric diseases. Biomolecules 10(1):12

McVey Neufeld KA, Kay S, Bienenstock J (2018) Mouse strain affects behavioral and neuroendocrine stress responses following administration of probiotic *Lactobacillus rhamnosus* JB-1 or traditional antidepressant fluoxetine. Front Neurosci 8(12):294

Messaoudi M, Lalonde R, Violle N, Javelot H, Desor D, Nejdi A, Bisson JF, Rougeot C, Pichelin M, Cazaubiel M, Cazaubiel JM (2011) Assessment of psychotropic-like properties of a probiotic formulation (*Lactobacillus helveticus* R0052 and *Bifidobacterium longum* R0175) in rats and human subjects. Br J Nutr 105(5):755–764

Misra S, Mohanty D (2019) Psychobiotics: a new approach for treating mental illness? Crit Rev Food Sci Nutr 59(8):1230–1236

Miyaoka T, Kanayama M, Wake R, Hashioka S, Hayashida M, Nagahama M, Okazaki S, Yamashita S, Miura S, Miki H, Matsuda H (2018) *Clostridium butyricum* MIYAIRI 588 as adjunctive therapy for treatment-resistant major depressive disorder: a prospective open-label trial. Clin Neuropharmacol 41(5):151–155

Nguyen TT, Kosciolek T, Eyler LT, Knight R, Jeste DV (2018) Overview and systematic review of studies of microbiome in schizophrenia and bipolar disorder. J Psychiatr Res 99:50–61

Novik G, Savich V (2020) Beneficial microbiota, probiotics and pharmaceutical products in functional nutrition and medicine. Microbes Infect 22(1):8–18

Ogawa N, Ito M, Yamaguchi H, Shiuchi T, Okamoto S, Wakitani K, Minokoshi Y, Nakazato M (2012) Intestinal fatty acid infusion modulates food preference as well as calorie intake via the vagal nerve and midbrain-hypothalamic neural pathways in rats. Metabolism 61(9):1312–1320

Oleskin AV, Zhilenkova OG, Shenderov BA, Amerhanova AM, Kudrin VS, Klodt PM (2014) Lactic-acid bacteria supplement fermented dairy products with human behavior-modifying neuroactive compounds. J Pharm Nutr Sci 4:199–206

Owada Y (2008) Fatty acid binding protein: localization and functional significance in the brain. Tohoku J Exp Med 214(3):213–220

Pompei A, Cordisco L, Amaretti A, Zanoni S, Matteuzzi D, Rossi M (2007) Folate production by bifidobacteria as a potential probiotic property. Appl Environ Microbiol 73(1):179–185

Rasheed N, Alghasham A (2012) Central dopaminergic system and its implications in stress-mediated neurological disorders and gastric ulcers: short review. Adv Pharmacol Sci 2012:1–11

Réus GZ, Jansen K, Titus S, Carvalho AF, Gabbay V, Quevedo J (2015) Kynurenine pathway dysfunction in the pathophysiology and treatment of depression: evidences from animal and human studies. J Psychiatr Res 68:316–328

Roy CC, Kien CL, Bouthillier L, Levy E (2006) Short-chain fatty acids: ready for prime time? Nutr Clin Pract 21(4):351–366

Saduakhasova S, Kushugulova A, Kozhakhmetov S, Shakhabayeva G, Tynybayeva I, Nurgozhin T, Zhumadilov Z (2013) Antioxidant activity of the probiotic consortium *in vitro*. Centr Asian J Glob Health 2:115

Sarkar A, Lehto SM, Harty S, Dinan TG, Cryan JF, Burnet PWJ (2016) Psychobiotics and the manipulation of bacteria-gut-brain signals. Trends Neurosci 39:763–781

Savignac HM, Kiely B, Dinan TG, Cryan JF (2014) *Bifidobacteria* exert strain-specific effects on stress-related behavior and physiology in BALB/c mice. Neurogastroenterol Motil 26(11):1615–1627

Selhub EM, Logan AC, Bested AC (2014) Fermented foods, microbiota, and mental health: ancient practice meets nutritional psychiatry. J Physiol Anthropol 33(1):1–12

Smith SM, Vale WW (2006) The role of the hypothalamic-pituitary-adrenal axis in neuroendocrine responses to stress. Dialogues Clin Neurosci 8(4):383–395

Soderholm JD, Perdue MH II (2001) Stress and intestinal barrier function. Am J Physiol Gastrointest Liv Physiol 280(1):G7–G13

Sudo N, Chida Y, Aiba Y, Sonoda J, Oyama N, Yu XN et al (2004) Postnatal microbial colonization programs the hypothalamic-pituitary-adrenal system for stress response in mice. J Physiol 558: 263–275

Sullivan RM, Dufresne MM (2006) Mesocortical dopamine and HPA axis regulation: role of laterality and early environment. Brain Res 1076(1):49–59

Sun J, Wang F, Hu X, Yang C, Xu H, Yao Y, Liu J (2018) *Clostridium butyricum* attenuates chronic unpredictable mild stress-induced depressive-like behavior in mice via the gut-brain axis. J Agric Food Chem 66(31):8415–8421

Tache Y, Bonaz B (2007) Corticotropin-releasing factor receptors and stress-related alterations of gut motor function. J Clin Invest 117(1):33–40

Taverniti V, Guglielmetti S (2011) The immunomodulatory properties of probiotic microorganisms beyond their viability (ghost probiotics: proposal of paraprobiotic concept). Genes Nutr 6(3): 261–274

Thomas CM, Hong T, Van Pijkeren JP, Hemarajata P, Trinh DV, Hu W, Britton RA, Kalkum M, Versalovic J (2012) Histamine derived from probiotic *Lactobacillus reuteri* suppresses TNF via modulation of PKA and ERK signaling. PLoS One 7(2):e31951

Tian P, Wang G, Zhao J, Zhang H, Chen W (2019) *Bifidobacterium* with the role of 5-hydroxytryptophan synthesis regulation alleviates the symptom of depression and related microbiota dysbiosis. J Nutr Biochem 66:43–51

Tsilingiri K, Rescigno M (2013) Postbiotics: what else? Benefic Microbes 4(1):101–107

Wei CL, Wang S, Yen JT, Cheng YF, Liao CL, Hsu CC, Wu CC, Tsai YC (2019) Antidepressant-like activities of live and heat-killed Lactobacillus *paracasei* PS23 in chronic corticosterone-treated mice and possible mechanisms. Brain Res 1711:202–213

Pathophysiological Role of Gut Microbiota Affecting Gut–Brain Axis and Intervention of Probiotics and Prebiotics in Autism Spectrum Disorder

4

Firdosh Shah and Mitesh Dwivedi ⓘ

Abstract

The human gut harbors up to 100 trillion of microorganisms out of which only a few microorganisms could be considered as probiotics. The alteration in the gut microbial composition can cause a considerable effect on central physiology as well as act on behavioral or the physiological phenotype of an individual. Presently, various studies on probiotics have increased our interest in the role played by the brain–gut–microbiota axis in diverse neurodevelopmental disorders such as obesity, depression, and autism. Although the exact etiology and pathology of autism spectrum disorder (ASD) are not understood well; several studies indicate the link between gut microbiota and neurodevelopmental disorders in offspring, including ASD. Certain intestinal bacteria such as *Clostridium* and *Sutterella* genera have been reported in abundance in ASD condition, suggesting their crucial role in manifestation of ASD. Similarly, there are also preclinical evidences which mark that the supplementation of probiotics and/or prebiotics results in an improvement in behavior of the children with ASD. The chapter will discuss different mechanisms involved in manifestation and amelioration of ASD by gut microbiota and probiotics–prebiotics, respectively. Therefore the chapter will provide all the available preclinical and animal studies for depicting the current clinical picture of autism by representing the relationship between gut microbiota, probiotics, prebiotics, and their role in ASD.

Keywords

Autism spectrum disorder (ASD) · Gut microbiota · Probiotics · Prebiotics · Gastrointestinal symptoms

F. Shah · M. Dwivedi (✉)
C. G. Bhakta Institute of Biotechnology, Faculty of Science, Uka Tarsadia University, Bardoli, Gujarat, India

© Springer Nature Singapore Pte Ltd. 2022
P. K. Deol, S. K. Sandhu (eds.), *Probiotic Research in Therapeutics*,
https://doi.org/10.1007/978-981-16-6760-2_4

Abbreviations

5-HT	5-hydroxytryptamine
AGEs	Advanced glycation end products
ASD	Autism spectrum disorder
ATEC	Autism treatment evaluation checklist
BDNF	Brain-derived neurotrophic factor
CHARGE	Childhood autism risks from genetics and environment
CLDN	Claudin
CNS	Central nervous system
DBC-P	Development behavior checklist
ENS	Enteric nervous system
FAST	Seizure-prone rats
FMT	Fecal microbiota transplantation
FOS	Fructooligosaccharides
GF	Germ-free
GI	Gastrointestinal
GOS	Galactooligosaccharides
HPA	Hypothalamic-pituitary axis
IL	Interleukin
ISAPP	International scientific association for probiotics and prebiotics
LPS	Lipopolysaccharide
MTT	Microbiota transfer therapy
PCR	Polymerase chain reaction
PDD-NOS	Pervasive developmental disorder-not otherwise specified
RAGE	Receptors for advanced glycation end products
SCFAs	Short chain fatty acids
SERT	Serotonin transporter variants
SLOW	Seizure-resistant rats
SPF	Specific pathogen free
TGF-β	Transforming growth factor-β
TLR	Toll-like receptor
TNF-α	Tumor necrosis factor-α

4.1   Introduction

Autism spectrum disorder (ASD) is a neurodevelopmental condition that causes communication and social interaction deficits in an individual. The symptom of this disorder starts to develop from the early age of 2–3 years of a child. Since 1996 Centers for Disease Control and Prevention (CDC) has been continuously keeping track on the prevalence of ASD. According to CDC, among 1000 children of age 8 years there are 18.5% of children who suffer from ASD (Maenner et al. 2020). It has been estimated that among every 68 children, one is affected by ASD and the rate of boys being affected by ASD is 4 times more compared to that of girls (Rosenfeld 2015). A few of the most common symptoms of ASD include repetitive body movements, challenges in verbal and nonverbal communication, uneven cognitive abilities, sensory impairments, and issue in expressing emotions. In addition to these symptoms, sleeping cycle dysfunction, immune dysregulation, gastrointestinal (GI) metabolic disorder, mitochondrial dysfunction, and neuroinflammation have also been suggested (Reddy and Saier 2015). Even though, the majority of the symptoms in autistic individuals remain similar but greatly differ in terms of the severity of the disease (Kanner 1943). There is an enormous amount of social, emotional as well as financial burdens on the families of such children.

Although various efforts have been put forward in past years in order to elucidate the exact cause behind the ASD still its pathophysiology remains largely obscure. So far, studies on ASD demonstrate that there are many factors including genetic predisposition and environmental factors that contribute to the etiology of ASD (Trottier et al. 1999; Persico and Napolioni 2013; Casanova 2007). There are many theories that have been proposed based on the cellular and molecular neurobiology in order to elucidate the pathogenesis of ASD but currently, it remains debatable in terms of its relevance with the condition (DiCicco-Bloom et al. 2006).

Several studies suggested the role of gut microbiota in the pathophysiology of ASD. The concept of microbial dysbiosis is not new for us, since it has been reported to be responsible for etiology and pathogenesis of various neurological conditions such as multiple sclerosis, Alzheimer's disease, Parkinson's disease, etc. (Adamczyk-Sowa et al. 2017; Mancuso and Santangelo 2018; Scheperjans et al. 2015). In all of these stated diseases, there is a major impact on all of the clinical entities due to the different compositions of microbiota. It has been shown that gut bacteria or their metabolic end products are considered to play an indispensable role in the physiological activities such as depression, mood, cognitive behavior, and brain development (Tuohy et al. 2015). Among all the symptoms of ASD, GI disorders such as stomachache, diarrhea, and constipation are the most occurring symptoms. It has been often seen that change within the microbial composition of GI results in the permeable gut which further ends up with septicemia or neurodevelopmental disorders such as ASD (Reddy and Saier 2015).

Despite the fact that research is carried forward in the direction of ASD, still the underlying mechanism for the cause of GI or the medical problem leading to the impairment in verbal communication within the individuals with ASD remained a major concern. In order to overcome or manage these health issues the alternative treatments such as probiotics, prebiotics, dietary supplements are brought into the picture (Buie 2015).

An important key component responsible for microbial activities and neurochemical processes is the diet of an individual. *Bifidobacterium*, *Lactobacillus* is the useful gut bacteria that produce short chain fatty acids (SCFAs) that affect mucosal integrity, immune function, and generation of bio-active compounds (Tuohy et al. 2015). Although there are very few positive shreds of evidence that supports the role of diet in ASD, nonetheless recent research is concentrated on the configuration of diet for the processes taking place in the gut microbiota–brain axis. Therefore, in the future, diet could be considered as a remedy in order to combat the neurodevelopmental disorders caused in ASD (Tuohy et al. 2015).

As far as treatments are considered for ASD there are various therapies such as applied behavior analysis, pivotal response treatment, cognitive behavior therapy, and dietary approaches which improve the symptoms of ASD. But in spite of various therapies, there is not a single treatment that can completely cure the disease. According to FDA, there are only two drugs, namely Risperidone and Aripiprazole which have been approved in order to use as a medication for symptoms associated with ASD (DeFilippis and Wagner 2016). Therefore, alternative treatment regimens are needed for ASD.

This chapter will discuss the role of gut microbiota in pathogenesis of ASD through animal and human clinical trial studies. In addition, the role of probiotics and prebiotics in amelioration of ASD and the different treatments which are under the pipelines or about to emerge in the future as an effective therapeutic intervention for ASD will also be discussed.

4.2   Relationship Link Between Brain–Gut–Microbiota Axis

The bidirectional neurohumoral communication system is also referred to as the gut–brain axis interaction and this communication between the brain and the gut is represented in Fig. 4.1. In most of the previous studies related to the impact of the gut–brain axis, the major focus was given to the GI syndromes (Sanger and Lee 2008). According to recent pieces of evidence it has been stated the gut bacteria have a potential to modulate brain development and bring about change in behavioral characteristics through the gut–brain axis (Diaz Heijtz et al. 2011). For instance, the central nervous system (CNS) controls the gut microbiota through peptides (Romijn et al. 2008; Cummings and Overduin 2007). It affects nutrient availability when peptides reach the level of satiation. Even in the case of the hypothalamic-pituitary-adrenal axis, the regulation of intestinal motility and integrity is controlled by the amount of cortisol released (Wang and Kasper 2014). Numerous studies so far suggest that the gut–brain axis plays an important role in the pathogenesis of ASD by impacting on neuroendocrine, neuroimmune, and autonomic nervous systems of the brain and thus modulating its function (Grenham et al. 2011; Mayer 2011). There are millions of neurons in the mucosa of the gastrointestinal tract, which constitute the enteric nervous system (ENS) and regulate gastrointestinal functions. Henceforth, the gut is also referred to as a "second brain."

The key factor underlying the relationships between the gut and the ASD is the increased permeability of the intestinal tract of ASD individuals, also called as "leaky gut" (Quigley 2016). Previous studies on ASD animal models depicted disruption in GI barrier effects on brain function which permits entry of the toxins and the bacterial products into the bloodstream (Onore et al. 2012; Hsiao et al. 2013). For example, study by Emanuele et al. (2010) depicts an association between gut microbiota and impaired social behavior due to an increase in the lipopolysaccharide

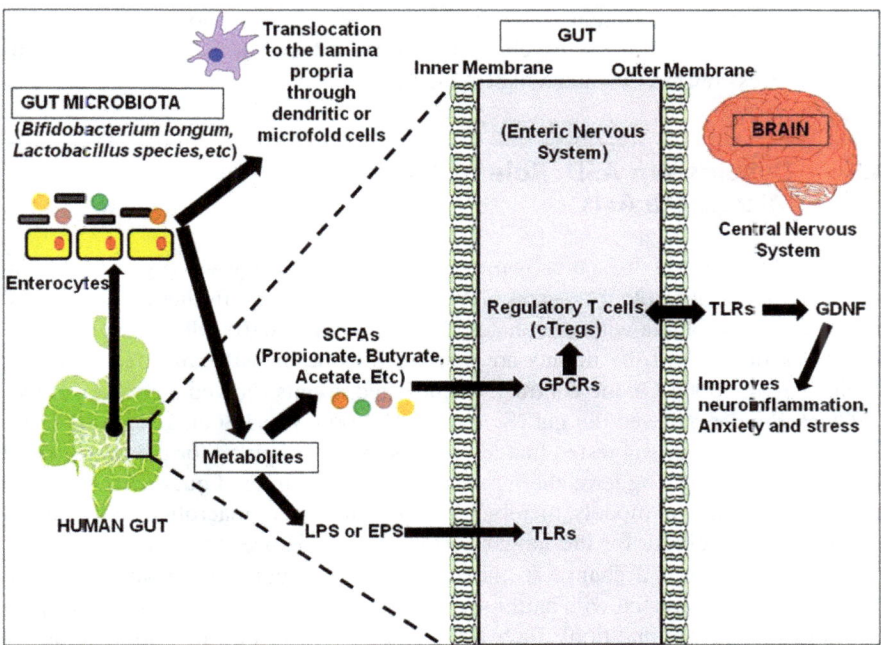

Fig. 4.1 The Relationship link between the Gut Microbiota, Gut, and Brain. There is direct communication between central nervous system (CNS) and enteric nervous system (ENS). Microfold cells or dendritic cells present with the intestinal lumen mediate translocation of gut microbiota through the lamina propria. The gut microbiota derived compounds such as short chain fatty acids (SCFAs), exopolysaccharides (EPS), and lipopolysaccharide (LPS) pass easily through the epithelium membrane and act upon ENS. These metabolites are also capable in interacting with specific receptors such as toll-like receptors (TLRs) or G-protein coupled receptors (GPCRs). SCFAs act on GPCRs and regulate homeostasis of colon regulatory T cells (cTreg) present within colon. In addition, gut microbiota also regulate ENS via TLR signaling which increases the number of enteric neurons and glial cells by stimulating expression of glial cell line-derived neurotrophic factor (GDNF)

(LPS) components of gram-negative bacteria in the serum of ASD individuals compared to that of healthy controls. Similarly, another report by Fiorentino et al. (2016) showed decreased levels of intestinal tight junction components such as claudin (CLDN)-1, OCLN, TRIC and increased levels of CLDN-5, CLDN-12, CLDN-3, and MMP-9 in the ASD individuals compared with controls suggesting impairment in the integrity of gut barrier and blood–brain barrier within ASD individuals. Moreover, lactulose: mannitol test which is used to determine intestinal permeability has shown an increase in gut permeability within autistic children in comparison with the healthy controls (de Magistris et al. 2010). According to a study by Braniste et al. (2014) germ-free mice demonstrate increased blood–brain barrier permeability. On contrast, SCFAs can enhance the integrity of the blood–brain barrier. Preclinical studies reported by O'Mahony et al. (2009, 2011) also support the connection between brain–gut–microbiota. These studies show changes in gut microbial composition due to psychological stress induced in the animal models of early life stress during the maternal separation.

4.3   Dysbiosis in ASD: Role of Altered Brain–Gut– Microbiota Axis

The ingestion of the high-fat diet by mothers during *pregnancy* was suggested as one of the factors responsible for causing ASD in children, *which* further resulted in the alteration of microbiota within neonates (Connolly et al. 2016). Moreover, there is a lower risk of ASD, if the infants are continued to be breastfeeding for 6 months (Schultz et al. 2006). On the contrary, formula-fed infants showed a greater number of *Clostridium difficile* in the gut (Schultz et al. 2006; Azad et al. 2013). Study by Korpela et al. (2016) suggested that antibiotics treatment given for the short period time may also bring long term alternations in the composition of gut microbiota both in humans and animal models, thereby indicating the role of macrolide antibiotics in altering gut microbiota for the longer period of time. However, study by Yassour et al. (2016) showed a change in the composition of gut microbiota among the children who were treated with antibiotics during the initial 3 years of their life span. In various medical conditions such as Crohn's disease, obesity, immunological defects, abnormal behaviors, and inflammatory bowel syndrome gut dysbiosis are reported among children (Cryan and Dinan 2012; Round and Mazmanian 2009; Ajslev et al. 2011; Jostins et al. 2012).

Furthermore, several studies on the animals and patients with pervasive developmental disorder-not otherwise specified (PDD-NOS) have shown to project dysbiosis in the composition of the microbiota and its metabolites (De Angelis et al. 2015; Borre et al. 2014a; Kushak et al. 2016; de Magistris et al. 2010). A study on female mice demonstrated differences within the *Firmicutes* and *Bacteroidetes* phyla in the offspring along with autism-like social behaviors upon administrating valproic acid during pregnancy to the female (De Theije et al. 2014a). It has been suggested that the increased level of microflora and reduction in microbial diversity result in the condition which may lead to the overgrowth of harmful bacteria which later on contributes in the severity of autistic symptoms (Finegold et al. 2012; Kang et al. 2013).

An anaerobic bacillus, *Desulfovibrio* is most commonly found in autistic children. This species of bacteria is considered to be resistant to antibiotic cephalosporin. But another crucial thing to point out is that these autistic children often have ear infections and consume a greater amount of antibiotics than healthy children, and this may result into excessive growth of *Desulfovibrio*, which contributes in the pathogenesis of autistic social behaviors due to production of virulent factors such as lipopolysaccharides (Finegold et al. 2010; Finegold 2011; Tomova et al. 2015; Emanuele et al. 2010). Further, *Bacteroidetes*, which produce propionic acid as a metabolite may modulate the gut–brain axis by influencing CNS and autism behavior (Finegold et al. 2010). A study conducted by De Angelis et al. (2013) suggests that though there is no difference in the abundance of *Firmicutes* in the fecal sample of PDD-NOS individuals on comparison with controls, in the case of autistic children *Firmicutes* is lower in the fecal samples compared to healthy ones. The higher proportion of bacterial species such as *Enterobacteriaceae*, *Clostridium*, *Lactobacillus*, *Alistipes*, *Sarcina*, *Akkermansia*, *Desulfovibrio*, *Suterellaceae*, and *Caloramator* have been reported among autistic children compared to the healthy ones (De Angelis et al. 2015, 2013; Finegold et al. 2012; Finegold 2011). The role of *Clostridium* in the etiopathogenesis of ASD was first reported by Ellen Bolte in 1998, who observed changes in neurobehavior and chronic diarrhea in her autistic child due to repetitive administration of antibiotics (Ding et al. 2017). In addition, there was another study that supported the hypothetical role of *Clostridium* in causing regressive autism in children. This study reported significant improvement in the gastrointestinal and neurobehavioral symptoms in the children upon treatment with vancomycin (Sandler et al. 2000). There are various other studies that report the abundance of *Clostridium* species including *Clostridium histolyticum* and *Clostridium perfringens* in the stools of autistic children (De Angelis et al. 2013; Song et al. 2004; Parracho et al. 2010). Unlike beneficial bacteria such as *Lactobacillus* and *Bifidobacterium*, the *Clostridium perfringens* and *Clostridium botulinum* are resistant to glyphosate. Therefore, glyphosate might cause an alteration in gut microbiota and contribute to the production of *Clostridium* toxins via influencing the gut–brain axis in ASD (Argou-Cardozo and Zeidán-Chuliá 2018).

Recently, studies have demonstrated the lower prevalence of *Prevotella* (a commensal microorganism) in dental samples of ASD individuals (Williams et al. 2012; Qiao et al. 2018). Individuals who are deficient with *Prevotella* experience impairment in carbohydrate digestion as *Prevotella* plays a crucial role in metabolizing saccharides and biosynthesis of vitamins (Arumugam et al. 2011).

The above mentioned several studies describe the alteration of gut microbiota in ASD individuals, but still, the exact composition of gut microbiota specific to autistic individuals is uncertain because some studies depict contrasting results. One of the reasons behind these contrasting results might be due to lack of homogeneity in terms of geographical area, treatment, age, diet, and severity of ASD among autistic individuals. Therefore, it is advisable to maintain homogeneity among the patient groups in order to correctly evaluate the findings.

4.3.1 Impact of Gut Microbiota on Brain and Behavior

Several studies have shown promising evidences to support the role of *the gut microbiota* in cognitive and metabolic activities (Dinan et al. 2015, Borre et al. 2014b). The ingestion of fructose and several other sugars by patients with depression was reported to cause abnormal excretion of hydrogen (Jones et al. 2011). This results in the alteration of microbiota and GI motility as fructose malabsorption gives substrate for rapid bacterial fermentation (Jones et al. 2011). A study on rats demonstrated an increase in plasma tryptophan levels after giving *Bifidobacterium infantis* continuously for 14 days. This further adds up to the role of gut microbiota in tryptophan metabolism (Desbonnet et al. 2008).

Study by Collins and Bercik (2009) on rats showed a change in the behavior of rats upon injecting *Helicobacter pylori*. This infection caused changes in the feeding habits of rats along with a change in gastric physiology. However, these symptoms were recovered upon curing *Helicobacter pylori*. The exact pathogenesis of *Helicobacter pylori* in causing abnormality in behavior is yet unknown but it is suggested to involve immune-mediated neural and humoral mechanisms (Collins and Bercik 2009). Though the activation of toll-like receptor (TLR)-2,4,5 increase during colonization but on the contrary, this study reveals that it gets reduced in germ-free rats. This, therefore, describes the interactions of receptors with microbiota. Moreover, the interaction between the epithelial layer and gut microbiota has also been implicated in the generation of immunoglobulins A and B (Collins and Bercik 2009).

4.3.2   Relationship Between Gut Microbiota and Central Nervous System

The gut functions may be regulated by both sympathetic and parasympathetic paths of the nervous system such as mucosal immune response, intestinal permeability, acid secretion, and mucus generation (Mayer 2000). There are reports which state that number of huge contractions in intestines decreases in the patients with consti-pation and further contribute to the symptoms of inflammatory intestinal disease (Lembo and Camilleri 2003; Chey et al. 2001). These contraction movements may be affected by the type of food consumed, sleep quality, and stress. An increase in parasympathetic activities in small and large intestines is associated with acute stress and a decrease in vagus activity within the stomach (Mayer 2000). The size and quality of the intestinal mucus layer of the enteric microbiota are affected by the autonomic nervous system-mediated modulation of the mucus secretion (Macfarlane and Dillon 2007). Along with this autonomic nervous system also affects the epithelial mechanism by activating the immune system by the gut. In several clinical studies, it has been reported that stress condition may increase the permeability of the gut epithelia, this may lead to the transmission of the microorganism in the lumen (Groot et al. 2000; Jacob et al. 2005; Keita and Soderholm 2010). Moreover, the metabolites produced by gut microbiota and cytokines give signals via the receptors in the gut (Mayer et al. 2015).

4.3.3   Gastrointestinal Symptoms of ASD

GI symptoms are considered as an important comorbidity in ASD (Coury et al. 2012). Although there are no reports on the mechanism of the cause of GI symptoms it could be said with conviction that if these symptoms are untreated may give rise to several behavioral challenges such as irritable and inattentive nature, impairment in communications (Erickson et al. 2005). It has been observed that more than 40% of ASD children experience an increased level of anxiety (van Steensel et al. 2011). According to this report, a model was proposed which links the connection between anxiety with a highly prevalent trait of sensory over-responsivity (Cermak et al. 2010; Green and Ben-Sasson 2010). According to a study, increase in stress responsiveness and anxiety due to alteration in the autonomic nervous system may lead to an increase in the epithelial permeability of the ASD patients (Keita and Soderholm 2010). However, there are several mechanisms put forth in explaining the alternative mechanisms of GI symptoms in ASD such as alteration in neural or endocrine elements of the enteric nervous system (Argyropoulos et al. 2013). This kind of dysbiosis is not only reported in ASD patients but also on several rodent models in form of mutation in the genes related to postsynaptic adhesion molecules affecting neurotransmission such as Neuroligin-3 and Shank-3 (Tabuchi et al. 2007) or serotonin transporter variants (SERT) (Veenstra-VanderWeele et al. 2012).

It is been assumed that the changes in the diet, medication, and overall hygiene are greatly responsible for developing various diseases including the diseases related to brain. Therefore, it has been speculated that such microbial alterations may also be associated with an increase in the prevalence of ASD (Collins et al. 2012; Wang and Kasper 2014). There are studies that suggest diet as a potential factor that may exert beneficial effects by altering the gut microbiome along with affecting the brain (de Theije et al. 2014b). Few reports represent abnormal levels of mammalian–microbial co-metabolites such as dimethylamine, phenylacetate glutamine, and hippurate (Shaw 2010; Yap et al. 2010). Hsiao et al. (2013) studied his hypothesis on dysbiosis of gut microbiota by using a rodent model of ASD where he points out the crucial role of serum metabolites in behavioral manifestations of autism-like behaviors and GI function. The study suggested that these conditions were further improved upon the ingestion of probiotics.

ASD individual experiences various GI abnormalities such as constipation, feeding problems, vomiting, diarrhea, and abdominal symptoms (Navarro et al. 2016). According to a meta-analysis report, GI disorders occur more often in ASD children compared with healthy individuals (McElhanon et al. 2014). It has been estimated that the GI symptoms significantly vary from 9% to 91% within ASD patients (Lefter et al. 2019; Coury et al. 2012). A study conducted by Childhood Autism Risks from Genetics and Environment (CHARGE) showed the association of GI abnormalities with the consistent autism behaviors, as measured by the Aberrant Behavior Checklist. On the other part, the severity of ASD was observed by the Autism Treatment Evaluation Checklist (ATEC) which was associated with a higher GI severity index (McElhanon et al. 2014). In addition, there are emerging evidences which mark the relation between gut dysbiosis due to an inflammatory state. Studies have reported abnormal fecal microbiota in both the conditions of patients either with inflammatory bowel diseases or in the autistic children due to impairment in GI (Walker et al. 2013; Fernell et al. 2007; Boso et al. 2006; Vuong and Hsiao 2017; Viggiano et al. 2015). Moreover, comparison of transcriptome profiles of ASD children gut biopsies, ileal, and colonic tissues were found similar to the individuals with inflammatory bowel disease (Walker et al. 2013).

Furthermore, advanced glycation end products (AGEs) have been reported to accumulate in excess amounts within the brain of autistic individuals (Junaid et al. 2004). AGEs can lead to oxidative stress, neuronal degeneration, and neuroinflammation by interacting with its receptor RAGE (receptors for advanced glycation end products) (Bierhaus et al. 2005). According to a study conducted by Boso et al. (2006) it has been suggested that systematic inflammation is the result of dysfunction due to a higher level of pro-inflammatory ligands S100A9 and the promoter ligand within the blood of autistic patients. This results in the alteration in the AGE-RAGE axis in ASD individuals. In addition, a study on astrocytes showed higher levels of pro-inflammatory cytokines in the neurons derived from ASD children compared to neurons from healthy children (Vuong and Hsiao 2017), which might led to alteration in the development of neurons and synapses in children with ASD (Doenyas 2018a).

The hyperactive cellular immune responses as well as increased plasma levels of pro-inflammatory cytokines such as tumor necrosis factor (TNF)-α, transforming growth factor (TGF)-β, and interleukins, IL-1β, IL-12, IL-6, and IL-8 have been reported in the children with ASD, suggesting that the elevated levels of immune or inflammatory molecules may contribute into the severity of the ASD-related neurobehavioral symptoms (Ashwood et al. 2011; Onore et al. 2012; Li et al. 2017). The gut microbiota has also been suggested to play a crucial role in modulating both innate and adaptive immune system, inflammation, as well as regulatory T cells (Treg-cells) (Umbrello and Esposito 2016).

4.3.4 Pathogenesis of *Clostridium*, *Sutterella*, and other Genera in ASD

In the regressive form of ASD, the onset of neurobehavioral symptoms and chronic diarrhea appears to occur after repetitive courses of antibiotics. The *Clostridium* species were suggested as one of the possible cause behind this (Barger et al. 2013). The hypothesis on the involvement of *Clostridium* species is based on the similarities of the behavior of the other toxin-mediated diseases (Voth and Ballard 2005). *Clostridium* species are a group of heterogeneous anaerobic, spore-forming gram-positive bacilli which are considered as beneficial members of a healthy intestinal microbiota. However, certain *Clostridium* species such as *Clostridium tetani* and *Clostridium perfringens* are among those groups which are potent producers of toxins and can cause a variety of human diseases (Bizzini 1984; Freeman 1979). As mentioned earlier, in certain cases, the trigger cause for the production of toxin release can lead to the disruption of the intestinal microbiota. *Clostridium difficile* is known for causing toxin-mediated diarrheal illness followed by antibiotic administration (Bizzini 1984). Similarly, toxins produced by *Clostridium botulinum* causes neuroparalytic illness due to disturbances caused in the gut microbiota due to antibiotics (Bolte 1998).

The study conducted by Sandler et al. (2000) clearly marks the role of *Clostridium* in children with regressive autism. The 6-week of oral vancomycin treatment given to Autistic children against clostridium gives out significant improvement in the neurobehavioral and GI symptoms. After discontinuation of vancomycin gradual regression in the bowel and behavioral symptoms was observed in all the subjects. Therefore, this fact further supports the involvement of the intestinal microbiota. However, this study needs to be further replicated in non-blind assessments and other subsequent controlled trials. Similarly, a study demonstrated by Finegold et al. (2012) shows a tenfold higher level of *Clostridium* species in feces of children with ASD and GI symptoms rather than healthy children. Another novel species of *Clostridium* found in abundance in ASD patients is *Clostridium bolteae* (Song et al. 2003). Another study conducted by these researchers also reported *clostridial* clusters I and XI in abundance within ASD individuals (Song et al. 2004). Moreover, the elevated levels of toxin-producing *Clostridium* species such as *Clostridium histolyticum* within the ASD group of individuals who are associated with GI problems have also been reported (Parracho et al. 2005).

Along with *Clostridium* species, there is another genus of anaerobic gram-negative called *Sutterella* which has also been hypothesized to play an important role in ASD. A comparative study between ASD children with GI disturbances and controls with only GI disturbances demonstrated higher prevalence of *Sutterella* species in ASD children as compared to controls (Williams et al. 2012). This study further provides evidence that *Sutterella* plays a crucial role in ASD children as it is the major part of the intestinal microbiota. Another study also demonstrated a higher prevalence of *Sutterella* as well as *Ruminococcus torques* in the feces of children with ASD as compared to the control groups (Wang et al. 2013). A study conducted by Kang et al. (2013) depicts reduction in the *Prevotella* and other *Bacteroidetes* who were associated with maintaining good colonic health. Therefore, this study points out that a lesser prevalence of beneficial bacteria leads to the manifestation of ASD.

Apart from the genera mentioned above, other microbes found in abundance in ASD children are *Desulfovibrio* (Finegold et al. 2010), *Akkermansia muciniphila* (De Angelis et al. 2013; Kang et al. 2013), *Faecalibacterium prausnitzii* (De Angelis et al. 2013). However, repeat studies on the same species did not show any significant results (Kang et al. 2013; Wang et al. 2013). Moreover, the contribution of these microorganisms in the manifestation of the neurodevelopmental disease is still not obvious, for instance, *Akkermansia muciniphila* and *Faecalibacterium prausnitzii* are considered as a biomarker of healthy gut microbiota. Overall, the study on compositional shifts between *Proteobacteria*, *Bacteroidetes*, *Firmicutes*, and *Actinobacteria* is still not meaningful as the mechanism behind the etiology is still unveiled. Overall, it can be said that the results vary greatly in terms of different analytical approaches, heterogeneous groups, and metrics to be used to consider the change significant. Moreover, the mechanistic role of microbes is still unclear; so it is difficult to verify whether microbiota truly plays a role in developing ASD. Hence, mechanistic studies are warranted for these microorganisms' role in the pathogenesis of ASD.

4.3.4.1 Role of Serotonin Pathway in Manifestation of Altered Gut–Brain Axis in ASD

Since the 1970s, it has been demonstrated that hyperserotonemia has a correlation with GI symptoms in children with ASD (Anderson et al. 1987; Hanley et al. 1977; Marler et al. 2016). A study on germ-free mice demonstrated a lower level of brain-derived neurotrophic factor (BDNF) and anxiety-like behaviors and a higher level of monoamines (such as noradrenaline, serotonin, and dopamine) compared to conventional mice (Diaz Heijtz et al. 2011; McVey Neufeld et al. 2013; O'Mahony et al. 2015). A recent study has shown an important role of neurotransmitter 5-hydroxytryptamine (5-HT) in building a connection between the gut–brain axis in ASD (Israelyan and Margolis 2018). 5-HT has been considered in the development of both CNS and ENS (Gaspar et al. 2003). The study conducted by Israelyan and Margolis (2018) was based on the SERT and its correlation with hyperserotonemia in ASD children which incorporated, the SERT Ala56 mutation, as the common variant in a mouse model. However, further studies are needed to elucidate whether the 5-HT agonism could act as a preventive measure for behavioral abnormalities and human studies as well.

It is believed that the elevated levels of serotonin in autistic individuals are observed not merely due to genetics but also due to the involvement of infections, GI disorders, and immune system impairment (Minderaa et al. 1987; Mulder et al. 2010). Studies on animals support a link between dysbiosis and enteric serotonin production. These studies found alteration in microbiome composition in the ASD mouse model (Meyza et al. 2013; Golubeva et al. 2017). According to one study, production of serotonin is stimulated by low-grade gut inflammation, mast cells, and platelets, which further leads to irregular intestinal motility and usage of tryptophan (De Theije et al. 2011). Since tryptophan acts as a precursor for various metabolites such as kynurenine and serotonin, it is suggested that reduction in tryptophan diet may lead to the impairment in cognitive autistic behavior in adults (Kraneveld et al. 2016). Overall, it could be said that dysbiosis among autistic individuals directly affects the availability of tryptophan due to a decrease in amino acids that are absorbed from the diet (Kraneveld et al. 2016). However, despite various studies on hyperserotonemia, there is no consistent clinical evidence on a link between hyperserotonemia and autistic behaviors (Kolevzon et al. 2010; Muller et al. 2016). Moreover, there is no enough data available to support the fact that the administration of tryptophan in the diet or serotonin reuptake inhibitor can be used as an effective therapy for ASD children. Hence, additional studies are warranted in these aspects.

4.3.4.2 Role of Gut Microbial Metabolites Pathway in Manifestation of Altered Gut–Brain Axis in ASD

Gut bacteria produce metabolites and co-metabolites which influence the brain, behavior, and gut through crossing the gut–blood and blood–brain barrier (Macfabe 2012). There are high levels of phenolic compounds such as p-cresol and its co-metabolite p-cresol sulfate within the fecal (De Angelis et al. 2015) and urine (Persico and Napolioni 2013) samples of ASD children. These phenolic compounds are produced by bacteria like *C. difficile* and *Bifidobacterium* which possess the ability to express *p*-cresol-synthesizing enzymes (Nicholson et al. 2012). According to one study, there are lower levels of butyric acid and higher levels of acetic acid and propionic acid within the ASD children (De Angelis et al. 2015). SCFAs are the fermented end products of non-digested carbohydrates and suggested to be the causative agent of ASD (Al-Lahham et al. 2010; MacFabe 2015). SCFAs influence CNS through epigenetic modulation of ASD-associated genes or by causing an alteration in mitochondrial functions (Yang et al. 2018). ASD-associated species such as *Clostridia, Bacteroides,* and *Desulfovibrio* produce propionic acid (Yang et al. 2018). This propionic acid has several functions such as anti-bacterial and anti-inflammatory effects, modulation of mitochondrial and lipid metabolism (Al-Lahham et al. 2010) and also modulation of neurotransmitter synthesis and release (DeCastro et al. 2005). Therefore, there are many supporting shreds of evidence which also suggest autistic-like behavior in rodents due to influence of propionic acid (MacFabe 2015; DeCastro et al. 2005; Shultz et al. 2009, 2008; Thomas et al. 2012; MacFabe et al. 2007). Apart from this, there are reports which show that propionic acid is also responsible for inducing neuroinflammation within the rodents along with the autistic behavior (Bhandari and Kuhad 2015, 2017; Bhandari et al. 2018). Although the exact mechanism behind the detrimental effect of propionic acid is not so far understood. Therefore, further studies are required for proper elucidation of the effect of propionic acid on neurodegenerative diseases including ASD.

Another crucial SCFA produced by gut microbiota is butyric acid which plays an important role in mitochondrial function, stimulating oxidative phosphorylation, modulates gut transepithelial transport and fatty acid oxidation (Hong et al. 2016). Butyric acid may serve as a therapeutic option in treating several neurological conditions such as dementia and depression (Govindarajan et al. 2011; Sun et al. 2016a, b). Moreover, animal model studies of ASD depict the beneficial role of butyric acid by modulating neurotransmitter gene expression through inhibiting histone deacetylation (Takuma et al. 2014; Kratsman et al. 2016). Recently, one study has demonstrated the role of butyric acid on the mitochondrial dysfunction within the ASD individual (Rose et al. 2018). In this study, a lymphoblastoid cell line model of ASD was developed which had mitochondrial dysfunction. After the administration of butyric acid, a positive effect on both healthy and ASD children was observed that are under physiological stress (Rose et al. 2018). However, more in vivo studies are required in order to elucidate the potential therapeutic effect of butyric acid on ASD-associated conditions.

It has been reported that ASD individuals have deregulated metabolism of free amino acids (Ming et al. 2012). These free amino acids are found in higher abundance in autistic children compared to healthy children suggesting the prevalence of protein degrading bacteria in children with autism (De Angelis et al. 2015). Autistic children also have a high level of tryptophan urinary excretion and this finding is related to other neuropsychiatric conditions (Noto et al. 2014). In addition to this, increased levels of 3-(3-hydroxyphenyl)-3-hydroxypropanoic acid (HPHPA) and 2-(4-hydroxyphenyl) propionate taurocholenate sulfate (Ming et al. 2012) and significantly lower level of 5-aminovalerate and 3-(3-hydroxyphenyl) propionate were also exhibited in the urine samples of ASD children (De Angelis et al. 2015; Noto et al. 2014; Shaw 2010). The HPHPA is phenylalanine metabolites of *Clostridium* which is reported to induce autistic-like behaviors in animal models (Shaw 2010).

4.4   Evidential In Vivo Animal and Human Clinical Trial Studies for Gut Dysbiosis and ASD-Like Behavioral Disturbances

There are several functions in animals that are regulated by intestinal microbes. Gut microbiota produce essential cofactors and vitamins such as vitamin B12 and folate which may affect DNA and histone protein methylation (LeBlanc et al. 2013; Le Galliard et al. 2008). Microbes metabolize complex proteins, lipids, and carbohydrates ingested by the host and through the fermentative process results in the production of SCFAs such as acetate, butyrate, and propionate (Hooper et al. 2002; Saulnier et al. 2008; Dai et al. 2012). Although these SCFAs are reported to show beneficial role, few reports suggest that these SCFAs may alter the intercellular spaces between the cells which leads to leaky gut, thus allowing more microbes and metabolites to pass across the epithelial barrier which subsequently results into detrimental neurological effects (Rosenfeld 2015). This alteration in the gut can affect not only host immunity but also neurobehavioral symptoms (Cryan and Dinan 2012; Douglas-Escobar et al. 2013; Ding and Schloss 2014; Galland 2014; Stilling et al. 2014; Sherman et al. 2015). The majority of the study till date has focused on the relationship between gut microbiota dysbiosis and ASD. So far it has been depicted clearly in the literature that how different organs and systems such as lung (Dickson et al. 2014), oral cavity (Ding and Schloss 2014), placenta (Amarasekara et al. 2015; Aagaard et al. 2014; Doyle et al. 2014; Antony et al. 2015), and vagina (Stumpf et al. 2013; Ding and Schloss 2014) consist of specific microbes that may affect or influence distal target systems. Therefore, it could be considered that disruption within the microbiomes may contribute to the etiology of ASD. Thus, there is pieces of evidence which proposes that these factors may aid in improving clinical signs upon exploring potential mechanism behind gut dysbiosis.

4.4.1   Animal Model Studies

The animal model studies linked to the alterations in the gut microbiota and neurobehavioral changes are summarized in Table 4.1.

The germ-free (GF), axenic mice were taken up for exploring the association of gut microbiota with ASD. These mice were devoid of any microorganism as it was obtained through the cesarean section which was further maintained in a sterile gnotobiotic environment. This model of mice helped indirectly interpreting the influence of the presence or absence of gut microbiota on behavioral patterns. The initial study reported hypersecretion of two commonly associated stress hormones,

Table 4.1 Animal model studies for relationship between the gut microbiota and ASD

Animal models	Study design	Type of treatments	Key findings	References
Seizure-prone (FAST) and seizure-resistant (SLOW) male rats	Nine week old male rats were assigned to each group	• FAST + PPA (4 mL, 0.26 M)	• FAST and SLOW rat strains showed social abnormalities upon administration of PPA	Shultz et al. (2015)
		• SLOW + PPA	• Neuroinflammation observed in the corpus callosum and cerebral cortex after PPA treatment	
		• FAST + PBS vehicle control (4 mL)	• Neuroinflammatory effects were more in FAST compared with SLOW rats	
		• SLOW + PBS		
Rats	Pregnant P0 rats were treated with the bacterial metabolite	• PPA (500 mg/kg b.wt. s.c. on GDs 12–16) or the LPS (50 mg/kg b.wt. s.c. on GD 15 or 16)	F1 male and female offspring derived from the treated rats showed impairments in the following symptoms:	Foley et al. (2015)
		• Control pregnant P0 rats received vehicle control on GDs 12–16 or 15–16	• Olfactory mediated social recognition	
			• Hyperlocomotion	
			• Disruptions in social behaviors	
GF and SPF mice	The gut of GF mice was reconstituted with *B. infantis* via oral administration to the parents with transmission to the offspring at the neonatal stage	GF mice received fecal transplantation from SPF animals (0.5 mL of 10^{-2} dilution of fresh SPF mouse feces 1 or 3 weeks prior to the stress protocol at 9 or 17 weeks of age)	• Adult GF mice were subjected to restraint stress which exhibited hypersecretion of ACTH and corticosterone	Sudo et al. (2004)
			• Suppression of neurotropin, BDNF, and NR2A in the cerebral cortex and hippocampus of adult GF mice	
			• Alleviated the exaggerated HPA responses upon reconstitution with *B. infantis*	

(continued)

Table 4.1 (continued)

Animal models	Study design	Type of treatments	Key findings	References
			• During early life, feces from SPF partially reversed the hormonal abnormalities in GF mice	
Rats	Rats (47–49 days of age) were exposed twice daily to intraventricular injection of PPA	PPA (4.0 mL of a 0.26 M solution) for an acute period (1 week)	Architectural and hyperlocomotion changes were observed in the brain after treatment	Thomas et al. (2012)
Mice	Pregnant P0 mice were treated on GD 11 with VPA	VPA (600 mg/kg body weight) or PBS (controls)	• Social behavioral deficits were demonstrated in F1 VPA-treated offspring	de Theije et al. (2014a, b)
			• The composition of SCFAs was changed in the F1 treated offspring	
			• Gut dysbiosis with changes in the *Bacteroidetes* and *Firmicutes* phyla and *Desulfovibrionales* resulted in F1 exposed offspring	
			• F1-exposed male offspring had alterations in *Alistipes*, *Enterorhabdus*, *Mollicutes*, and *Erysipelotrichalis* genera	
			• Gut microbiome disturbances were more pronounced in F1 male than female offspring	
Rats	1 M PPA was infused over 10 min in rats for inducing ASD-like symptoms. Resveratrol was	• 1 M PPA (4 µL) was infused over 10 min into the anterior portion of the lateral ventricle	• Treatment with resveratrol for 4 weeks significantly restored all the neurological,	Bhandari and Kuhad (2017)

(continued)

Table 4.1 (continued)

Animal models	Study design	Type of treatments	Key findings	References
	administered from the second day of the surgery up to 28th day		sensory, behavioral, biochemical and molecular deficits in PPA-induced autistic phenotype in rats	
		• Resveratrol (5, 10 and 15 mg/kg) was administered starting from the second day of the surgery and continued up to 28th day	• Suppression of oxidative-nitrosative stress, mitochondrial dysfunction, TNF-α and MMP-9 expression in PPA-induced ASD in rats	
Rats	1 M PPA was infused over 10 min in rats in order to induce ASD-like symptoms within the rat. Curcumin was administered from the second day of the surgery up to 28th day	• 1 M PPA (4 µL) was infused over 10 min into the anterior portion of the caudoputamen to induce autistic behavior in rats	• Curcumin restored the core and associated symptoms of autistic phenotype by suppressing oxidative-nitrosative stress, mitochondrial dysfunction, TNF-α and MMP-9 in PPA-induced autism in rats	Bhandari and Kuhad (2015)
		• Curcumin (50, 100, and 200 mg/kg) was administered per orally starting from second day of surgery and continued up to 28th day		
Rats	ASD-like phenotype was induced by infusion of 1 M PPA. Naringenin uncoated and coated (glutathione and tween 80) naringenin loaded PLGA nanoparticles and minocycline were administered orally	• 1 M PPA was infused into anterior portion of lateral ventricle in rats	• Naringenin and its nanoparticles significantly restored behavioral and biochemical deficits in ASD phenotype	Bhandari et al. (2018)
		• Naringenin (25, 50, and 100 mg/kg), uncoated and coated naringenin loaded PLGA nanoparticles (25 mg/kg) and minocycline (50 mg/kg) were	• Glutathione and tween 80 coated nanoparticles enhanced brain delivery of NGN by inhibition of P-glycoprotein	
			• Naringenin (100 mg/kg) and its nanocarriers	

(continued)

Table 4.1 (continued)

Animal models	Study design	Type of treatments	Key findings	References
		administered orally for 29 days	(25 mg/kg) demonstrated pharmacological efficacy comparable to minocycline (50 mg/kg)	

PPA propionic acid, *GD* gestational days, *ACTH* adrenocorticotropic hormone, *GF* germ-free, *SPF* specific pathogen free, *VPA* valproic acid, *BDNF* brain-derived neurotrophic factor, *PBS* phosphate buffered saline, *MMP* matrix metalloproteinases, *PLGA* polylactic-co-glycolic acid

namely adrenocorticotropic and corticosterone upon inducing stress as compared to specific pathogen free (SPF) controls (Sudo et al. 2004). However, the hypothalamic-pituitary axis (HPA) responses were ameliorated upon the administration of *Bifidobacterium infantis* within GF mice. This study also suggests that in the initial period of life, if the feces are transplanted from SPF to GF animals, it results in reversing the hormonal abnormalities partially. In addition, the study showed suppression of crucial neurotrophins, NR2A and BDNF proteins found in the hippocampus and cerebral cortex of GF mice (Sudo et al. 2004).

Studies have shown impairments of social behavior, cognition, and sensorimotor ability along with the neuroinflammation in Long-Evans rats upon intracerebroventricular injection of 4 mL of 0.26 M solution of propionic acid (Shultz et al. 2008, 2009, 2015). In another study, 500 mg/kg of propionic acid along with 50 mg/kg of LPS was injected daily into pregnant rats which resulted in impairment in hyperlocomotion, social behavior, olfactory mediated social recognition abnormalities in male and female offspring (Foley et al. 2015). A study conducted by Bhandari and Kuhad (2017) suggested decrease in oxidative-nitrosative stress, mitochondrial dysfunction, and TNF-α and MMP-9 expression in PPA-induced ASD in rats upon administration of resveratrol for continuously 4 weeks. Similarly, another study on curcumin also showed restoration in the autistic associated symptoms like neurological and behavioral deficits caused upon administration of propionic acid in rats (Bhandari and Kuhad 2015). Moreover, the treatment with naringenin and its nanoparticles showed a significant improvement in biochemical and behavioral deficiency within the propionic induced ASD rat (Bhandari et al. 2018).

According to an animal study conducted by Thomas et al. (2012), LPS treatment led to hypersensitive males but females who were administered with propionic acid exhibited sensitization in acoustic startle testing during both pre- and postnatal treatment. In prepulse inhibition testing, animals are initially exposed under weak acoustic prepulse. These animals decrease reflexive flinching startle response upon receiving exposure to a higher pulse. Normally healthy animals are capable of filtering out irrelevant auditory information but on the contrary, the animals with neurobehavioral deficiencies are incapable to do so. During the prenatal period, females who were exposed to propionic acid showed lower prepulse inhibition threshold, whereas there was not any similar effect on the males. It was observed that males and females who were treated with propionic acid during the prenatal period has an increase in anxiety-like behaviors among them as they used to spend less time in the center of the open-field maze. Females who were exposed to propionic acid during the prenatal and postnatal periods also exhibited anxiogenic behaviors. There were hyperlocomotion and architectural changes in the brain of rats after the intraventricular injection of propionic acid for a period of 8 days which resembles the pathobiology of the ASD cases (Thomas et al. 2012).

The most common drug used in the treatment of epilepsy and other neurological disorders is valproic acid. There is a link seen between the usages of this drug by mother during pregnancy which later on depicts characteristics related to ASD in the upcoming progeny (Christensen et al. 2013). Mice that were exposed to valproic acid demonstrated autistic-like social behavioral deficiencies gut dysbiosis and alteration in SCFAs (de Theije et al. 2014b). Animals that were exposed to valproic acid showed bacterial shifts of the phyla *Bacteroidetes* and *Firmicutes* and along with shift in the order of *Desulfovibrionales* which is quite similar to that observed in the ASD patients. Conversely, the male offspring upon receiving valproic acid showed alteration in genera *Alistipes*, *Enterorhabdus*, *Mollicutes*, and *Erysipelotrichalis* (de Theije et al. 2014a, b).

So far, all these studies bestow strong evidence suggesting that cascade effects of ASD-like behavioral symptoms are result of gut dysbiosis. Therefore, it is suggested that probiotics treatments in this animal model may ameliorate pathological changes in the gut such as bacterial metabolic disruptions and neurobehavioral abnormalities.

4.4.2 Human Clinical Trials

In ASD patients, GI symptoms are the most common comorbidity responsible for developing clinical severity (Buie et al. 2010; Mayer et al. 2014). Disruption of gut microbiota may be one of the reasons behind this disorder. Studies carried out on the impact of microbial alteration in the ASD children with GI symptoms have been shown in Table 4.2. There is only a single report so far, suggesting no connection between ASD symptoms and gut microbiota composition (Gondalia et al. 2012). The rest of the studies showed an increase in *Clostridial* groups such as *Clostridium bolteae* in ASD children (Song et al. 2004). Further, one study suggests that there is a suppression of transcripts that encodes for hexose transporters, transcription factor CDX2, and disaccharides (Williams et al. 2011). It is believed that these transcriptomic changes within the host are a result of gut dysbiosis as there is a downfall in the ratio of *Bacteroidetes* to *Firmicutes* and elevation in the *Betaproteobacteria* (Williams et al. 2011).

In another study, there was an increase in the *Sutterella* spp in the fecal sample of the ASD children with or without GI disorders (Wang et al. 2013). Moreover, the stool sample of children with ASD and GI symptoms showed a greater amount of *Ruminococcus torques* compared to those without such disorders (Wang et al. 2013).

Table 4.2 Human clinical studies for relationship between the gut microbiota and ASD

Study population	Study design	Method of therapy or analysis	Key findings	References
ASD children; nonrelated controls	• ASD children ($n = 15$) • Nonrelated controls ($n = 8$)	Group and species-specific primers were designed to target the 16S rRNA genes for qRT-PCR analysis of the stool samples	*C. bolteae* and clusters I and XI were elevated in the stool of ASD children	Song et al. (2004)
ASD children with GI dysfunction; nonrelated controls with only GI symptoms	• ASD children with GI dysfunction ($n = 15$) • Nonrelated controls with GI symptoms ($n = 7$)	• qRT-PCR with human mRNA samples for SI, MGAM, Lactase, *SGLT1, GLUT2*, Vili, and *CDX2* • Pyrosequencing of intestinal microbiota	• Decreased expression of disaccharidases, hexose transporters, and the transcription factor CDX2 was exhibited in ASD children	Williams et al. (2011)
ASD children with GI dysfunction; nonrelated controls only with GI symptoms	• ASD children with GI dysfunction ($n = 15$) • Nonrelated controls with GI symptoms only ($n = 7$)	Pyrosequencing and qPCR of ileal and cecal biopsies	*Sutterella* spp. (*wadsworthensis* and *stercoicanis*) was found in abundance in the gut microbiota of ASD children with GI dysfunction, but these species were absent in children with only GI symptoms	Williams et al. (2012)
ASD children; nonrelated controls	• ASD children ($n = 58$) • Nonrelated controls ($n = 39$)	• Bacterial culture of the stool samples • Concentrations of lysozyme, lactoferrin, secretory IgA, elastase, SCFAs were measured in the stool	• Decreased number of SCFAs, specifically acetate, propionate, and valerate were identified in ASD children, especially those consuming a daily probiotic • Stool of ASD children contained greater amounts of *Lactobacillus* but less *Bifidobacter*	Adams et al. (2011)
ASD children without GI	• ASD children	bEFAP was performed on the	• *Firmicutes* (70%),	

(continued)

Table 4.2 (continued)

Study population	Study design	Method of therapy or analysis	Key findings	References
dysfunction; ASD children with GI dysfunction; neurotypical siblings	without GI dysfunction ($n = 23$)	stool samples from all three groups	*Bacteroidetes* (20%), and *Proteobacteria* (4%) comprised the major microbiota present in the stool	Gondalia et al. (2012)
	• ASD children with GI dysfunction ($n = 28$) • Neurotypical siblings ($n = 53$)		• No evidence was found linking gut disease, dysbiosis, and ASD symptoms	

bEFAP bacterial tag encoded FLX amplicon pyrosequencing, *qRT-PCR* quantitative real-time polymerase chain reaction, *SI* sucrose isomaltase, *MGAM* maltase glucoamylase, *SGLT1* sodium dependent glucose cotransporter, *GLUT2* glucose transporter 2, *CDX2* caudal type home box 2

Study by De Angelis et al. (2013) has indicated microbial disruptions and changes in microbial phyla such as *Firmicutes*, *Fusobacteria*, *Bacteroidetes*, and *Verrucomicrobia* along with alteration in the *Lachnospiraceae* family of ASD children compared with healthy children. ASD children also demonstrated a rise in genera such as *Caloramator*, *Sarcina*, *Clostridium*, *Alistipes*, *Akkermansia* spp, and *Sutterellaceae*, whereas a decrease in *Enterobacteriaceae*, *Eubacteriaceae*, and *Bifidobacterium* spp (De Angelis et al. 2013). Another study also found lower amount of *Bifidobacteria* species but showed elevated levels of the *Mucolytic* and *Akkermansia muciniphila* in children with ASD (Wang et al. 2011a, b). Moreover, *Wadsworthensis* and *stercoicanis*, which are part of *Sutterella* spp. highlight relationship between the microbiota–gut–brain axis as it is absent in children with GI symptoms and present within ASD children with GI dysfunction (Williams et al. 2012).

Study by Adams et al. (2011) showed that there is a link between ASD clinical severity and GI symptoms. The stool sample of ASD children exhibited increase in amount of *Lactobacillus* and decrease in the level of *Bifidobacteria* along with the decrease in fecal SCFAs. In contrast, another study reported fecal SCFAs such as acetic, butyric, isobutyric, valeric, propionic, and isovaleric acids to be elevated, whereas there was a fall down in the levels of caproic acid in the stool of ASD children (Wang et al. 2012). The concentration of ammonia was relatively higher in ASD patients compared to the controls (Wang et al. 2012).

Overall, the above mentioned studies suggest that *Clostridia*, *Desulfovibrio*, *Sutterella*, and *Bacteroidetes* are higher in the stool of ASD children. On the contrary, *Firmicutes*, *Prevotella*, and *Bifidobacteria* tend to be in less proportion in these patients. So far, there exist conflicting reports on whether probiotics can be used in treating bacterial imbalances observed in ASD patients. Therefore, further studies are needed with larger data sets in order to make better comparisons within the same group with the different analysis criteria. For example, determination of bacterial SCFAs through fecal and urinary analysis, establishing the amount of ammonia in ASD patients with or without GI symptoms, and while addressing ammonia concentration it is also necessary to consider patients receiving probiotics treatment separately in different group from the ones who are not receiving. In addition, each patient should be monitored and repeatedly analyzed over time in order to elucidate the metabolic parameters associated with disease symptoms.

4.5   Physiological Effects of Dietary Supplements and Prebiotics on the Gut–Brain Axis

Dietary interventions are hugely popular when it comes in terms of improving the health of ASD children but despite its beneficial effects it could also be potentially harmful. As ASD children are already reported to exhibit picky eating behaviors, so restrictive diets may further lead to limit the variety of food intake as well as macronutrient and micronutrient deficiencies (Santocchi et al. 2016; Doenyas 2018b). On the contrary, the Mediterranean diet has been demonstrated to have an impact on gut microbiota, neurobehavioral and cardiovascular diseases (Liu et al. 2017; Fallon 2005). Therefore, in order to understand different potential therapeutic diets, it is necessary to investigate the physiological effects of dietary interventions on the gut–brain axis. The proposed mechanism for the role of prebiotics in the amelioration of ASD symptoms is represented in Fig. 4.2.

Children with ASD are considered to have impaired protein digestion, increased gut permeability, and decreased digestive enzyme activity which could be a possible reason for the elevated levels of urinary dietary peptides as well altered level of plasma amino acids profiles and fecal metabolites. Study has shown that casein and gluten, agonists of opioid receptors exert effects on the neurobehaviors and CNS (Grimaldi et al. 2017). Dietary peptides which show opioid activity pass through intestinal membrane abnormally and further enter into the CNS to exert effects on neurotransmission and other physiology associated symptoms (Grimaldi et al. 2017). On implementation of gluten and casein deprived diet within patients with ASD seems to decrease the level of urinary peptides and improve behavior. Nevertheless, the evidences projecting such outcomes are still limited (Schmidt et al. 2015; Lavasani et al. 2010). Gluten and casein deprived diet can only decrease the levels of gluteomorphins and caseomorphins toxins but cannot restore gut microbiota composition or heal the mucosa (Kwon et al. 2013). Even though few promising evidences of the beneficial effect of gluten and casein deprived diet have been shown, but the mechanistic aspects of the gluten and casein deprived diet are still unclear (De Magistris et al. 2010; Chae et al. 2012).

Furthermore, probiotic is considered as a better alternative therapeutic option than restrictive dietary supplements because it contributes in healing the intestinal

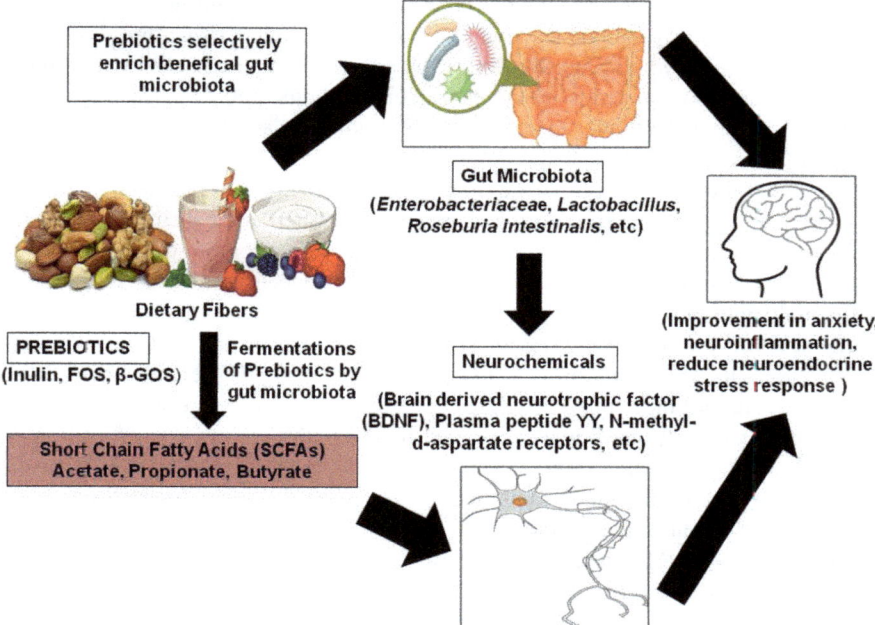

Fig. 4.2 Proposed mechanism of modulation caused in gut–brain axis by prebiotics. Ingestion of balanced amount of dietary fiber is necessary in order to ensure the right balance of beneficial bacteria within the gut which further improves health. Prebiotics such as inulin, FOS, β-GOS play an essential role in selectively enriching the beneficial gut microbiota such as *Enterobacteriaceae*, *Lactobacillus*, *Anaerostipes caccae*, *Roseburia intestinalis*, etc. The prebiotics get fermented by the gut microbiota and release short chain fatty acids (SCFAs). These SCFAs are important in improving neurological symptoms associated with ASD such as anxiety, neuroinflammation, and also reduction in neuroendocrine stress response through influencing neurochemical factors and signaling pathways via various neurological receptors such as BDNF, Plasma peptide YY, *N*-methyl-ᴅ-aspartate receptors, etc. These receptors play an essential role in maintaining synaptic plasticity and optimal memory function

mucosa, protects the epithelial barrier through the production of mucin and fortifying tight junction, increases the production of digestive enzymes and antioxidants, and also modulates immune functions (Doenyas 2018a).

4.5.1   Effects of Novel Personalized Dietary Treatment

Dietary interventions have been proven to affect ASD symptoms through the gut–brain axis. The low doses of narcotics are reported to induce behaviors which resemble with ASD due to unusual activity in brain opioid systems of children with ASD (Panksepp 1979). This brought up to an opioid-excess theory of autism which states that individuals with opioid activity are formed from dietary sources (i.e. casein and gluten). These peptides enter through CNS and influence neurotransmission and other physiological symptoms due to abnormal passage through the gut membrane (Whiteley and Shattock 2002). Moreover, the altered activity in ASD of the DPP4 enzyme is related to an increase in endogenous opioids in the body fluids of ASD individuals (Whiteley et al. 2010). This enormous amount of opioid peptides are said to be passed through the blood–brain barrier (BBB) due to binding with opioid receptors which results in ASD symptoms. Therefore, it has been suggested that removal of opioid peptides precursors such as gluten and casein may improve ASD by positively influencing neurotransmission in the brain (Ciéslinska et al. 2017). However, it is still not confirmed that this mechanism will work in all individuals with ASD. A randomized, double-blind, placebo-controlled trial on the administration of digestive enzyme supplement did not reveal any significant clinical improvement in ASD symptoms (Munasinghe et al. 2010). Therefore, these studies suggest that not all individuals with ASD possess improper digestion of gluten and casein, hence not all individuals may lead to generation of opioid peptides upon administrating gluten and casein. In another study, there was an elevation in the immune response to gliadin or cow milk protein in the children who responded to a gluten-free or casein-free diet (Jyonouchi et al. 2002). These studies suggest that not all individuals with ASD have digestive issues with immune reactivity to cow milk protein or gliadin. Therefore, before recommending a gluten-free or casein-free diet, individual measurements should be taken, so that it can be ensured that there is no negative factor aggravating (Sponheim 1991). Besides inflammatory reaction to dietary proteins in some individuals with ASD, it was also found that subgroup of children with ASD also produced antibodies against Purkinje cells and gliadin peptides (Vojdani et al. 2004). The production of antibodies upon administration of gluten-free or casein-free diets results in the improvement of behavioral symptoms of some ASD patients but not all of them (Vojdani et al. 2004). Hence, these findings suggest that there is a need for developing individual dietary treatment regimens for ASD. Table 4.3 lists preclinical and clinical studies involving Ketogenic diet.

4.5.2 Effects of Prebiotics

Prebiotics are defined as substrates selectively utilized by host microorganisms, conferring a health benefit by The International Scientific Association for Probiotics and Prebiotics (ISAPP) (Gibson et al. 2017). The benefits of probiotics are not only limited to gut homeostasis but can also be extended in improving metabolic, endocrine, immune, and nervous functions. Prebiotics include carbohydrates such as fructans, fructooligosaccharides (FOS), inulin, galactans, galactooligosaccharides (GOS), and other food ingredients indigestible by the host. These ingestible compounds promote the growth of bacteria which are capable of fermenting SCFAs that can affect distal target organs and gut (Saulnier et al. 2008; Fond et al. 2014). No study so far has been reported to improve ASD symptoms solely by prebiotic treatment. One of the studies demonstrates that two prebiotic treatments,

Table 4.3 Experimental studies on animal models and ASD patients receiving Ketogenic Diet

Sr. No.	Sample	Dietary treatment	Tests	Inferences	References
Animal models of ASD					
1.	Males inbred B6.129S-Mecp2^{tm1Hzo}/J (Mecp2$^{2308/y}$) Rett mice ($n = 12$, control-18)	KD: 3.3% carbohydrate, 80% fat, 16.7% protein, 7.2 kcal/g	Behavioral tests sensitive to motor and sensory function	KD improved motor behavior and reduced anxiety	Mantis et al. (2009)
	Control: wild-type Mecp2$^{+/y}$	SD: 65% carbohydrate, 6.9% fat, 28.1% protein, 4.1 kcal/g			
2.	Pregnant Wistar Han rats ($n = 6$, control-6)	KD (6:1 fat to carbohydrates plus proteins) or the standard diet (SD) for 10 days beginning on postnatal Day 21	Social behavioral testing; bioenergetic analysis; bioenergetic profiling, blood chemistry	The KD induced a significant increase in β-hydroxybutyrate levels compared to SD. KD was able to restore aspects of bioenergetic dysfunction	Ahn et al. (2014)
3.	Male BTBR T+ tf/Animal strain to control: C57Bl/6 ($n = 28$, control-30)	KD: 3.2% carbohydrate, 75.1% fat, 8.6% protein, 7.24 kcal/g; SD: 57.9% fat, 23.9% protein, 3.02 kcal/g	Sociability test; intrahippocampal EEG, seizure threshold, blood chemistry	The KD improved sociability and communication and reduced self-directed repetitive behavior in BTBR mouse	Ruskin et al. (2013)
ASD patients					
1.	ASD children with clinical or subclinical seizures	Administration of KD. On-line surveys reporting seizures and other clinical factors	Sleep, communication, behavior, attention, and mood	The ketogenic diet was perceived to significantly improve seizures more than the gluten-free casein-free diet	Frye et al. (2011)
2.	A girl with adenylosuccinate lyase deficiency type II, ASD	KD (2.5:1-lipids: carbohydrates)	–	After introduction of the diet, reduction of seizure was achieved up to 95%	Jurecka et al. (2014)
3.	30 children with autistic behavior (16 boys and 14 girls)	30% of energy as medium-chain triglyceride oil, 30% as fresh cream, 11% as saturated	Childhood autism rating scale	– Only 18 patients tolerated the ketogenic diet – Improvements in learning, social behavior, speech,	Evangeliou et al. (2003)

(continued)

Table 4.3 (continued)

Sr. No.	Sample	Dietary treatment	Tests	Inferences	References
		fat, 19% as carbohydrates, and 10% as protein		cooperation, stereotypy and hyperactivity were observed	
4.	Six patients (2.7% of total sample) who had demonstrated a pathological increase in serum b-OH-b associated with glucose loading	Ketogenic diet	Childhood autism rating scale	16 patients who had demonstrated increased in serum b-OH-b associated with glucose loading	Spilioti et al. (2013)
	Total sample: 187 Greek children (105 males, 82 females)			– Only patient 1 showed remarkable improvement according to the CARS scale	
5.	One girl with sub domains of Autism	Gluten-free, casein-free diet and after the introduced KD (1.5:1 medium-chain triglycerides rather than butter and cream as its primary source of fat)	CARS. Intelligence quotient. Electroencephalogram	Improved cognitive and language function, marked improvement in social skills, increased calmness, and complete resolution of stereotypies	Herbert and Buckley (2013)

FOS and Bimuno GOS reduce salivary cortisol secretion, suppress neuroendocrine stress response, and increase attention span within the subjects (Schmidt et al. 2015). The major concern with prebiotics is to selectively enrich *Lactobacillus* and *Bifidobacterium* spp which could be used for the treatment of ASD. There are various mechanisms through which probiotics can influence gut energy metabolism, elongation of microvilli, increase in mucus layer thickness, protections of the gut epithelium, water fluxes, motility and barrier function (Gibson et al. 1995). Few animal studies have marked improvement in prominent clinical features of ASD such as anxiety and neuroinflammation through normalization induced in the LPS-mediated anxiety and elevation in BDNF levels upon oral administration of prebiotics (Savignac et al. 2013, 2016). So far, only three clinical trials have been reported for the use of prebiotics in the children with ASD (NCT02086110, NCT04261595, and NCT02720900) which is mentioned in Table 4.4 (https://www.clinicaltrials.gov as on March 28, 2020). In the first clinical study (NCT02086110), GOS was combined with bovine colostrum which improved GI symptoms as well as irritability scores. The second study (NCT04261595) carried out the examination of dates as a functional food for autism via its prebiotic effect, modulation of anti-inflammatory and anti-oxidative activity but still no results have been reported so far. The third preclinical study (NCT02720900) reported the supports improvement in anti-social behavior. Therefore, it could be suggested that combined intervention therapies may have a better therapeutic output for ASD compared to the single dietary approach. The proposed mechanism for the role of prebiotics in the amelioration of ASD symptoms is represented in Fig. 4.2. However, more studies are required with multiple prebiotics to confirm their role in ASD treatment or amelioration.

4.6 Role of Probiotics in Amelioration of ASD

Probiotics are defined as living bacteria that, when administered in adequate amounts, confer a health benefit on the host (FAO/WHO 2001). Probiotics have been considered as an emerging therapeutic approach in the ASD. Several studies suggest that probiotics may decrease the inflammatory state and ameliorate behavioral symptoms in children with ASD (Liu et al. 2010; Jonkers et al. 2012; Zhang et al. 2016; Giannetti and Staiano 2016; Navarro et al. 2016). A study conducted on the maternal-activation (MLA) model of ASD showed improvement in gut permeability and autistic-like behaviors upon oral administration of *Bacteroides fragilis* (Hsiao et al. 2013). Although *Bacteroidetes* has shown to improve ASD symptoms in the study mentioned by Hsiao et al. (2013) but it also indicates that the probiotics treatments may require some specificity in terms of gut permeability and neurobehavioral responses.

In a recent study, autistic-like behavior was induced in hamsters by propionic acid and clindamycin (El-Ansary et al. 2018). Later, these hamsters were treated with an oral mixture of *Bifidobacteria* and *Lactobacilli* strains (ProtexinR) for 3 weeks. The findings showed clindamycin and propionic acid-induced depletion in Mg^{2+} and

Table 4.4 Clinical studies of prebiotics in the treatment of ASD

Prebiotic	Title of study	Study design	Clinical trials gov. identifiers	Recruitment status[a]	Key findings
Milk oligosaccharides from bovine colostrum	A pilot study examining microbiota composition in children with autism and gastrointestinal symptoms after use of Bifidobacterium Infantis and milk oligosaccharides	Randomized ($n = 11$)	NCT02086110	Completed	• Reduced frequency of certain GI symptoms • Significant improvement in irritability scores • Improvements may be due to a reduction in IL-13 and TNF-α production
Dates as a functional food for autism	Evaluating dates as a functional food for autism via its prebiotic effect, modulation of anti-inflammatory and anti-oxidative activity	Non-randomized ($n = 120$)	NCT04261595	Active, not recruiting	No results declared so far
Prebiotic intervention for autism spectrum disorders	Effect of a prebiotic (B-GOS) supplementation on microbiota and gastrointestinal (GI) symptoms in children with autism spectrum disorders (ASD)	Randomized ($n = 41$)	NCT02720900	Completed	• Metabolic shifts were observed in urine spectra profile and fecal samples after B-GOS® intervention • Reduced gastrointestinal (GI) discomfort but no significant difference in GI symptoms or sleep (volunteer diaries)

Gov government
[a]Status is based on the review of https://www.clinicaltrials.gov as on March 28, 2020

γ-aminobutyric acid receptors (GABA) due to an increase in brain glutamate excitotoxicity in hamsters (El-Ansary et al. 2018). Previous studies suggested that Mg^{2+} deficiency results in the elevation of Ca^{2+} and glutamate. Therefore, this could be one of the possible reasons for repetitive and impaired social behaviors in ASD (MacFabe et al. 2007, 2011).

There are shreds of evidence that depicts that maternal obesity and diabetes share a connection with autism (Li et al. 2016; Nahum Sacks et al. 2016). It has been reported that maternal high-fat diet is responsible to induce alteration in gut microbiome composition and social withdrawal in offspring (Buffington et al. 2016). However, the administration of *Lactobacillus reuteri* reverses abnormalities related to social behaviors but not repetitive behaviors and anxiety suggesting that the specificity of probiotics strains is responsible for its effect (Buffington et al. 2016). Moreover, *Lactobacillus reuteri* (NCT03337035) also increased the levels of oxytocin involved in the mesolimbic dopamine reward system, which is thought to be deregulated in ASD (Buffington et al. 2016; Donaldson and Young 2008).

Various clinical trials have been performed on children in order to demonstrate the potent role of probiotics in ASD (Table 4.5). A study was conducted on 62 children with ASD within 3–16 years of age group in which *Lactobacillus plantarum* WCFS1 was continuously administered at a daily dose of 4.5×10^{10} CFU for 3 weeks. Children were divided into two groups: group one received a placebo during the first feeding period (3 weeks) and a probiotic during the second (3 weeks); and vice versa for the second group. Through a standardized Development Behavior Checklist (DBC-P), impact of both the food regimes was measured on the behavioral and emotional characteristics of an individual.

Table 4.5 reports the main ongoing and completed clinical trials performed to evaluate the effects of probiotics strains on individuals with ASD (https://www.clinicaltrials.gov as on March 28, 2020). Overall, it can be said that the positive effect of probiotics therapy on neurobehavioral symptoms of children with ASD is still a controversial subject. There are various drawbacks in the clinical trials mentioned above, such as small sample sizes, different study periods, and lack of homogeneity in patients enrolled in terms of age, dietary habits, and severity of GI symptoms.

4.6.1 Beneficial Effects of Probiotics on Gastrointestinal Symptoms in ASD

Along with social-emotional impairments and nonverbal communication, the GI symptoms have also been reported in ASD. Few studies described below have shown varied GI symptoms upon uptake of probiotics and are presented in Table 4.6. A cohort study was conducted among 22 children who were orally supplemented with *Lactobacillus acidophilus* twice daily for a period of 2 months but unfortunately, there was no improvement seen within the behavior and emotional state of the subjects (Kaluzna-Czaplinska and Blaszczyk 2012). Another study examined the fecal microbiota of ten children with ASD through real-time

Table 4.5 Clinical studies of probiotics in the treatment of ASD

Probiotics	Title of study	Study design	Clinical trials Gov. identifiers	Recruitment status[a]	Key findings
L. Reuteri Supplementation	Efficacy of Lactobacillus Reuteri in managing social deficits in children with autistic spectrum disorder: a randomized clinical trial with evaluation of gut microbiota and metabolomics profiles	Randomized, double-blind (n = 80)	NCT04293783	Recruiting	Prophylactic use of L reuteri DSM 17938 during the first 3 months of life reduced the onset of functional gastrointestinal disorders and reduced private and public costs for the management of this condition
Oral L. reuteri probiotics and oxytocin nasal spray	The effects of probiotics and oxytocin nasal spray on social behaviors of ASD children—a pilot study	Randomized (n = 35)	NCT03337035	Completed	No results reported so far
Combination probiotic: BB-12 with LGG (different doses)	Road to discovery for combination of probiotic BB-12 With LGG (different doses) in treating autism spectrum disorder	Randomized (n = 70)	NCT03514784	Recruiting	No results reported so far
The probiotic mix (VISBIOME) will be mainly Bifidobacteria and Lactobacilli	Probiotics for quality of life in autism spectrum disorder	Randomized (n = 13)	NCT02903030	Completed	No results reported so far
Multistrain probiotics Vivomixx (The probiotic strains contained in the intervention are Streptococcus thermophilus, Bifidobacterium breve,	The efficacy of the multistrain Probiotic, Vivomixx, on behavior and gastrointestinal symptoms in children with autism spectrum disorder (ASD)	Randomized (n = 82)	NCT03369431	Recruiting	No results reported so far

Bifidobacterium longum, Bifidobacterium infantis, Lactobacillus acidophilus, Lactobacillus plantarum, Lactobacillus paracasei, Lactobacillus delbrueckii subsp. Bulgaricus)					
Lactobacillus Plantarum PS128	Psychophysiological effects of Lactobacillus Plantarum PS128 in preschool children with autism spectrum disorder: a randomized, placebo-controlled trial	Double-blinded, randomized, placebo-controlled (n = 250)	NCT03982290	Recruiting	No results reported so far
Mixture probiotics capsule	Efficacy of multiple strain probiotics reduces the neurobehavioral disorder in premature very low birth weight infants	Randomized (n = 250)	NCT03858816	Recruiting	No results reported so far

Gov government

[a]Status is based on the review of https://www.clinicaltrials.gov as on March 28, 2020

Table 4.6 Role of probiotics in modulating gastrointestinal dysfunction in ASD children

Strain of probiotic	Dose and duration of probiotic therapy	Study design	Measurement of GI symptoms	Key findings	References
Lactobacillus plantarum WCFS1	Dose: 4.5×10^{10} CFUs Duration: Daily for 3 weeks	Double-blind, placebo-controlled, crossover trial ($n = 22$)	Parents recorded GI function and symptoms in a diary throughout the study	Probiotic therapy significantly increased the amount of Lactobacilli/ Enterococci ($p < 0.05$) and decreased the Clostridium coccoides ($p < 0.05$) found in the stool samples of children with ASD as compared with placebo	Parracho et al. (2010)
L. acidophilus	Dose: 5×10^{9} CFUs Duration: 2 times/day for 2 months	Noncontrolled trial ($n = 22$)	No	Urine DA (a metabolite of Candida species) level was higher in children before probiotics supplementation (160.04 ± 22.88 μmol/mmol creatinine, $p < 0.05$)	Kaluzna-Czaplinska and Blaszczyk (2012)
Delpro capsule: 2 billion CFUs of each of the following: Lactobacillus delbrueckii, L. acidophilus, Lactobacillus casei, B. longum, Bifidobacteria bifidum, with an additional 8 mg Del-immune V powder	Dose: 10 billion CFUs total Duration: 1 capsule, 3 times/day for 21 days	Noncontrolled trial ($n = 33$)	21-Day stool frequency diary before and after intervention, fourth domain in ATEC includes two questions measuring severity of diarrhea and constipation	52% of the total respondents reported severe constipation at baseline with a decline to 20% reporting severe constipation after treatment, 20% of the total respondents reported severe diarrhea at baseline with a decline to none after treatment	West et al. (2013)
"Children Dophilus": Blend of three strains of Lactobacillus (60%), two strains of Bifidumbacteria	Dose: Not provided Duration: 3 times/day for 4 months	Noncontrolled trial ($n = 29$)	Parent questionnaire at baseline only	Children with ASD and their siblings had more GI dysfunction than did controls ($p < 0.05$) with GI	Tomova et al. (2015)

(25%), one strain of *Streptococcus* (15%)			symptom severity and autism severity being strongly, positively correlated ($R = 0.78$, $p = 0.01$)-Bacteroidetes-to-Firmicutes ratio was lower ($p < 0.05$) and *Lactobacillus* was higher ($p < 0.05$) in children with ASD as compared with controls	
Three strains: *Lactobacillus acidophilus*, *Lactobacillus rhamnosus*, *Bifidobacteria longum*	Dose: 5 g of powder/day (each gram contained 100×10^6 CFUs of each strain) Duration: 1 time/day for 3 months	Prospective, open-label noncontrolled trial ($n = 60$) 6-GSI used to assess GI symptoms (lower score = fewer GI symptoms)	Children with ASD had lower *Bifidobacteria* levels than did age and sex-matched controls at baseline and fecal *Lactobacilli* and *Bifidobacteria* concentrations increased as a result of supplementation ($p < 0.0001$)	Shaaban et al. (2017)

polymerase chain reaction (PCR) (Tomova et al. 2015). The study showed stabilized ratio of *Bacteroidetes/Firmicutes* in ASD children upon supplementation of one capsule of "Children Dophilus" (which comprises three strains of *Lactobacillus*, two strains of *Bifidobacterium*, and one strain of *Streptococcus*), and it also exhibited depletion of TNF-α levels in their stools upon thrice a day supplementation of capsule for continuous 4 months (Tomova et al. 2015).

Furthermore, in a study *Lactobacillus rhamnosus* GG was given to 40 subjects out of 75 infants in a clinical trial for a period of 6 months (Pärtty et al. 2015). The infants were examined after attaining the age of 13 years, among which 6 of the 35 were diagnosed with Asperger syndrome/hyperactivity disorder in the placebo group, whereas there were no reports of such conditions in the probiotics group. This study suggests that if *Lactobacillus* is administered in early life, it may reduce the risk of developing ASD (Pärtty et al. 2015). However, further studies in this domain are recommended for confirming the results. In addition, De Simone Formulation (a multistrain mixture of ten probiotics) showed remission in GI symptoms and neurobehavioral symptoms of ASD upon daily administration in a boy (Blades 2000; Grossi et al. 2016). Similarly, another probiotics therapy, namely Delpro (a mixture of five probiotic strains) was administered along with the immunomodulator Del-Immune V (*Lactobacillus rhamnosus* V lysate) in ASD children for 21 days (West et al. 2013). This study reported improvement after evaluating multiple domains such as speech, language, communication, sociability, cognitive behavior through the ACTEC (West et al. 2013). Moreover, a significant decrease in the severity of speech/communication categories of ATEC was also observed in a study involving 30 patients of ASD (age group of 5–9 years) who were administered with a formula containing *Lactobacillus acidophilus*, *Lactobacillus rhamnosus*, and *Bifidobacteria longum* for continuous 3 months (Shaaban et al. 2017). An ongoing clinical trial (NCT03369431) is utilizing Vivomixx® (a probiotic mixture of one strain of *Streptococcus thermophilus* DSM 24731 and three strains of *Bifidobacterium* and four strains of *Lactobacillus*) on children with ASD. The aim of this clinical trial is to evaluate improvements in cognitive, behavior, and GI symptoms. Along with these, the changes in fecal and urinary biomarkers would also be analyzed to observe neurological patterns after probiotics therapy.

Although there are only few evidences available for probiotics to prove them as a promising therapeutic approach for ASD, however, various mechanisms behind their involvement in neural, metabolic, and immune signaling pathways assure the benefits in improving the neural health of ASD individuals. These proposed probiotic mechanisms in ASD are presented in Fig. 4.3.

4.7   Microbial Reconstitution: A Potent Future Therapeutic Intervention in ASD

A study documented that changes can be induced in the onset of autism-like behaviors and serum metabolites via initiating a shift in microbial flora within the gut of the mouse (Hsiao et al. 2013). *Bacteroides fragilis* has shown to cause

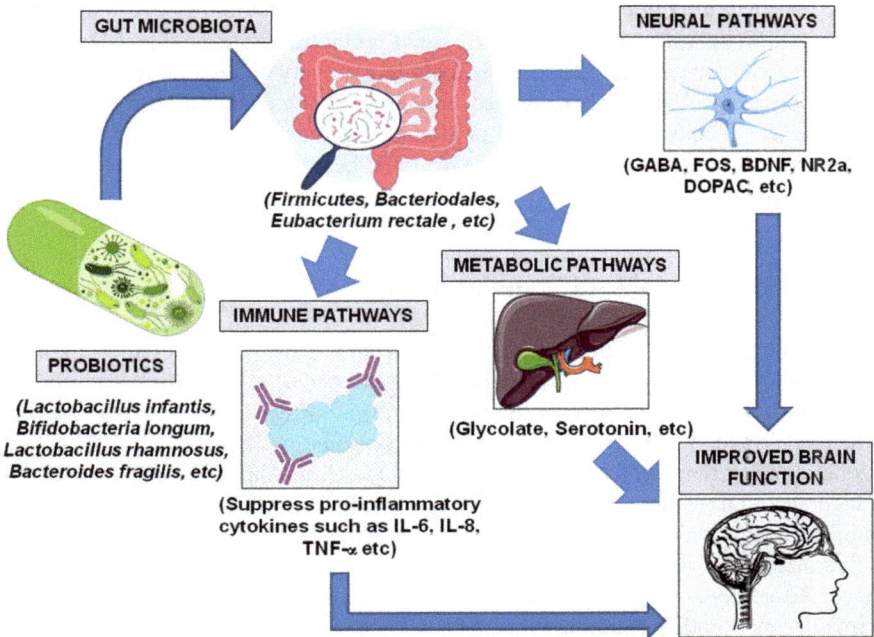

Fig. 4.3 Proposed mechanism of modulation caused in gut–brain axis through Probiotics. Administration of selected probiotics (*Lactobacillus infantis*, *Bifidobacteria longum*, *Lactobacillus rhamnosus*, *Bacteroides fragilis*, etc.) exhibit positive effects on the brain activity of humans by modulating brain function through altering the composition of those bacteria which are responsible for inducing ASD-like symptoms in an individual, i.e. *Firmicutes*, *Bacteriodales*, *Eubacterium rectale*, etc. These probiotics regulate several signaling pathways including neural pathways by improving cognitive functions such as memory and learning in ASD patients which provide protection against oxidative stress through fatty acids such as GABA, FOS, DOPAC, etc. Moreover, it also influences immune pathways by inducing immune signals through downregulating expression of pro-inflammatory cytokines such as IL-6, IL-8, TNF-α, etc. which further can modulate the CNS and contribute to the amelioration of autism by influencing the early stages of brain development. The probiotics also showed an increase in the anti-inflammatory cytokine IL-10 which shows immune response-dependent antidepressant properties of gut bacteria. Metabolic pathways are altered in the presence of glycolate and serotonin which also indicates the role of probiotics in enhancing brain function by modulating the metabolism of the host. The anxiolytic actions of *Bifidobacterium infantis* and *Bifidobacterium longum* can be modulated by a BDNF or cytokine-independent pathway. Therefore, it is considered that *Lactobacillus* and *Bifidobacterium* significantly decrease anxiety symptoms in ASD individuals

improvement within sensory motor, anxiety behaviors, and abnormal communicative symptoms along with amelioration in intestinal permeability defects and cytokine alterations (Hsiao et al. 2013). These findings provide pieces of evidence suggesting that microbial substitution of beneficial bacteria whether it is through bacteriotherapy, probiotics, or fecal microbiota transplantation (FMT) can serve as an effective therapeutic option in improving symptoms of ASD.

So far, there are not enough clinical studies carried out with respect to microbial reconstitution in ASD. Similarly, in the case of probiotics limited evidence exists for improving other psychiatric conditions including anxiety and depressive symptoms (Pirbaglou et al. 2016). There are several ongoing clinical trials on developing strategies for reconstituting microbiota including probiotics which may be considered as an alternative treatment in improving the health of ASD individuals.

4.7.1   Fecal Microbiota Transfer (FMT)

FMT is the interposition of fecal microbiota from a healthy individual to a patient with dysbiotic gut microbiota. The concept of FMT can be tracked back from fourth-century China, where the doctor prescribed an oral human fecal suspension which effectively treated severe diarrhea and food poisoning (Zhang et al. 2012). Few studies suggest that FMT has a high impact on the treatment of *Clostridium difficile* infections (CDI) (Vrieze et al. 2013; Lessa et al. 2015). FMT has also been reported in improving constipation symptoms by normalizing gut microbiota in patients with inflammatory bowel disease and irritable bowel syndrome (Aroniadis and Brandt 2013; Rossen et al. 2015). Therefore, research is increasing in the field of FMT in terms of ameliorating children with ASD.

The growing evidence towards the role of FMT in improving gut microbiota disruption and symptoms of ASD has paved the way towards clinical trials using FMT to treat children with ASD. FMT has already been proven to modify gut microbiota composition through a significantly increasing diversity of *Bifidobacteria* and *Prevotella* species (Kang et al. 2017). Re-evaluation of patients even after 2 years of completion of this clinical trial (NCT02504554) showed improved GI and ASD symptoms (Kang et al. 2019). According to this study, FMT intervention could be considered as an effective and well-tolerated approach in treating children with ASD. However, there are some adverse effects of FMT which include abdominal cramps, short term belching, mild abdominal floating (Kelly et al. 2015). Therefore, the safety of FMT has to be considered as well as the focus should also be made in the direction of reducing opportunistic pathogens to recipients before FMT can be adopted widely.

4.7.2   Microbiota Transfer Therapy (MTT)

Microbiota transfer therapy (MTT) is a therapy similar to that of FMT but in the case of MTT individuals are treated with antibiotic vancomycin for 14 days followed by administration of standardized human gut microbiota and bowel cleaning for 7–8 weeks. A study involving MTT has shown improvement in GI symptoms such as diarrhea, abdominal pain, constipation, etc. and also normalized the microbiota of ASD patients (Kang et al. 2017).

4.8 Conclusions

Several studies have demonstrated the clinical importance of microbes in different diseases. As far as neurobehavioral disorder ASD is concerned it has been evident that intestinal microbiota controls the health and well-being of ASD individuals. So far, it is not clear whether dysbiosis is the primary reason for impacting brain function and development or it is the secondary aberration that is responsible for causing impairment in gut function through alteration in neural regulation. We also suggest that the studies in the future should also focus on the confounding factors of the study group such as high-fat diet, GI symptoms, SCFAs which may be helpful in finding the relationship between gut microbiota and autism. The significant link between GI symptoms and ASD has been established and it has reported increased levels of *Clostridium* spp. and *Desulfovibrio* and reduced levels of *Bifidobacterium*. However, there are not sufficient data for defining GI as a peculiar characteristic or unique profile for ASD. If it is confirmed anyhow that GI is an exaggerating factor in ASD, then these will open doors for various therapeutic approaches such as probiotics, prebiotics, FMT, and other nutritional strategies in the treatment of ASD patients. Among all these therapeutic options for ASD, probiotics are mostly investigated as a potential treatment in reducing autism symptoms. Although clinical trials with children have shown promising results, it is still considered to be limited in terms of lack in safety, tolerability evaluations, and in the method of analysis of microbiota. Therefore, there is a requirement of well-designed, randomized, placebo-controlled clinical trials in order to identify accurate dose and time of treatment along with a more suitable strain of microbe. The emerging growth in gaining insights into pathophysiology and therapeutic aspects of ASD may bring novel approaches in managing the clinical severity of the disease and help in reducing the burden of this disorder.

Acknowledgments We are grateful to Uka Tarsadia University, Maliba Campus, Tarsadi, Gujarat, India for providing the facilities needed for the preparation of this chapter.

Conflict of Interest The authors declare no conflict of interest.

References

Aagaard K, Ma J, Antony KM et al (2014) The placenta harbors a unique microbiome. Sci Transl Med 6:237ra265

Adamczyk-Sowa M, Medrek A, Madej P et al (2017) Does the gut microbiota influence immunity and inflammation in multiple sclerosis pathophysiology? J Immunol Res 2017:7904821

Adams JB, Johansen LJ, Powell LD et al (2011) Gastrointestinal flora and gastrointestinal status in children with autism-comparisons to typical children and correlation with autism severity. BMC Gastroenterol 11:22

Ahn Y, Narous M, Tobias R et al (2014) The ketogenic diet modifies social and metabolic alterations identified in the prenatal valproic acid model of autism spectrum disorder. Dev Neurosci 36(5):371–380

Ajslev TA, Andersen CS, Gamborg M et al (2011) Childhood overweight after establishment of the gut microbiota: the role of delivery mode, pre-pregnancy weight and early administration of antibiotics. Int J Obes 35(4):522–529

Al-Lahham SH, Peppelenbosch MP, Roelofsen H (2010) Biological effects of propionic acid in humans, metabolism, potential applications and underlying mechanisms. Biochim Biophys Acta 1801:1175–1183

Amarasekara R, Jayasekara RW, Senanayake H et al (2015) Microbiome of the placenta in pre-eclampsia supports the role of bacteria in the multifactorial cause of pre-eclampsia. J Obstet Gynaecol Res 41(5):662–669

Anderson GM, Freedman DX, Cohen DJ et al (1987) Whole blood serotonin in autistic and normal subjects. J Child Psychol Psychiatry 28:885–900

Antony KM, Ma J, Mitchell KB et al (2015) The preterm placental microbiome varies in association with excess maternal gestational weight gain. Am J Obstet Gynecol 212(5):653.e1–653.16

Argou-Cardozo I, Zeidán-Chuliá F (2018) Clostridium bacteria and autism spectrum conditions: a systematic review and hypothetical contribution of environmental glyphosate levels. Med Sci 6(2):E29

Argyropoulos A, Gilby KL, Hill-Yardin EL (2013) Studying autism in rodent models: reconciling endophenotypes with comorbidities. Front Hum Neurosci 7:417

Aroniadis OC, Brandt LJ (2013) Fecal microbiota transplantation: past, present and future. Curr Opin Gastroenterol 29:79–84

Arumugam M, Raes J, Pelletier E et al (2011) Enterotypes of the human gut microbiome. Nature 473(7346):174–180

Ashwood P, Krakowiak P, Hertz-Picciotto I (2011) Elevated plasma cytokines in autism spectrum disorders provide evidence of immune dysfunction and are associated with impaired behavioral outcome. Brain Behav Immun 25:40–45

Azad MB, Konya T, Maughan H et al (2013) CHILD Study Investigators. Gut microbiota of healthy Canadian infants: profiles by mode of delivery and infant diet at 4 months. CMAJ 185(5):385–394

Barger BD, Campbell JM, McDonough JD (2013) Prevalence and onset of regression within autism spectrum disorders: a meta-analytic review. J Autism Dev Disord 43(4):817–828

Bhandari R, Kuhad A (2015) Neuropsychopharmacotherapeutic efficacy of curcumin in experimental paradigm of autism spectrum disorders. Life Sci 141:156–169

Bhandari R, Kuhad A (2017) Resveratrol suppresses neuroinflammation in the experimental paradigm of autism spectrum disorders. Neurochem Int 103:8–23

Bhandari R, Paliwal JK, Kuhad A (2018) Naringenin and its nanocarriers as potential phytotherapy for autism spectrum disorders. J Funct Foods 47:361–375

Bierhaus A, Humpert PM, Morcos M et al (2005) Understanding RAGE, the receptor for advanced glycation end products. J Mol Med 83:876–886

Bizzini B (1984) Tetanus: bacterial vaccines. Academic Press, San Diego, CA, pp 37–68. Chapter 2

Blades M (2000) Autism an interesting dietary case history. Nutr Food Sci 30:137–139

Bolte ER (1998) Autism and *Clostridium tetani*. Med Hypotheses 51(2):133–144

Borre YE, O'Keeffe GW, Clarke G et al (2014a) Microbiota and neurodevelopmental windows: implications for brain disorders. Trends Mol Med 20(9):509–518

Borre YE, Moloney RD, Clarke G et al (2014b) The impact of Microbiota on brain and behavior: mechanisms & therapeutic potential. Adv Exp Med Biol 817:373–403

Boso M, Emanuele E, Minoretti P et al (2006) Alterations of circulating endogenous secretory RAGE and S100A9 levels indicating dysfunction of the AGE-RAGE axis in autism. Neurosci Lett 410:169–173

Braniste V, Al-Asmakh M, Kowal C et al (2014) The gut microbiota influences blood-brain barrier permeability in mice. Sci Transl Med 6(263):263ra158

Buffington SA, Di Prisco GV, Auchtung TA et al (2016) Microbial reconstitution reverses maternal diet-induced social and synaptic deficits in offspring. Cell 165:1762–1775

Buie T (2015) Potential etiologic factors of microbiome disruption in autism. Clin Ther 37:976–983

Buie T, Campbell DB, Fuchs GJ et al (2010) Evaluation, diagnosis, and treatment of gastrointestinal disorders in individuals with ASDs: a consensus report. Pediatrics 125(Suppl 1) S1–S18

Casanova MF (2007) The neuropathology of autism. Brain Pathol 17:422–433

Cermak SA, Curtin C, Bandini LG (2010) Food selectivity and sensory sensitivity in children with autism spectrum disorders. J Am Diet Assoc 110:238–246

Chae CS, Kwon HK, Hwang JS et al (2012) Prophylactic effect of probiotics on the development of experimental autoimmune myasthenia gravis. PLoS One 7:e52119

Chey WY, Jin HO, Lee MH et al (2001) Colonic motility abnormality in patients with irritable bowel syndrome exhibiting abdominal pain and diarrhea. Am J Gastroenterol 96(5):1499–1506

Christensen J, Gronborg TK, Sorensen MJ et al (2013) Prenatal valproate exposure and risk of autism spectrum disorders and childhood autism. JAMA 309:1696–1703

Ciéslinska A, Kostyra E, Savelkoul HFJ (2017) Treating autism spectrum disorder with gluten-free and casein-free diet: the underlying microbiota-gut-brain axis mechanisms. HSOA J Clin Immunol Immunother 3:1–11

Collins SM, Bercik P (2009) The relationship between intestinal microbiota and the central nervous system in normal gastrointestinal function and disease. Gastroenterology 136(6) 2003–2014

Collins SM, Surette M, Bercik P (2012) The interplay between the intestinal microbiota and the brain. Nat Rev Microbiol 10:735–742

Connolly N, Anixt J, Manning P et al (2016) Maternal metabolic risk factors for autism spectrum disorder-an analysis of electronic medical records and linked birth data. Autism Res 8:829–837

Coury DL, Ashwood P, Fasano A et al (2012) Gastrointestinal conditions in children with autism spectrum disorder: developing a research agenda. Pediatrics 130(Suppl 2):S160–S168

Cryan JF, Dinan TG (2012) Mind-altering microorganisms: the impact of the gut microbiota on brain and behaviour. Nat Rev Neurosci 13(10):701–712

Cummings DE, Overduin J (2007) Gastrointestinal regulation of food intake. J Clin Invest 117:13–23

Dai ZL, Li XL, Xi PB et al (2012) Metabolism of select amino acids in bacteria from the pig small intestine. Amino Acids 42:1597–1608

De Angelis M, Piccolo M, Vannini L et al (2013) Fecal microbiota and metabolome of children with autism and pervasive developmental disorder not otherwise specified. PLoS One 8(10):e76993

De Angelis M, Francavilla R, Piccolo M et al (2015) Autism spectrum disorders and intestinal microbiota. Gut Microbes 6(3):207–213

De Theije CG, Wu J, da Silva SL (2011) Pathways underlying the gut-to-brain connection in autism spectrum disorders as future targets for disease management. Eur J Pharmacol 668(Suppl. 1): S70–S80

DeCastro M, Nankova BB, Shah P (2005) Short chain fatty acids regulate tyrosine hydroxylase gene expression through a cAMP-dependent signaling pathway. Brain Res Mol Brain Res 142: 28–38

DeFilippis M, Wagner KD (2016) Treatment of autism spectrum disorder in children and adolescents. Psychopharmacol Bull 46(2):18–41

Desbonnet L, Garrett L, Clarke G et al (2008) The probiotics Bifidobacteria infantis: an assessment of potential antidepressant properties in the rat. J Psychiatr Res 43(2):164–174

Diaz Heijz R, Wang S, Anuar F et al (2011) Normal gut microbiota modulates brain development and behavior. Proc Natl Acad Sci U S A 108:3047–3052

DiCicco-Bloom E, Lord C, Zwaigenbaum L et al (2006) The developmental neurobiology of autism spectrum disorder. J Neurosci 26(26):6897–6906

Dickson RP, Erb-Downward JR, Prescott HC et al (2014) Cell-associated bacteria in the human lung microbiome. Microbiome 2:28

Dinan TG, Stilling RM, Stanton C et al (2015) Collective unconscious: how gut microbes shape human behavior. J Psychiatr Res 63:1–9

Ding T, Schloss PD (2014) Dynamics and associations of microbial community types across the human body. Nature 509:357–360

Ding HT, Taur Y, Walkup JT (2017) Gut microbiota and autism: key concepts and findings. J Autism Dev Disord 47(2):480–489

Doenyas C (2018a) Gut microbiota, inflammation, and probiotics on neural development in autism spectrum disorder. Neuroscience 374:271–286

Doenyas C (2018b) Dietary interventions for autism spectrum disorder: new perspectives from the gut-brain axis. Physiol Behav 194:577–582

Donaldson ZR, Young LJ (2008) Oxytocin, vasopressin and the neurogenetics of sociality. Science 322:900–904

Douglas-Escobar M, Elliott E, Neu J (2013) Effect of intestinal microbial ecology on the developing brain. JAMA Pediatr 167:374–379

Doyle RM, Alber DG, Jones HE et al (2014) Term and preterm labour are associated with distinct microbial community structures in placental membranes which are independent of mode of delivery. Placenta 35:1099–1101

El-Ansary A, Bacha AB, Bjørklund G et al (2018) Probiotic treatment reduces the autistic-like excitation/inhibition imbalance in juvenile hamsters induced by orally administered propionic acid and clindamycin. Metab Brain Dis 33:1155–1164

Emanuele E, Orsi P, Boso M et al (2010) Low-grade endotoxemia in patients with severe autism. Neurosci Lett 471(3):162–165

Erickson CA, Stigler KA, Corkins MR et al (2005) Gastrointestinal factors in autistic disorder: a critical review. J Autism Dev Disord 35(6):713–727

Evangeliou A, Vlachonikolis I, Mihailidou H et al (2003) Application of a ketogenic diet in children with autistic behavior: pilot study. J Child Neurol 18(2):113–118

Fallon J (2005) Could one of the most widely prescribed antibiotics amoxicillin/clavulanate "Augmentin" be a risk factor for autism? Med Hypotheses 64:312–315

FAO/WHO (2001) Health and nutritional properties of probiotics in food including powder milk with live lactic acid bacteria. FAO, Rome

Fernell E, Fagerberg UL, Hellström PM (2007) No evidence for a clear link between active intestinal inflammation and autism based on analyses of faecal calprotectin and rectal nitric oxide. Acta Paediatr 96:1076–1079

Finegold SM (2011) Desulfovibrio species are potentially important in regressive autism. Med Hypotheses 77(2):270–274

Finegold SM, Dowd SE, Gontcharova V et al (2010) Pyrosequencing study of fecal microflora of autistic and control children. Anaerobe 16(4):444–453

Finegold SM, Downes J, Summanen PH (2012) Microbiology of regressive autism. Anaerobe 18(2):260–262

Fiorentino M, Sapone A, Senger S et al (2016) Blood-brain barrier and intestinal epithelial barrier alterations in autism spectrum disorders. Mol Autism 7:49

Foley KA, MacFabe DF, Kavaliers M et al (2015) Sexually dimorphic effects of prenatal exposure to lipopolysaccharide, and prenatal and postnatal exposure to propionic acid, on acoustic startle response and prepulse inhibition in adolescent rats: relevance to autism spectrum disorders. Behav Brain Res 278:244–256

Fond G, Boukouaci W, Chevalier G et al (2014) The "psychomicrobiotic": targeting microbiota in major psychiatric disorders: a systematic review. Pathol Biol 63:35–42

Freeman BA (1979) Clostridium the spore-forming anaerobes. In: Burrows textbook of microbiology, 21st edn. Saunders, Philadelphia, PA. Chapter 27

Frye RE, Sreenivasula S, Adams JB et al (2011) Traditional and non-traditional treatments for autism spectrum disorder with seizures: an on-line survey. BMC Pediatr 11:37

Galland L (2014) The gut microbiome and the brain. J Med Food 17:1261–1272

Gaspar P, Cases O, Maroteaux L (2003) The developmental role of serotonin: news from mouse molecular genetics. Nat Rev Neurosci 4:1002–1012

Giannetti E, Staiano A (2016) Probiotics for irritable bowel syndrome: clinical data in children. J Pediatr Gastroenterol Nutr 63(Suppl. 1):S25–S26

Gibson G, Roberfroid M, De Louuain C (1995) Dietary modulation of the human colonic microbiota: introducing the concept of prebiotics. J Nutr 125:1401–1412

Gibson GR, Hutkins R, Sanders ME et al (2017) The International Scientific Association for Probiotics and Prebiotics (ISAPP) consensus statement on the definition and scope of prebiotics. Nat Rev Gastroenterol Hepatol 14:491–502

Golubeva AV, Joyce SA, Moloney G et al (2017) Microbiota-related changes in bile acid & tryptophan metabolism are associated with gastrointestinal dysfunction in a mouse model of autism. EBioMedicine 24:166–178

Gondalia SV, Palombo EA, Knowles SR et al (2012) Molecular characterisation of gastrointestinal microbiota of children with autism (with and without gastrointestinal dysfunction) and their neurotypical siblings. Autism Res 5:419–427

Govindarajan N, Agis-Balboa RC, Walter J et al (2011) Sodium butyrate improves memory function in an Alzheimer's disease mouse model when administered at an advanced stage of disease progression. J Alzheimers Dis 26:187–197

Green SA, Ben-Sasson A (2010) Anxiety disorders and sensory overresponsivity in children with autism spectrum disorders: is there a causal relationship? J Autism Dev Disord 40:1495–1504

Grenham S, Clarke G, Cryan JF et al (2011) Brain gut-microbe communication in health and disease. Front Physiol 2:94

Grimaldi R, Cela D, Swann JR et al (2017) In vitro fermentation of B-GOS: impact on faecal bacterial populations and metabolic activity in autistic and non-autistic children. FEMS Microbiol Ecol 93:fiw233

Groot J, Bijlsma P, Van Kalkeren A et al (2000) Stress-induced decrease of the intestinal barrier function. The role of muscarinic receptor activation. Ann N Y Acad Sci 915:237–246

Grossi E, Melli S, Dunca D et al (2016) Unexpected improvement in core autism spectrum disorder symptoms after long-term treatment with probiotics. SAGE Open Med Case Rep 4 2050313X16666231

Hanley HG, Stahl SM, Freedman DX (1977) Hyperserotonemia and amine metabolites in autistic and retarded children. Arch Gen Psychiatry 34:521–531

Herbert MR, Buckley JA (2013) Autism and dietary therapy: case report and review of the literature. J Child Neurol 28(8):975–982

Hong J, Jia Y, Pan S et al (2016) Butyrate alleviates high fat diet-induced obesity through activation of adiponectin-mediated pathway and stimulation of mitochondrial function in the skeletal muscle of mice. Oncotarget 7:56071–56082

Hooper LV, Midtvedt T, Gordon JI (2002) How host-microbial interactions shape the nutrient environment of the mammalian intestine. Annu Rev Nutr 22:283–307

Hsiao EY, McBride SW, Hsien S (2013) Microbiota modulate behavioral and physiological abnormalities associated with neurodevelopmental disorders. Cell 155(7):1451–1463

Israelyar N, Margolis KG (2018) Serotonin as a link between the gut-brain microbiome axis in autism spectrum disorders. Pharmacol Res 132:1–6

Jacob C, Jacob C, Yang PC, Darmoul D et al (2005) Mast cell tryptase controls paracellular permeability of the intestine. Role of protease-activated receptor 2 and beta-arrestins. J Biol Chem 280(36):31936–31948

Jones HF, Butler RN, Brooks DA (2011) Intestinal fructose transport and malabsorption in humans. Am J Physiol Gastrointest Liver Physiol 300(2):G202–G206

Jonkers D, Penders J, Masclee A et al (2012) Probiotics in the management of inflammatory bowel disease: a systematic review of intervention studies in adult patients. Drugs 72:803–823

Jostins L, Ripke S, Weersma RK et al (2012) Host-microbe interactions have shaped the genetic architecture of inflammatory bowel disease. Nature 491(7422):119–124

Junaid MA, Kowal D, Barua M et al (2004) Proteomic studies identified a single nucleotide polymorphism in glyoxalase I as autism susceptibility factor. Am J Med Genet A 131:11–17

Jurecka A, Zikanova M, Jurkiewicz E et al (2014) Attenuated adenylosuccinate lyase deficiency: a report of one case and a review of the literature. Neuropediatrics 45(1):50–55

Jyonouchi H, Sun S, Itokazu N (2002) Innate immunity associated with inflammatory responses and cytokine production against common dietary proteins in patients with autism spectrum disorder. Neuropsychobiology 46:76–84

Kaluzna-Czaplinska J, Blaszczyk S (2012) The level of arabinitol in autistic children after probiotic therapy. Nutrition 28:124–126

Kang DW, Park JG, Ilhan ZE et al (2013) Reduced incidence of Prevotella and other fermenters in intestinal microflora of autistic children. PLoS One 8(7):e68322

Kang DW, Adams JB, Gregory AC et al (2017) Microbiota transfer therapy alters gut ecosystem and improves gastrointestinal and autism symptoms: an open-label study. Microbiome 5:1–16

Kang D, Adams JB, Co DM et al (2019) Long-term benefit of microbiota transfer therapy on autism symptoms and gut microbiota. Sci Rep 9:5821

Kanner L (1943) Autistic disturbances of affective contact. Nervous Child 2:217–250

Keita AV, Soderholm JD (2010) The intestinal barrier and its regulation by neuroimmune factors. Neurogastroenterol Motil 22:718–733

Kelly CR, Kahn S, Kashyap P et al (2015) Update on fecal microbiota transplantation 2015: indications, methodologies, mechanisms, and outlook. Gastroenterology 149:223–237

Kolevzon A, Newcorn JH, Kryzak L et al (2010) Relationship between whole blood serotonin and repetitive behaviors in autism. Psychiatry Res 175:274–276

Korpela K, Salonen A, Virta LJ et al (2016) Intestinal microbiome is related to lifetime antibiotic use in Finnish pre-school children. Nat Commun 7:10410

Kraneveld AD, Szklany K, de Theije CG et al (2016) Gut-to-brain axis in autism spectrum disorders: central role for the microbiome. Int Rev Neurobiol 131:263–287

Kratsman N, Getselter D, Elliott E (2016) Sodium butyrate attenuates social behavior deficits and modifies the transcription of inhibitory/excitatory genes in the frontal cortex of an autism model. Neuropharmacology 102:136–145

Kushak RI, Buie TM, Murray KF et al (2016) Evaluation of intestinal function in children with autism and gastrointestinal symptoms. J Pediatr Gastroenterol Nutr 62(5):687–691

Kwon HK, Kim GC, Kim Y et al (2013) Amelioration of experimental autoimmune encephalomyelitis by probiotic mixture is mediated by a shift in T helper cell immune response. Clin Immunol 146:217–227

Lavasani S, Dzhambazov B, Nouri M et al (2010) A novel probiotic mixture exerts a therapeutic effect on experimental autoimmune encephalomyelitis mediated by IL-10 producing regulatory T cells. PLoS One 5:e9009

Le Galliard JF, Cote J, Fitze PS (2008) Lifetime and intergenerational fitness consequences of harmful male interactions for female lizards. Ecology 89:56–64

LeBlanc JG, Milani C, de Giori GS et al (2013) Bacteria as vitamin suppliers to their host: a gut microbiota perspective. Curr Opin Biotechnol 24:160–168

Lefter R, Ciobica A, Timofte D et al (2019) A descriptive review on the prevalence of gastrointestinal disturbances and their multiple associations in autism spectrum disorder. Medicina 56(1):11

Lembo A, Camilleri M (2003) Chronic constipation. New Engl J Med 349(14):1360–1368

Lessa FC, Mu Y, Bamberg WM et al (2015) Burden of Clostridium difficile infection in the United States. N Engl J Med 372:825–834

Li M, Fallin MD, Riley A et al (2016) The association of maternal obesity and diabetes with autism and other developmental disabilities. Pediatrics 137:e20152206

Li Q, Han Y, Dy ABC (2017) Hagerman, R. The gut microbiota and autism spectrum disorders. Front Cell Neurosci 11:120

Liu Y, Fatheree NY, Mangalat N et al (2010) Human-derived probiotic Lactobacillus reuteri strains differentially reduce intestinal inflammation. Am J Physiol Gastrointest Liver Physiol 299:G1087–G1096

Liu J, Liu X, Xiong XQ et al (2017) Effect of vitamin A supplementation on gut microbiota in children with autism spectrum disorders—a pilot study. BMC Microbiol 17:204

Macfabe DF (2012) Short-chain fatty acid fermentation products of the gut microbiome: implications in autism spectrum disorders. Microb Ecol Health Dis 23:19260

MacFabe DF (2015) Enteric short-chain fatty acids: microbial messengers of metabolism, mitochondria, and mind: implications in autism spectrum disorders. Microb Ecol Health Dis 26:28177

MacFabe DF, Cain DP, Rodriguez-Capote K et al (2007) Neurobiological effects of intraventricular propionic acid in rats: possible role of short chain fatty acids on the pathogenesis and characteristics of autism spectrum disorders. Behav Brain Res 176:149–169

MacFabe DF, Cain NE, Boon F et al (2011) Effects of the enteric bacterial metabolic product propionic acid on object-directed behavior, social behavior, cognition, and neuroinflammation in adolescent rats: relevance to autism spectrum disorder. Behav Brain Res 217 47–54

Macfarlane S, Dillon JF (2007) Microbial biofilms in the human gastrointestinal tract. J Appl Microbiol 102(5):1187–1196

Maenner MJ, Shaw KA, Baio J et al (2020) Prevalence of autism spectrum disorder among children aged 8 years — autism and developmental disabilities monitoring network, 11 Sites, United States, 2016. MMWR Surveill Summ 69:1–12

de Magistris L, Familiari V, Pascotto A et al (2010) Alterations of the intestinal barrier in patients with autism spectrum disorders and in their first-degree relatives. J Pediatr Gastroenterol Nutr 51:418–424

Mancuso C, Santangelo R (2018) Alzheimer's disease and gut microbiota modifications: the long way between preclinical studies and clinical evidence. Pharmacol Res 129:329–336

Mantis JG, Fritz CL, Marsh J et al (2009) Improvement in motor and exploratory behavior in Rett syndrome mice with restricted ketogenic and standard diets. Epilepsy Behav 15 133–141

Marler S, Ferguson BJ, Lee EB et al (2016) Brief report: whole blood serotonin levels and gastrointestinal symptoms in autism spectrum disorder. J Autism Dev Disord 46:1124–1130

Mayer EA (2000) The neurobiology of stress and gastrointestinal disease. Gut 47(6):861–869

Mayer EA (2011) Gut feelings: the emerging biology of gut-brain communication. Nat Rev Neurosci 12:453–466

Mayer EA, Padua D, Tillisch K (2014) Altered brain-gut axis in autism: comorbidity or causative mechanisms? BioEssays 36:933–939

Mayer EA, Tillisch K, Gupta A (2015) Gut/brain axis and the microbiota. J Clin Invest 125(3): 926–938

McElhanon BO, McCracken C, Karpen S (2014) Gastrointestinal symptoms in autism spectrum disorder: a meta-analysis. Pediatrics 133:872–883

McVey Neufeld KA, Mao YK, Bienenstock J et al (2013) The microbiome is essential for normal gut intrinsic primary afferent neuron excitability in the mouse. Neurogastroenterol Motil 25: 183–188

Meyza KZ, Defensor EB, Jensen AL et al (2013) The BTBR T+ tf/J mouse model for autism spectrum disorders-in search of biomarkers. Behav Brain Res 251:25–34

Minderaa RB, Anderson GM, Volkmar FR et al (1987) Urinary 5-hydroxyindoleacetic acid and whole blood serotonin and tryptophan in autistic and normal subjects. Biol Psychiatry 22:933–940

Ming X, Stein TP, Barnes V et al (2012) Metabolic perturbance in autism spectrum disorders: a metabolomics study. J Proteome Res 11:5856–5862

Mulder EJ, Anderson GM, Kemperman RF et al (2010) Urinary excretion of 5-hydroxyindoleacetic acid, serotonin and 6-sulphatoxymelatonin in normoserotonemic and hyperserotonemic autistic individuals. Neuropsychobiology 61:27–32

Muller CL, Anacker AMJ, Veenstra-VanderWeele J (2016) The serotonin system in autism spectrum disorder: from biomarker to animal models. Neuroscience 321:24–41

Munasinghe SA, Oliff C, Finn J et al (2010) Digestive enzyme supplementation for autism spectrum disorders: a double-blind randomized controlled trial. J Autism Dev Disord 40: 1131–1138

Nahum Sacks K, Friger M, Shoham-Vardi I et al (2016) Prenatal exposure to gestational diabetes mellitus as an independent risk factor for long-term neurologic morbidity of the offspring. Am J Obstet Gynecol 214:S48–S49

Navarro F, Liu Y, Rhoads JM (2016) Can probiotics benefit children with autism spectrum disorders? World J Gastroenterol 22:10093–10102

Nicholson JK, Holmes E, Kinross J et al (2012) Host-gut microbiota metabolic interactions. Science 336:1262–1267

Noto A, Fanos V, Barberini L et al (2014) The urinary metabolomics profile of an Italian autistic children population and their unaffected siblings. J Matern Fetal Neonatal Med 27(Suppl. 2): 46–52

O'Mahony SM, Marchesi JR, Scully P et al (2009) Early life stress alters behavior, immunity, and microbiota in rats: implications for irritable bowel syndrome and psychiatric illnesses. Biol Psychiatry 65(3):263–267

O'Mahony SM, Hyland NP, Dinan TG et al (2011) Maternal separation as a model of brain–gut axis dysfunction. Psychopharmacology 214(1):71–88

O'Mahony SM, Clarke G, Borre YE et al (2015) Serotonin, tryptophan metabolism and the brain-gut-microbiome axis. Behav Brain Res 277:32–48

Onore C, Careaga M, Ashwood P (2012) The role of immune dysfunction in the pathophysiology of autism. Brain Behav Immun 26:383–392

Panksepp J (1979) A neurochemical theory of autism. Trends Neurosci 2:174–177

Parracho HM, Bingham MO, Gibson GR et al (2005) Differences between the gut microflora of children with autistic spectrum disorders and that of healthy children. J Med Microbiol 54 (Pt 10):987–991

Parracho HM, Gibson GR, Knott F et al (2010) A double-blind, placebo-controlled, crossover-designed probiotic feeding study in children diagnosed with autistic spectrum disorders. Int J Probiot Prebiot 5(2):69–74

Pärtty A, Kalliomäki M, Wacklin P (2015) A possible link between early probiotics intervention and the risk of neuropsychiatric disorders later in childhood: a randomized trial. Pediatr Res 77: 823–828

Persico AM, Napolioni V (2013) Urinary p-cresol in autism spectrum disorder. Neurotoxicol Tertol 36:82–90

Pirbaglou M, Katz J, de Souza RJ et al (2016) Probiotic supplementation can positively affect anxiety and depressive symptoms: a systematic review of randomized controlled trials. Nutr Res 36(9):889–898

Qiao Y, Wu M, Feng Y et al (2018) Alterations of oral Microbiota distinguish children with autism spectrum disorders from healthy controls. Sci Rep 8(1):1597

Quigley EM (2016) Leaky gut-concept or clinical entity? Curr Opin Gastroenterol 32:74–79

Reddy BL, Saier MH (2015) Autism and our intestinal microbiota. J Mol Microbiol Biotechnol 25: 51–55

Romijn JA, Corssmit EP, Havekes LM et al (2008) Gut-brain axis. Curr Opin Clin Nutr Metab Care 11:518–521

Rose S, Bennuri SC, Davis JE et al (2018) Butyrate enhances mitochondrial function during oxidative stress in cell lines from boys with autism. Transl Psychiatry 8:42

Rosenfeld CS (2015) Microbiome disturbances and autism spectrum disorders. Drug Metab Dispos 43:1557–1571

Rossen NG, MacDonald JK, de Vries EM et al (2015) Fecal microbiota transplantation as novel therapy in gastroenterology: a systematic review. World J Gastroenterol 21:5359–5371

Round JL, Mazmanian SK (2009) The gut microbiota shapes intestinal immune responses during health and disease. Nat Rev Immunol 9(5):313–323

Ruskin DN, Svedova J, Cote JL et al (2013) Ketogenic diet improves core symptoms of autism in BTBR mice. PLoS One 8(6):e65021

Sandler RH, Finegold SM, Bolte ER et al (2000) Short-term benefit from oral vancomycin treatment of regressive-onset autism. J Child Neurol 15(7):429–435

Sanger GJ, Lee K (2008) Hormones of the gut-brain axis as targets for the treatment of upper gastrointestinal disorders. Nat Rev Drug Discov 7:241–254

Santocchi E, Guiducci L, Fulceri F et al (2016) Gut to brain interaction in autism spectrum disorders: a randomized controlled trial on the role of probiotics on clinical, biochemical and neurophysiological parameters. BMC Psychiatry 16:183

Saulnier DM, Gibson GR, Kolida S (2008) In vitro effects of selected synbiotics on the human faecal microbiota composition. FEMS Microbiol Ecol 66:516–527

Savignac HM, Corona G, Mills H et al (2013) Prebiotic feeding elevates central brain derived neurotrophic factor, N-methyl-D-aspartate receptor subunits and D-serine. Neurochem Int 63: 756–764

Savignac HM, Couch Y, Stratford M et al (2016) Prebiotic administration normalizes lipopolysaccharide (LPS)-induced anxiety and cortical 5-HT2A receptor and IL1-b levels in male mice. Brain Behav Immun 52:120–131

Scheperjans F, Aho V, Pereira PA et al (2015) Gut microbiota are related to Parkinson's disease and clinical phenotype. Mov Disord 30(3):350–358

Schmidt K, Cowen PJ, Harmer CJ et al (2015) Prebiotic intake reduces the waking cortisol response and alters emotional bias in healthy volunteers. Psychopharmacology 232:1793–1801

Schultz ST, Klonoff-Cohen HS, Wingard DL et al (2006) Breastfeeding, infant formula supplementation, and Autistic Disorder: the results of a parent survey. Int Breastfeed J 15(1):16

Shaaban SY, ElGendy YG, Mehanna NS et al (2017) The role of probiotics in children with autism spectrum disorder: a prospective, open-label study. Nutr Neurosci 7:1–6

Shaw W (2010) Increased urinary excretion of a 3-(3-hydroxyphenyl)-3-hydroxypropionic acid (HPHPA), an abnormal phenylalanine metabolite of Clostridia spp. in the gastrointestinal tract, in urine samples from patients with autism and schizophrenia. Nutr Neurosci 13:135–143

Sherman MP, Zaghouani H, Niklas V (2015) Gut microbiota, the immune system, and diet influence the neonatal gut-brain axis. Pediatr Res 77:127–135

Shultz SR, MacFabe DF, Ossenkopp KP et al (2008) Intracerebroventricular injection of propionic acid, an enteric bacterial metabolic end-product, impairs social behavior in the rat: implications for an animal model of autism. Neuropharmacology 54:901–911

Shultz SR, Macfabe DF, Martin S et al (2009) Intracerebroventricular injections of the enteric bacterial metabolic product propionic acid impair cognition and sensorimotor ability in the Long-Evans rat: further development of a rodent model of autism. Behav Brain Res 200:33–41

Shultz SR, Aziz NA, Yang L et al (2015) Intracerebroventricular injection of propionic acid, an enteric metabolite implicated in autism, induces social abnormalities that do not differ between seizure-prone (FAST) and seizure resistant (SLOW) rats. Behav Brain Res 278 542–548

Song Y, Liu C, Molitoris DR et al (2003) Clostridium bolteae sp. nov., isolated from human sources. Syst Appl Microbiol 26(1):84–89

Song Y, Liu C, Finegold SM (2004) Real-time PCR quantitation of clostridia in feces of autistic children. Appl Environ Microbiol 70:6459–6465

Spilioti M, Evangeliou AE, Tramma D et al (2013) Evidence for treatable inborn errors of metabolism in a cohort of 187 Greek patients with autism spectrum disorder (ASD). Front Hum Neurosci 24(7):858

Sponheim E (1991) Gluten-free diet in infantile autism. A therapeutic trial. Tidsskr Nor Laegeforen 111:704–707

van Steensel FJ, Bögels SM, Perrin S (2011) Anxiety disorders in children and adolescents with autistic spectrum disorders: a meta-analysis. Clin Child Fam Psychol Rev 14(3):302–317

Stilling RM, Dinan TG, Cryan JF (2014) Microbial genes, brain & behaviour – epigenetic regulation of the gut-brain axis. Genes Brain Behav 13:69–86

Stumpf RM, Wilson BA, Rivera A et al (2013) The primate vaginal microbiome: comparative context and implications for human health and disease. Am J Phys Anthropol 152(Suppl 57): 119–134

Sudo N, Chida Y, Aiba Y et al (2004) Postnatal microbial colonization programs the hypothalamic-pituitary-adrenal system for stress response in mice. J Physiol 558:263–275

Sun J, Ling Z, Wang F et al (2016a) Clostridium butyricum pretreatment attenuates cerebral ischemia/reperfusion injury in mice via anti-oxidation and anti-apoptosis. Neurosci Lett 613: 30–35

Sun J, Wang F, Hong G et al (2016b) Antidepressant-like effects of sodium butyrate and its possible mechanisms of action in mice exposed to chronic unpredictable mild stress. Neurosci Lett 618: 159–166

Tabuchi K, Blundell J, Etherton MR et al (2007) A neuroligin-3 mutation implicated in autism increases inhibitory synaptic transmission in mice. Science 318:71–76

Takuma K, Hara Y, Kataoka S et al (2014) Chronic treatment with valproic acid or sodium butyrate attenuates novel object recognition deficits and hippocampal dendritic spine loss in a mouse model of autism. Pharmacol Biochem Behav 126:43–49

de Theije CG, Wopereis H, Ramadan M et al (2014a) Altered gut microbiota and activity in a murine model of autism spectrum disorders. Brain Behav Immun 37:197–206

de Theije CG, Wu J, Koelink PJ et al (2014b) Autistic like behavioural and neurochemical changes in a mouse model of food allergy. Behav Brain Res 261:265–274

Thomas RH, Meeking MM, Mepham JR et al (2012) The enteric bacterial metabolite propionic acid alters brain and plasma phospholipid molecular species: further development of a rodent model of autism spectrum disorders. J Neuroinflammation 9:153

Tomova A, Husarova V, Lakatosova S et al (2015) Gastrointestinal microbiota in children with autism in Slovakia. Physiol Behav 138:179–187

Trottier G, Srivastava L, Walker CD (1999) Etiology of infantile autism: a review of recent advances in genetic and neurobiological research. J Psychiatry Neurosci 24:103–115

Tuohy KM, Venuti P, Cuva S et al (2015) Diet and the gut microbiota-how the gut:brain axis impacts on autism. In: Tuohy K, Del Rio D (eds) Diet-microbe interactions in the gut. Academic Press, London, pp 225–245

Umbrello G, Esposito S (2016) Microbiota and neurologic diseases: potential effects of probiotics. J Transl Med 14:298

Veenstra-VanderWeele J, Muller CL et al (2012) Autism gene variant causes hyperserotonemia, serotonin receptor hypersensitivity, social impairment and repetitive behavior. Proc Natl Acad Sci U S A 109:5469–5474

Viggiano D, Ianiro G, Vanella G et al (2015) Gut barrier in health and disease: focus on childhood. Eur Rev Med Pharmacol Sci 9:1077–1085

Vojdani A, O'Bryan T, Green JA et al (2004) Immune response to dietary proteins, gliadin and cerebellar peptides in children with autism. Nutr Neurosci 7:151–161

Voth DE, Ballard JD (2005) Clostridium difficile toxins: mechanism of action and role in disease. Clin Microbiol Rev 2:247–263

Vrieze A, de Groot PF, Kootte RS (2013) Fecal transplant: a safe and sustainable clinical therapy for restoring intestinal microbial balance in human disease? Best Pract Res Clin Gastroenterol 27: 127–137

Vuong HE, Hsiao EY (2017) Emerging roles for the gut microbiome in autism spectrum disorder. Biol Psychiatry 81:411–423

Walker SJ, Fortunato J, Gonzalez LG et al (2013) Identification of unique gene expression profile in children with regressive autism spectrum disorder (ASD) and ileocolitis. PLoS One 8:e58058

Wang Y, Kasper LH (2014) The role of microbiome in central nervous system disorders. Brain Behav Immun 38:1–12

Wang L, Christophersen CT, Sorich MJ et al (2011a) Low relative abundances of the mucolytic bacterium Akkermansia muciniphila and Bifidobacterium spp. in feces of children with autism. Appl Environ Microbiol 77:6718–6721

Wang LW, Tancredi DJ, Thomas DW (2011b) The prevalence of gastrointestinal problems in children across the United States with autism spectrum disorders from families with multiple affected members. J Dev Behav Pediatr 32:351–360

Wang L, Christophersen CT, Sorich MJ et al (2012) Elevated fecal short chain fatty acid and ammonia concentrations in children with autism spectrum disorder. Dig Dis Sci 57:2096–2102

Wang L. Christophersen CT, Sorich MJ et al (2013) Increased abundance of Sutterella spp. and Ruminococcus torques in feces of children with autism spectrum disorder. Mol Autism 4:42

West R, Roberts E, Sichel LS et al (2013) Improvements in gastrointestinal symptoms among children with autism spectrum disorder receiving the Delpro® probiotic and immunomodulator formulation. J Prob Health 1(1):102

Whiteley P, Shattock P (2002) Biochemical aspects in autism spectrum disorders: updating the opioid-excess theory and presenting new opportunities for biomedical intervention. Expert Opin Ther Targets 6:175–183

Whiteley P, Shattock P, Carr K et al (2010) How could a gluten and casein-free diet ameliorate symptoms associated with autism spectrum conditions? Autism Insights 2010:39–53

Williams BL, Hornig M, Buie T et al (2011) Impaired carbohydrate digestion and transport and mucosal dysbiosis in the intestines of children with autism and gastrointestinal disturbances. PLoS One 6:e24585

Williams BL, Hornig M, Parekh T (2012) Application of novel PCR-based methods for detection, quantitation, and phylogenetic characterization of Sutterella species in intestinal biopsy samples from children with autism and gastrointestinal disturbances. MBio 3(1):e00261–e00211

Yang Y, Tian J, Yang B (2018) Targeting gut microbiome: a novel and potential therapy for autism. Life Sci 194:111–119

Yap Y, Angley M, Veselkov KA et al (2010) Urinary metabolic phenotyping differentiates children with autism from their unaffected siblings and age-matched controls. J Proteome Res 9:2996–3004

Yassour M, Vatanen T, Siljander H et al (2016) Natural history of the infant gut microbiome and impact of antibiotic treatment on bacterial strain diversity and stability. Sci Transl Med 8(343):343ra81

Zhang F, Luo W, Shi Y et al (2012) Should we standardize the 1,700-year-old fecal microbiota transplantation? Am J Gastroenterol 107:1755–1756

Zhang Y, Li L, Guo C et al (2016) Effects of probiotic type, dose and treatment duration on irritable bowel syndrome diagnosed by Rome III criteria: a meta-analysis. BMC Gastroenterol 16:62

...tics Ameliorate Gut–Brain Dysbiosis ...ism Spectrum Disorder by ...lating Nrf2-Keap1 Signaling Pathway

5

... Bhandari and Anurag Kuhad

Abstract

Autism spectrum disorder (ASD) involves deficient ability to socially interact and communicate, pervasive, and stereotypic behavior. It also involves co-morbidities such as anxious and aggressive nature and epilepsy. Apart from above, this disorder also involves physiological symptoms which co-exist with behavioral symptoms, such as dysfunction of immune system and mitochondria as well as gastrointestinal complications, which lead to oxidative stress and neuroinflammation, further worsening the behavioral complications. Twenty-three percent to 70% of patients suffering from ASD account for gastrointestinal complications and these correlate with behaviors which are relevant to autistic endophenotype. A strong gut–brain dysbiosis occurs in ASD patients in consequence to the enormous production of short-chain fatty acids such as propanoic acid (PPA) by abnormal gut flora which worsens the behavioral, neurochemical, and mitochondrial dysfunction. This further leads to the generation of free radical species responsible for the synthesis of pro-inflammatory cytokines, which cause microglia activation. The Nrf2-Keap1 signaling pathway in the brain might be a plausible therapeutic target in autism for targeting gut–brain dysbiosis responsible for worsening behavioral and biochemical alterations. Its activation can serve as a putative link in targeting oxidative stress as the delicate balance of the Nrf2-NFkB pathway is responsible for either protection by antioxidant genes and phase II enzymes or generation of pro-inflammatory cytokines. Probiotics are potential neurotherapeutics which can target the Nrf2-Keap1/ARE pathway and mitigate the neuroinflammatory cascade. Hence, it can prove to be utilized as an adjunct therapy for targeting the gut–brain dysbiosis generated inflammatory cascade.

R. Bhandari · A. Kuhad (✉)
Pharmacology Research Laboratory, University Institute of Pharmaceutical Sciences, UGC-Centre of Advanced Study, Panjab University, Chandigarh, India
e-mail: anurag.kuhad@pu.ac.in

© Springer Nature Singapore Pte Ltd. 2022
P. K. Deol, S. K. Sandhu (eds.), *Probiotic Research in Therapeutics*,
https://doi.org/10.1007/978-981-16-6760-2_5

Keywords

Autism spectrum disorders · Nrf2-Keap1 · Gut–brain dysbiosis · Nrf2-NFk
cross-talk · Probiotics

5.1   Introduction

Autism spectrum disorder (ASD) is an intricate neurodevelopmental disorder which is typically characterized by deficient social interaction and communication skills, restrictive, rigid, pervasive, as well as stereotypic behavior (American Psychiatric Association 2013; Estabillo et al. 2018). The prevalence of ASD is 1 in 59 as reported by the US Center for Disease Control and Prevention (CDC) (Baio et al. 2018). ASD involves complex etiology where there is an amalgamation of genetic factors, epigenetic, and environmental factors, as suggested by current research reports (Santangelo and Tsatsanis 2005). Patients who have ASD are also associated with co-morbidities such as anxiety, depression, aggression, epilepsy, and the most commonly seen gastrointestinal complications (American Psychiatric Association 2013; Chaidez et al. 2014; Estabillo et al. 2018). Research and clinical studies have indicated that these gastrointestinal complications are further responsible for immune system deregulation, mitochondrial dysfunction leading to oxidative stress and neuroinflammation, which worsens the behavioral complexities of ASD patients (Hsiao et al. 2013).

The prevalence of gastrointestinal complications such as diarrhea, bloated stomach, abdominal pain, and constipation range from 23% to 70% in autistic patients as these patients have abnormal gut flora (Chaidez et al. 2014). Abnormal growth of gut micro-flora includes bacteria such as Clostridia, Desulfovibrio, Sutterella, and Ruminococcus species, and these produce short-chain fatty acids (SCFAs) such as propanoic acid (PPA). These are a result of the metabolism of carbohydrates as well as amino acids obtained from the diet (Hooper et al. 2002; Finegold et al. 2002; Macfabe 2012; Wang et al. 2013; Frye et al. 2016). Release of pro-inflammatory cytokines and the reduction in the levels of endogenous antioxidants such as glutathione, superoxide dismutase (SOD), and an abnormal increase in the levels of lipid peroxidase was a result of excess production of PPA by the gut bacteria (MacFabe et al. 2007; Shultz et al. 2009).

There are no available treatments for these physiological co-morbidities such as gastrointestinal complications resulting in gut–brain dysbiosis, which is also responsible for the worsening of core behavioral symptoms of autism. Pharmacological agents such as antipsychotics (risperidone and aripiprazole), anticonvulsants, antidepressants, mood stabilizers, medications for ADHD, NMDA receptor antagonists, melatonin, oxytocin, and omega-3 fatty acids provide relief only for associated symptoms such as anxiety, depression, irritability, and epilepsy (Benvenuto et al. 2013; Young and Findling 2015).

5.2 Gut–Brain Dysbiosis in ASD

An intense gut–brain dysbiosis exists in autism, which leads to the worsening of behavioral as well as gastrointestinal symptoms after consumption of a diet rich in carbohydrates or food in which short-chain fatty acids such as PPA which is a by-product of gut microbiota are present as a preservative (Horvath et al. 1999; Jyonouchi et al. 2002). PPA can stimulate the release of serotonin from enteric neurons and cause severe contractions of abdominal muscles (MacFabe et al. 2007). Studies have been done to understand the effect of CNS exposure of PPA on the behavioral, biochemical, and molecular alterations and to understand its similarity with the ASD endophenotype (MacFabe et al. 2007, 2008). Gut microbiota plays an integral part in the modulation of gut–brain cross-talk, and dysbiosis can have adverse effects on the gastrointestinal tract as well as behaviors of autistic children by either regulating neuroimmune system via hypothalamic-pituitary (HPA) axis or production of chemicals such as short-chain fatty acids (Macfabe 2012; Sudo 2016). Liu et al. (2019) have investigated in their systematic review the critical role of gut microbiota in gastrointestinal symptoms and behavioral alterations occurring in ASD children. Sixteen studies include a total sample size of 381 ASD patients and 283 healthy controls (HCs) in this systematic review. They have indicated that 381 ASD patients demonstrated an increase in Desulfovibrio, Clostridia, Bacteroides, and Lactobacillus species as compared to 283 healthy controls, and these results correlated with their behavioral changes.

Ng et al. (2019) indicated that open-label trials found that probiotic supplementation resulted in significant improvement in gastrointestinal symptoms as well as mitigated behavioral alterations in autistic children. Prebiotics in a combination of gluten and casein-free diet showed a considerable increase in the sociability index in autistic children. However, they showed limited efficacy in double-blind placebo-controlled randomized clinical trials indicating that further mechanistic studies and clinical trials with large sample size are required for some final pieces of evidence. Hence, gut microbiota has a critical link with dysbiosis and immune system deregulation, which also affects the behaviors of ASD patients.

Gut–brain dysbiosis results in activation of the immune system and can cause mitochondrial dysfunction as PPA inhibits the activity of mitochondrial complexes I, II, III, and IV because it will enter Kreb's cycle as propionyl-CoA. Inhibition of mitochondrial complex activity will cause inhibition of proximal part of Kreb's cycle resulting in a reduction in the ratio of NADH:FADH$_2$ from 3:1 to 1:1, which is further responsible for disruption of ETC by reducing the production of ATP (Thomas et al. 2012; Frye et al. 2013; Macfabe 2015). Disruption of electron transport chain (ETC) of mitochondria leads to damage of mitochondrial DNA. This further leads to the generation of reactive oxygen species (ROS) as a result of oxidative stress. ROS will further lead to the production of pro-inflammatory cytokines responsible for activation of NF-kB leading to neuroinflammation and further behavioral changes which are characteristic of ASD (Macfabe 2015; Frye et al. 2016). These observations are in line with the clinical reports on immune system activation and mitochondrial dysfunction occurring in ASD patients.

Dysregulation of immune system, as well as enhanced levels of pro-inflammatory cytokines such as TNF-α, IL-6, IL-β, IL-2, IL-4, IL-13, has been found in the brain of ASD patients (Bjorkland et al. 2016). Patients who had autism have enhanced levels of pro-inflammatory cytokines such as TNF-α, reduced INF-γ, and increased activation of glial cells in the cerebrospinal, plasma as well as amniotic fluid (Croonenberghs et al. 2002; Nakagawa and Chiba 2016). These inflammatory cytokines might be responsible for the generation of ASD-like behaviors (Masi et al. 2017).

ASD involves various physiological co-morbidities, and hence, mitochondrial dysfunction, as well as oxidative stress, is considered to be a major part of these (Chauhan and Chauhan 2006; Frye et al. 2016). Electron transport chain dysfunction is one of the key mechanisms involved in the pathogenesis of autism. Rossignol and Frey (Rossignol and Frye 2012) have revealed through their meta-analysis that 30% of children suffering from autism show specific biomarkers indicating that these children suffer from mitochondrial dysfunction and the rate of its occurrence in autism was 5.0% which is markedly greater in comparison to those found in healthy population indicating that there is a close connection between autism and dysfunction of the mitochondrial respiratory chain (Skladal et al. 2003). Immune system dysregulation and oxidative stress as a result of gut–brain cross-talk can release pro-inflammatory cytokines and generate neuroinflammatory cascade which might be responsible for mitochondrial dysfunction. Another plausible reason for acquired mitochondrial dysfunction could be a result of exposure to environmental pollutants such as heavy metals, pesticides, insecticides, and biphenyls which can result in epigenetic modifications. Hence, gut–brain dysbiosis, genetic susceptibility as well as various environmental factors lead to immune system activation, oxidative stress, and mitochondrial dysfunction in autism spectrum disorder patients (Gottfried and Savino 2015; Sealey et al. 2016; Cristiano et al. 2018). All of these physiological co-morbidities with autism can worsen their behavioral complications.

5.3   The Putative Role of Nrf2-Keap1 as a Therapeutic Target in ASD

In general, Nrf2, a transcription factor protects the cellular damage by oxidative-nitrosative stress. Protection from oxidative stress by Nrf2 occurs due to activation of various antioxidant genes and phase II enzymes such as NADPH, GSH, SOD, catalase, HO-1, and NQO1 expressed on antioxidant response elements (ARE). Nrf2 binds to ARE and maintains homeostasis in the cell by balancing redox pathways. Reduction in its activity can lead to apoptosis. Under basal conditions, Nrf2 binds covalently to cysteine residues on Keap1 (Kelch-like ECH associated protein 1), which is its repressor in the plasma membrane. It sequesters Nrf2 inside the cell. Binding of Nrf2 with Keap1 will eventually result in the ubiquitination of Nrf2 and proteasomal degradation responsible for inhibition of antioxidant response. Keap1 acts as a linker responsible for the interaction of Nrf2 via CuI3-Rbx E3 ubiquitin ligase complex (Itoh et al. 1997; Zhang and Hannink 2003; Nguyen et al. 2009; Lu et al. 2016). Children who have ASD have leaky gut and have associated gastrointestinal complications which are responsible for enhanced production of short-chain fatty acids such as PPA by abnormal gut flora such as Clostridia, Desulfovibrio, and Sutterella species (Macfabe 2015). PPA can cross the BBB and can cause mitochondrial dysfunction as a result of the disruption of the electron transport chain. There will be a generation of reactive oxygen species (ROS) and reactive nitrogen species (RNS) as a result of oxidative stress (Macfabe 2012; Vuong and Hsiao 2017; Cristiano et al. 2018). In the case of mild-moderate oxidative stress, there is the activation of the Nrf2/Keap1/ARE pathway for protecting the cell. ROS will modify cysteine residues of Keap1, disrupting the Nrf2-Keap1 complex and hence, leading to translocation of Nrf2 into the nucleus. In the nucleus, Nrf2 will get associated with Maf proteins. After that it binds to the promoter region of the antioxidant response element (ARE) initiating the transcription of antioxidant genes and phases II enzymes, which will inhibit ROS (Zhang and Hannink 2003; Nguyen et al. 2009; Lu et al. 2016). NF-kB is a key transcription factor responsible for regulating neuroinflammation in response to immune system dysregulation as a result of enhanced oxidative stress. It suppresses the transcription of antioxidant genes and phases II enzymes. It consists of p65, p50, p52, and RelB subunits. Under physiological conditions, NF-kB is in an inactive state as it is sequestered inside the cell by its inhibitor IkB-α. Enhanced oxidative stress causes activation of IKK-β, which is responsible for the phosphorylation of IkB-α (Liu et al. 2017). This subsequently leads to the translocation of NF-kB into the nucleus as a result of the proteasomal degradation of its inhibitor. NF-kB binds in the nucleus with the genome at the k region which is the binding site for NF-kB. This leads to the transcription of various pro-inflammatory cytokines and other genes such as TNF-α, Il-β, Il-6, iNOS, COX-2 (Sandberg et al. 2014; Wardyn et al. 2015; Liu et al. 2017). NF-kB binding sites have been found in the promoter region of the Nrf2 gene, indicating that there might be cross-talk between Nrf2 and NF-kB pathway (Nair et al. 2008). Nrf2/Keap1/ARE pathway inhibits NF-kB pathway activation as a result of enhanced expression of HO-1 and other antioxidant genes. There is another process by which the Nrf2/Keap1/ARE pathway inhibits NF-kB pathway activation. After translocation of Nrf2 inside the nucleus, free Keap1 inhibits phosphorylation of IkB-α by binding to IKK-β and will prevent the activation of the NF-kB pathway (Nair et al. 2008). The p65 subunit of NF-kB acts as a negative regulator of the Nrf2/Keap1 pathway by facilitating the binding of HDAC3, which is a corepressor to the ARE region via interaction of Maf proteins resulting in hypoacetylation of local histone proteins and hence leading to downregulation of Nrf2 signaling (Liu et al. 2008). This can also occur via deprivation of CBP, which is a binding protein of CREB. Literature reports have indicated that p65 also interacts with Keap1 and inhibits Nrf2 signaling as well as it can enhance the ubiquitination of Nrf2 and prevents binding of Nrf2 to ARE for the facilitation of transcription (Yu et al. 2011). Hence, Nrf2-Keap1 pathway and its cross-talk with NF-kB can be a potential therapeutic target by various dietary phytochemicals as well as probiotics (Fig. 5.1).

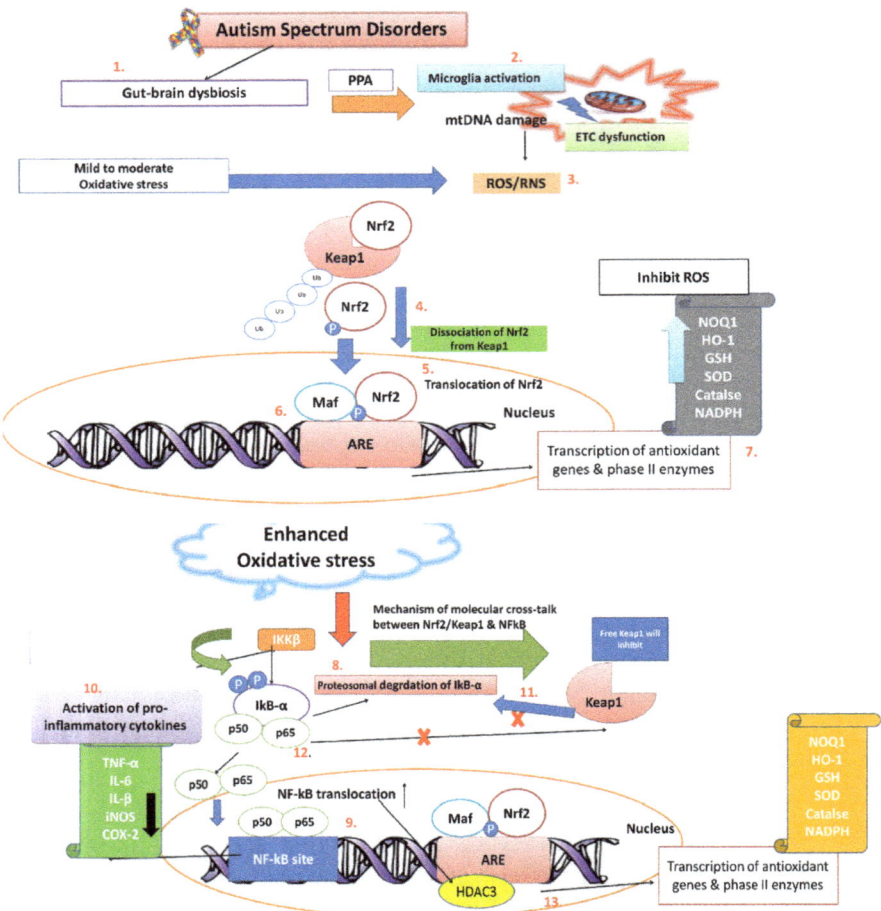

Fig. 5.1 Molecular mechanism of the Nrf2-Keap1 pathway as a putative link in targeting oxidative stress and neuroinflammation occurring as a result of gut–brain dysbiosis in ASD (1) Gastrointestinal complications occur in ASD which results in enhanced production of PPA by abnormal gut bacteria such as Clostridia, Desulfovibrio, and Sutterella species responsible for gut–brain dysbiosis. (2) Electron transport chain is disrupted by PPA after crossing the BBB leading to mitochondrial dysfunction. (3) Oxidative stress leads to generation of reactive oxygen species (ROS). (4) Mild-moderate oxidative stress leads to the activation of the Nrf2/ARE signaling pathway resulting in dissociation of Nrf2 from its inhibitor Keap1. (5) It will then translocate in the nucleus. (6) Nrf2 will associate with Maf protein in the nucleus and bind to antioxidant response element (ARE). (7) Antioxidant genes and phase II enzymes such as NADPH, GSH, SOD, catalase, heme-oxygenase-1, and NQO1 will undergo transcription. These will inhibit ROS. (8) Enhanced oxidative stress will activate IKKβ, which causes phosphorylation of IkB-α, a NF-kB inhibitor, and results in proteasomal degradation of IkB-α. (9) NF-kB, transcription factor regulating immune system activation, will bind to its region after migrating into the nucleus (10) This will result in transcription of pro-inflammatory cytokines and other genes such as TNF-α, Il-β, Il-6, iNOS, COX-2. (11) There is molecular cross-talk between the Nrf2/ARE signaling pathway and the NF-kB pathway as free Keap1 will prevent degradation of IkB-α, leading to inhibition of the NF-kB pathway. (12) Keap1 is inhibited by p65 subunit of NF-kB. Hence, interfering with the transcription caused by Nrf2. (13) NF-kB also facilitates the binding of HDAC3

5.4   Probiotics in Autism

Several pre-clinical and clinical studies have suggested the involvement of gut–brain dysbiosis in ASD. Probiotics can help in the restoration of the altered microbiota and can cause modulation of gut–brain cross-talk by influencing both behaviors and neurochemistry in ASD patients (Slattery et al. 2016). International Scientific Association for Probiotics and Prebiotics (ISAPP) has defined probiotics as living organisms which when consumed in adequate amounts result in improvement of the health of the host. These are useful in maintaining the composition of the intestinal as well as gut-microflora and result in improved resistance of the host against various infections (Li and Zhou 2016; Gibson et al. 2017). There are several research reports which indicate the beneficial effect of probiotics as a therapeutic in the improvement of social interaction, communication deficits, and gastrointestinal complications in ASD patients.

The culture of probiotic microbes is available in the market in the form of lyophilized powders or as highly concentrated frozen cultures. Probiotic microorganisms which are used are various strains of lactobacillus bacteria such as *L. rhamnosus*, *L. plantarum*, *L. casei*, *L. reuteri*, and *L. acidophilus*. Other microbes such as Bifidobacteria species, *Bacillus coagulans*, and *Enterococcus faecium* as well as yeast are also used. Its effect is highly specific to the strain used and many times commercially available products contain a mixture of two or more strains (Pandey et al. 2015; Fowlie et al. 2018; Ng et al. 2019).

Hsiao et al. (2013) have indicated gastrointestinal complications as a result of altered microbiota as demonstrated in the maternal immune activation model of ASD. It was observed that per-oral treatment with the probiotic microbe, Bacteroides fragilis resulted in amelioration of social interaction and communication deficits, improvement in repetitive and pervasive behavior as well as sensorimotor dysfunction. There was also a significant improvement in the gut-permeability. Several clinical studies have indicated the beneficial effect of probiotics in amelioration of behavioral and biochemical abnormalities in autistic children (Table 5.1).

Hence, alteration of microbiota as a result of probiotic treatment can be explored as an adjunct therapy for the amelioration of symptoms associated with ASD. It has also been found that short-chain fatty acids produced as a result of fermentation by gut bacteria play a major role in the modulation of the gut–brain axis and amelioration of behavioral symptoms associated with ASD. These short-chain fatty acids can also cause alteration of neurotransmitter levels such as serotonin and dopamine. Administration of probiotic produces neurotransmitters such as serotonin, dopamine, GABA, acetylcholine (MacFabe et al. 2011; Macfabe 2015; Hughes et al. 2018; Pulikkan et al. 2019).

Fig. 5.1 (continued) (histone deacetylase3) to the ARE region because of its binding to Maf proteins This will lead to repression of transcription caused by Nrf2. Nrf2-Keap1 pathway and its cross-talk with NF-kB can be a potential therapeutic target by various dietary phytochemicals as well as probiotics

Table 5.1 Clinical Study indicating the beneficial effect of probiotic treatment in ASD patients

S. No.	Clinical study	Probiotic treatment	Result	References
1.	Children were administered probiotic supplement for 6 months	DelPro® contained various strains of Lactobacilli such as *acidophilus*, *casei*, and *delbrueckii*, Bifidobacterium *longum*, and *bifidum* (10 billion colony forming units). It also included 8 mg powder containing peptidoglycan, peptides, and various nucleosides containing components obtained from the strain of *L. rhamnosus*	The effect of probiotic treatment was evaluated based on ATEC score (Autism Treatment Evaluation Checklist score). It was observed that there was a significant improvement in the gastrointestinal complications as well as sociability	West et al. (2013)
2.	A randomized, double-blind controlled clinical study was carried out in children suffering from autism. These children were in the age group of 3–16 years. These children either received a placebo or treatment	Chronic treatment was started with *Lactobacillus plantarum* WCFS1 for 6 weeks (4.5×10^{10} CFU)	Treatment with probiotics led to an improvement in communication deficit, social interaction ability, and a reduction in anxiety	Parracho et al. (2010)
3.	The abnormal gut flora as a result of overproduction of Clostridia and Bacteroidia has resulted in changes in the permeability of the gut, gut–brain dysbiosis, and immune system activation as indicated by the production of inflammatory cytokines. This was observed in the maternal immune system activated model of autism. Changes were also	Probiotic treatment was started with Bacteroides fragilis	Treatment resulted in the improvement of gut-permeability, communication deficits, sensorimotor and locomotion dysfunction, anxiety of the offspring of MIA model mothers. Treatment with probiotics also resulted in the restoration of the microbiome. Hence, it was inferred that the treatment of autistic children with probiotics resulted in the restoration of gut dysbiosis caused by	Doenyas (2018)

(continued)

Table 5.1 (continued)

S. No.	Clinical study	Probiotic treatment	Result	References
	observed in core behavioral symptoms specific to autism		abnormal gut flora. Hence, leading to a reduction in immune system activation, improvement of permeability of gut, and amelioration of behavioral symptoms of autism	
4.	This clinical study involved autistic patients suffering from gastrointestinal complications	Lactobacillus acidophilus probiotic capsules were given to patients as a treatment, BID for 2 months orally. These capsules contained $(5 \times 10^9 \text{ CFU/g})$ of *L. acidophilus*	It was observed that after treatment with probiotic for 2 months there was a significant reduction in the levels of D-arabinitol in the urine of autistic children post-treatment. There was also improvement in behavioral symptoms	Kałużna-Czaplińska and Błaszczyk (2012)
5.	A clinical trial was conducted in autistic patients where they were administered with vancomycin 500 mg/day, TID for 8 weeks	Following the Vancomycin treatment patients were also administered probiotic orally for 4 weeks consisting of *Lactobacillus acidophilus*, *Lactobacillus bulgaricus*, and *B. bifidum*	Results indicated that probiotic therapy led to an improvement in the social interaction and communication deficit as well as reduction in repetitive and pervasive behavior	Sandler et al. (2000)
6.	Twenty-two autistic children within the age group of 4–10 years were given probiotic treatment	These patients were administered a sugar-free diet and as a treatment was given Lactobacillus acidophilus capsules $(5 \times 10^9 \text{ CFU/g})$ for 2 months, BID	Results indicated that there was a significant improvement in some of the autistic behaviors such as rigid and pervasive behavior and the ability to concentrate on different tasks. However, no improvement was observed in other behavior as well as eye-contact ability	Romeo et al. (2011)

(continued)

Table 5.1 (continued)

S. No.	Clinical study	Probiotic treatment	Result	References
7.	This clinical study involved 26 autistic children and 24 healthy children. This study was divided into two stages. In the first stage 16S rRNA, gene sequencing analysis was used to compare the gut microbiota profile of autistic and normal children. Apart from this, the fecal concentration of short-chain fatty acids (SCFA) utilizing GC-MS as well as estimation of levels of plasma neurotransmitters and measurement of metabolites was also done using UHPLC-MS/MS. Gut-permeability was accessed by the estimation of zonulin as a marker by ELISA. In the second stage, the subjects were divided into two groups. 16 subjects received probiotic supplement along with fructo-oligosaccharide (FOS) as an intervention. While ten subjects were given a placebo. Then again they were tested for gut microbiota analysis, SCFA content, and neurotransmitter analysis	Children were administered probiotics along with fructo-oligosaccharide (FOS). The probiotic supplement included a mixture of strains such as Bifidobacterium longum, *B. infantis*, *L. rhamnosus*, *L. paracasei*, *B. lactis* (10^{10} CFU/day). FOS was added as it is a growth factor for Bifidobacterium and can stimulate the growth of probiotics	Results indicated that gut dysbiosis occurred in autistic children as indicated by significantly lower levels of good bacteria such as Bifidobacterium longum and Bifidobacteriales. Treatment with probiotic and FOS resulted in the suppression of Clostridia. This resulted in a significant reduction in the behavioral as well as gastrointestinal complications. Treatment with probiotics resulted in a reduction in the levels of serotonin and an increase in HVA levels. Hence, treatment with probiotics resulted in modulation of the microbiome, short-chain fatty acid levels as well as improvement in ASD symptoms	Wang et al. (2020)

5.5   Modulation of the Nrf2-Keap1 Pathway by Probiotics: Plausible Mechanism

Nrf2-Keap1 pathway as also explained in the above section can be explored as a potential therapeutic target in ASD for targeting the gut–brain dysbiosis and the neuroinflammation as well as oxidative stress associated with it. Activation of the Nrf2 pathway upregulates various genes that are involved in detoxification of reactive oxygen species (ROS) (Wardyn et al. 2015). Hence, inducers of this pathway can protect the cellular function against the environmental stressors (Kraft et al. 2004). It has been reported by Jones et al. (2013) that lactobacilli can induce physiological levels of ROS in cells of the intestinal epithelium and can consequently trigger the induction of cellular proliferation. Gut bacteria can stimulate ROS production in epithelial cells. This cellular ROS is generated as a result of the catalytic action of NADPH oxidases. Here, Nox1 is an isozyme in non-phagocytic tissues that generate ROS. Nox1 expression is very high in enterocytes of the colon (Jones et al. 2013; Lambeth and Neish 2014). Hence, this non-pathogenic generation of ROS can also be done by symbiotic gut bacteria via Nox1 especially lactobacilli strain and helps in the promotion of cellular proliferation and can also cause modification of NF-kB signaling (Kumar et al. 2007). Nrf2-Keap1 signaling pathway responds to reactive oxygen species which is responsible for causing oxidative stress. Hence, at low levels of ROS as generated in the case of gut bacteria Nrf2 is bound to its inhibitor Keap1, which is its repressor in the plasma membrane. It sequesters Nrf2 inside the cell. Binding of Nrf2 with Keap1 will eventually result in the ubiquitination of Nrf2 and proteasomal degradation responsible for inhibition of antioxidant response. Keap1 acts as a linker responsible for the interaction of Nrf2 via Cul3-Rbx E3 ubiquitin ligase complex (Itoh et al. 1997; Zhang and Hannink 2003; Nguyen et al. 2009; Lu et al. 2016). ROS will modify cysteine residues of Keap1, disrupting the Nrf2-Keap1 complex and hence, leading to translocation of Nrf2 into the nucleus. In the nucleus, Nrf2 will get associated with Maf proteins. After that, it binds to the promoter region of the antioxidant response element (ARE) initiating the transcription of antioxidant genes and phases II enzymes, which will inhibit ROS (Zhang and Hannink 2003; Nguyen et al. 2009; Lu et al. 2016). Hence, it has been indicated by Jones et al. (2015) through their study that the physiological level of ROS is generated by this symbiotic bacteria such as Lactobacilli via Nox1 which can activate Nrf2 signaling pathway and therefore are responsible for mechanistic mediation of beneficial effects of gut microbiota. The results have indicated that L. plantarum induces cellular protection against oxidative stress and causes upregulation of CncC pathway genes in Drosophila. This signaling pathway is responsible for the mediation of L. plantarum induced cellular protection against oxidative stress in Drosophila. Nox1 is essential for ROS generation by Lactobacilli and hence is required for optimal lactobacilli induced cellular protection. This will further activate the Nrf2-Keap1 signaling pathway and will cause cellular protection by the generation of phase 2 detoxification response. Wang et al. (2017) have indicated in their study on Bacillus amyloliquefaciens SC06 that oxidative stress is primarily involved in gastrointestinal complications and this could be ameliorated after treatment with probiotics. Hence, they deciphered the mechanism behind the antioxidant capacity of various Bacillus strains utilizing hydrogen peroxide to induce intestinal porcine epithelial cell1 oxidative stress model. Various bacillus strains were co-cultured with IPEC-1 for 3 h. Out of all the strains *Bacillus amyloliquefaciens* SC06, mediated IPEC-1 showed maximum antioxidant capacity. The various antioxidant as well as apoptotic genes were studied such as the effects of Bacillus SC06 on the Nrf2/Keap1 pathway. Apart from this, production of reactive oxygen species, membrane potential, apoptosis as well as necrosis were also looked into. It was observed that HO-1 gene expression had increased. Further, it was observed that gene expression of phase II genes had altered as well as there was an alteration in the expression of SOD-1, GPX-2, TRX-1. Even the levels of reduced glutathione had increased after co-culturing with SC06. Hence, the changes in the expression of phase II genes and detoxification enzymes indicated the involvement of the Nrf2/Keap1 signaling pathway. It was also observed that the Bacillus SC06 strain could also phosphorylate Nrf2. There are some other research reports as well which have revealed that during oxidative stress administration of probiotic could induce the Nrf2/Keap1 pathway and its antioxidant systems (Gao et al. 2013). *Lactobacillus gasseri* (LG2055) is a probiotic bacteria and it was demonstrated by Kobatake et al. (2017) that when *C. elegans* were fed with *L. gasseri* their lifespan was extended and it was related to the regulation of reactive oxygen species. It was observed that after treatment with paraquat mouse embryonic fibroblast cells and NIH-3T3 cells showed high levels of ROS and low proliferation of cells. Treatment with probiotics suppressed ROS and increased proliferation of cells. The mRNA expression as indicated by qRT-PCR of genes related to oxidative stress and Nrf2 protein levels was significantly enhanced after treatment with probiotic. Translocation of Nrf2 in the nucleus after dissociation from Keap1 was also enhanced after treatment with *L. gasseri*. Hence, there was the activation of the Nrf2-Keap1/ARE signaling pathway. It was also observed that the Nrf2-Keap1 signaling pathway was activated by *L. gasseri* via the MAPK pathway. This pathway includes ERK, JNK, and p38 MAPK which are responsible for activation of Nrf2-signaling. All the three MAPKs were studied utilizing various inhibitors and it was observed that activation of Nrf2-Keap1/ARE signaling pathway involved JNK activation. After inhibition of JNK, it was observed that there was significant suppression of Nrf2 induction and the expression of its target genes in comparison to ERK or p38 MAPK. However, further studies are required to elucidate the mechanism of activation of JNK by *L. gasseri* (Kobatake et al. 2017). Saeedi et al. (2020) have also demonstrated that the gut microbiota can induce Nrf2-Keap1 signaling pathway in the liver. Lactobacilli can induce this pathway and it has been demonstrated in both Drosophila and mice. Oral administration of *L. rhamnosus* resulted in protection against liver injury as a result of oxidative stress via activation of Nrf2 pathway by 5-methoxyindoleacetic acid, produced by lactobacilli. Hence, probiotics can cause modulation of Nrf2-Keap1 signaling pathway leading to activation of phase II genes and detoxifying enzymes via Nox1 or by activation of MAPK/ERK pathway. Still, further pre-clinical and clinical research is required to elucidate the mechanistic pathway at the molecular

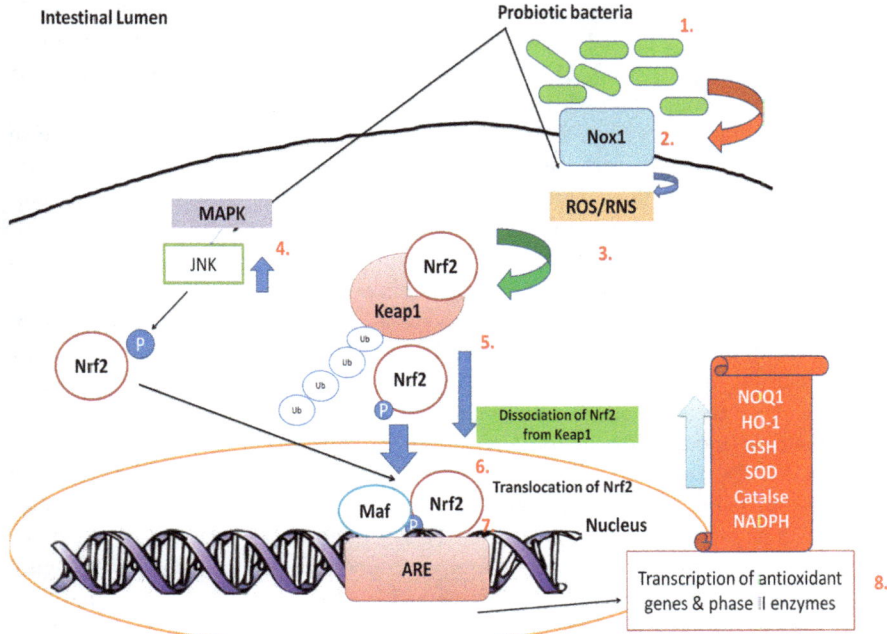

Fig. 5.2 Plausible mechanism of action of probiotics via modulation of the Nrf2-Keap1 pathway. (1) Probiotic bacteria such as *Lactobacilli/Bacillus* can stimulate ROS production via catalytic action of Nox1. (2) Nox1 is an isozyme and it belongs to the category of NADPH oxidases. (3) Once, ROS/RNS is generated it leads to activation of the Nrf2/Keap1 pathway. (4) Probiotic bacteria also activate JNK signaling and subsequently leads to activation of the Nrf2/Keap1 pathway. (5) After activation, Nrf2 dissociates from Keap1. (6) Translocation of Nrf2 takes place inside the nucleus. (7) In the nucleus, Nrf2 gets associated with Maf protein and binds to antioxidant response element (ARE). (8) This leads to the inhibition of ROS by transcription of antioxidant genes and phase II enzymes such as NADPH, GSH, SOD, catalase, heme-oxygenase-1, and NQO1

level to understand the intricacies. Figure 5.2 highlights the plausible mechanism of action of probiotics via modulation of the Nrf2-Keap1 pathway.

5.6   Perspectives

Autism spectrum disorder is an intricate disorder with the heterogeneous nature and multifactorial origin. Complex genetic mutations as well as epigenetic changes are involved in this. It is a complex disorder involving behavioral complications and several co-morbidities such as anxiety, epilepsy, gut–brain dysbiosis, oxidative stress, mitochondrial dysfunction, and biochemical changes. The physiological co-morbidities further worsen behavioral and biochemical complications. Till date, no neurotherapeutic agent exists which reduces the occurrence of physiological co-morbidities such as gastrointestinal complications, oxidative stress, mitochondrial dysfunction, and neuroinflammation arising as a result of these. Nrf2-Keap1/ARE signaling cascade can serve as a putative link in regulating the gut–brain dysbiosis occurring in ASD. This dysbiosis is also one of the causes of oxidative stress, mitochondrial dysfunction, and immune system deregulation leading to neuroinflammation (Lu et al. 2016). Probiotics can serve as a safer alternative in comparison to existing therapeutic agents discovered to date for autism. Several clinical studies have indicated the beneficial effect of probiotics such as *Lactobacilli*, *Bacillus*, and *Bifidobacteria* strain in amelioration of gastrointestinal complications, improvement in behavioral symptoms as well as biochemical and mitochondrial dysfunction. Research reports have indicated that probiotics hold enormous potential as a neurotherapeutic agent for autism spectrum disorders as they are activators of Nrf2 pathway either via activation of ERK and PI3K/Akt signaling pathway or by the generation of ROS via Nox1. However, we further require some more pre-clinical as well as clinical studies of these probiotics, which elucidate the mechanism of modulation of the Nrf2-Keap1 pathway in autism utilizing either Nox1 or MAPK pathways. Enormous cost of healthcare causes a burden on the caregivers of patients suffering from autism spectrum disorders. Hence, probiotics represent a safe and inexpensive alternative to mitigate the oxidative stress, mitochondrial dysfunction, and neuroinflammation occurring in consequence of gut–brain dysbiosis in autistic children which also worsens their behavioral complications. Therefore, the current unmet requirement is to formulate brain targeted delivery systems of probiotics and to explore their clinical potential as neurotherapeutics in ASD.

5.7   Conclusions

Therefore, probiotics ameliorate gut–brain dysbiosis in autism spectrum disorders by modulating the Nrf2-Keap1 signaling pathway and can be utilized as potential neurotherapeutics in ASD.

Conflict of Interest The authors declare no conflict of interest.

References

American Psychiatric Association (2013) Diagnostic and statistical manual of mental disorders. American Psychiatric Association, Washington, DC

Baio J, Wiggins L, Christensen DL et al (2018) Prevalence of autism spectrum disorder among children aged 8 years - autism and developmental disabilities monitoring network, 11 Sites, United States, 2014. MMWR Surveill Summ 67:1. https://doi.org/10.15585/mmwr.ss6706a1

Benvenuto A, Battan B, Porfirio MC, Curatolo P (2013) Pharmacotherapy of autism spectrum disorders. Brain and Development 35:119–127. https://doi.org/10.1016/j.braindev.2012.03.015

Bjorkland G, Saad K, Chirumbolo S et al (2016) Immune dysfunction and neuroinflammation in autism spectrum disorder. Acta Neurobiol Exp 76:257–268. https://doi.org/10.5772/22318

Chaidez V, Hansen RL, Hertz-Picciotto I (2014) Gastrointestinal problems in children with autism, developmental delays or typical development. J Autism Dev Disord 44:1117–1127. https://doi.org/10.1007/s10803-013-1973-x

Chauhan A, Chauhan V (2006) Oxidative stress in autism. Pathophysiology 13:171–181. https://doi.org/10.1016/j.pathophys.2006.05.007

Cristiano C, Lama A, Lembo F et al (2018) Interplay between peripheral and central inflammation in autism spectrum disorders: possible nutritional and therapeutic strategies. Front Physiol 9:184. https://doi.org/10.3389/fphys.2018.00184

Croonenberghs J, Bosmans E, Deboutte D et al (2002) Activation of the inflammatory response system in autism. Neuropsychobiology 45:1–6

Doenyas C (2018) Gut microbiota, inflammation, and probiotics on neural development in autism spectrum disorder. Neuroscience 374:271–286

Estabillo JA, Matson JL, Cervantes PE (2018) Autism symptoms and problem behaviors in children with and without developmental regression. J Dev Phys Disabil 30:17–26. https://doi.org/10.1007/s10882-017-9573-x

Finegold SM, Molitoris D, Song Y et al (2002) Gastrointestinal microflora studies in late-onset autism. Clin Infect Dis 35:S6–S16. https://doi.org/10.1086/341914

Fowlie G, Cohen N, Ming X (2018) The perturbance of microbiome and gut-brain axis in autism spectrum disorders. Int J Mol Sci 19:2251

Frye RE, Melnyk S, Macfabe DF (2013) Unique acyl-carnitine profiles are potential biomarkers for acquired mitochondrial disease in autism spectrum disorder. Transl Psychiatry 3:e220. https://doi.org/10.1038/tp.2012.143

Frye RE, Rose S, Chacko J et al (2016) Modulation of mitochondrial function by the microbiome metabolite propionic acid in autism and control cell lines. Transl Psychiatry 6:e927. https://doi.org/10.1038/tp.2016.189

Gao D, Gao Z, Zhu G (2013) Antioxidant effects of Lactobacillus plantarum via activation of transcription factor Nrf2. Food Funct 4:982–989. https://doi.org/10.1039/c3fo30316k

Gibson GR, Hutkins R, Sanders ME et al (2017) Expert consensus document: the International Scientific Association for Probiotics and Prebiotics (ISAPP) consensus statement on the definition and scope of prebiotics. Nat Rev Gastroenterol Hepatol 14:491–502

Gottfried C, Savino W (2015) The impact of neuroimmune alterations in autism spectrum disorder. Front Psychiatry 6:1–16. https://doi.org/10.3389/fpsyt.2015.00121

Hooper LV, Midtvedt T, Gordon JI (2002) How host-microbial interactions shape the nutrient environment of the mammalian intestine. Annu Rev Nutr 22:283–307

Horvath K, Papadimitriou JC, Rabsztyn A et al (1999) Gastrointestinal abnormalities in children with autistic disorder. J Pediatr 135:559–563

Hsiao EY, McBride SW, Hsien S et al (2013) Microbiota modulate behavioral and physiological abnormalities associated with neurodevelopmental disorders. Cell 155:1451–1463. https://doi.org/10.1016/j.cell.2013.11.024

Hughes HK, Rose D, Ashwood P (2018) The gut microbiota and dysbiosis in autism spectrum disorders. Curr Neurol Neurosci Rep 18:81

Itoh K, Chiba T, Takahashi S et al (1997) An Nrf2/small Maf heterodimer mediates the induction of phase II detoxifying enzyme genes through antioxidant response elements. Biochem Biophys Res Commun 236:313–322. https://doi.org/10.1006/BBRC.1997.6943

Jones RM, Luo L, Ardita CS et al (2013) Symbiotic lactobacilli stimulate gut epithelial proliferation via Nox-mediated generation of reactive oxygen species. EMBO J 32:3017–3028. https://doi.org/10.1038/emboj.2013.224

Jones RM, Desai C, Darby TM et al (2015) Lactobacilli modulate epithelial cytoprotection through the Nrf2 pathway. Cell Rep 12:1217–1225. https://doi.org/10.1016/j.celrep.2015.07.042

Jyonouchi H, Sun S, Itokazu N (2002) Innate immunity associated with inflammatory responses and cytokine production against common dietary proteins in patients with autism spectrum disorder. Neuropsychobiology 46:76–84. PMID 65416

Kałuzna-Czaplińska J, Błaszczyk S (2012) The level of arabinitol in autistic children after probiotic therapy. Nutrition 28:124–126. https://doi.org/10.1016/j.nut.2011.08.002

Kobatake E, Nakagawa H, Seki T, Miyazaki T (2017) Protective effects and functional mechanisms of Lactobacillus gasseri SBT2055 against oxidative stress. PLoS One 12:1–17. https://doi.org/10.1371/journal.pone.0177106

Kraft AD, Johnson DA, Johnson JA (2004) Nuclear factor E2-related factor 2-dependent antioxidant response element activation by tert-butylhydroquinone and sulforaphane occurring preferentially in astrocytes conditions neurons against oxidative insult. J Neurosci 24:1101–1112. https://doi.org/10.1523/JNEUROSCI.3817-03.2004

Kumar A, Wu H, Collier-Hyams LS et al (2007) Commensal bacteria modulate cullin-dependent signaling via generation of reactive oxygen species. EMBO J 26:4457–4466. https://doi.org/10.1038/sj.emboj.7601867

Lambeth JD, Neish AS (2014) Nox enzymes and new thinking on reactive oxygen: a double-edged sword revisited. Annu Rev Pathol Mech Dis 9:119–145. https://doi.org/10.1146/annurev-pathol-012513-104651

Li Q, Zhou JM (2016) The microbiota-gut-brain axis and its potential therapeutic role in autism spectrum disorder. Neuroscience 324:131–139

Liu G-H, Qu J, Shen X (2008) NF-κB/p65 antagonizes Nrf2-ARE pathway by depriving CBP from Nrf2 and facilitating recruitment of HDAC3 to MafK. Biochim Biophys Acta, Mol Cell Res 1783:713–727. https://doi.org/10.1016/J.BBAMCR.2008.01.002

Liu T, Zhang L, Joo D, Sun SC (2017) NF-κB signaling in inflammation. Signal Transduct Target Ther 2:17023

Liu F, Li J, Wu F et al (2019) Altered composition and function of intestinal microbiota in autism spectrum disorders: a systematic review. Transl Psychiatry 9:43. https://doi.org/10.1038/s41398-019-0389-6

Lu M-C, Ji J-A, Jiang Z-Y, You Q-D (2016) The Keap1-Nrf2-ARE pathway as a potential preventive and therapeutic target: an update. Med Res Rev 36:924–963. https://doi.org/10.1002/med.21396

Macfabe DF (2012) Short-chain fatty acid fermentation products of the gut microbiome: implications in autism spectrum disorders. Microb Ecol Health Dis 23. https://doi.org/10.3402/mehd.v23i0.19260

Macfabe DF (2015) Enteric short-chain fatty acids: microbial messengers of metabolism, mitochondria, and mind: implications in autism spectrum disorders. Microb Ecol Health Dis 26:28177. https://doi.org/10.3402/mehd.v26.28177

MacFabe DF, Cain DP, Rodriguez-Capote K et al (2007) Neurobiological effects of intraventricular propionic acid in rats: possible role of short chain fatty acids on the pathogenesis and characteristics of autism spectrum disorders. Behav Brain Res 176:149–169. https://doi.org/10.1016/j.bbr.2006.07.025

MacFabe DF, Rodríguez-Capote K, Hoffman JE et al (2008) A novel rodent model of autism: intraventricular infusions of propionic acid increase locomotor activity and induce neuroinflammation and oxidative stress in discrete regions of adult rat brain. Am J Biochem Biotechnol 4:146–166. https://doi.org/10.3844/ajbbsp.2008.146.166

MacFabe DF, Cain NE, Boon F et al (2011) Effects of the enteric bacterial metabolic product propionic acid on object-directed behavior, social behavior, cognition, and neuroinflammation in adolescent rats: relevance to autism spectrum disorder. Behav Brain Res 217:47–54. https://doi.org/10.1016/j.bbr.2010.10.005

Masi A, Glozier N, Dale R, Guastella AJ (2017) The immune system, cytokines, and biomarkers in autism spectrum disorder. Neurosci Bull 33:194–204. https://doi.org/10.1007/s12264-017-0103-8

Nair S, Doh ST, Chan JY et al (2008) Regulatory potential for concerted modulation of Nrf2- and Nfkb1-mediated gene expression in inflammation and carcinogenesis. Br J Cancer 99:2070–2082. https://doi.org/10.1038/sj.bjc.6604703

Nakagawa Y, Chiba K (2016) Minireviews involvement of neuroinflammation during brain development in social cognitive deficits in autism spectrum disorder and schizophrenia. J Pharmacol Exp Ther 358:504–515

Ng QX, Loke W, Venkatanarayanan N et al (2019) A systematic review of the role of prebiotics and probiotics in autism spectrum disorders. Medicina 55:129. https://doi.org/10.3390/medicina55050129

Nguyen T, Nioi P, Pickett CB (2009) The Nrf2-antioxidant response element signaling pathway and its activation by oxidative stress. J Biol Chem 284:13291–13295. https://doi.org/10.1074/jbc.R900010200

Pandey KR, Naik SR, Vakil BV (2015) Probiotics, prebiotics and synbiotics - a review. J Food Sci Technol 52:7577–7587

Parracho HMRT, Gibson GR, Knott F, Bosscher D, Kleerebezem M, McCartney AL (2010) A double-blind, placebo-controlled, crossover designed probiotic feeding study in children diagnosed with autism spectrum disorders. Int J Probiot Prebiot 5:69–74

Pulikkan J, Mazumder A, Grace T (2019) Role of the gut microbiome in autism spectrum disorders. In: Advances in experimental medicine and biology. Springer, New York, NY, pp 253–259

Romeo MG, Romeo DM, Trovato L et al (2011) Role of probiotics in the prevention of the enteric colonization by Candida in preterm newborns: incidence of late-onset sepsis and neurological outcome. J Perinatol 31:63–69. https://doi.org/10.1038/jp.2010.57

Rossignol DA, Frye RE (2012) Mitochondrial dysfunction in autism spectrum disorders: a systematic review and meta-analysis. Mol Psychiatry 17:290–314. https://doi.org/10.1038/mp.2010.136

Saeedi BJ, Liu KH, Owens JA et al (2020) Gut-resident Lactobacilli activate hepatic Nrf2 and protect against oxidative liver injury. Cell Metab 31:956–968.e5. https://doi.org/10.1016/j.cmet.2020.03.006

Sandberg M, Patil J, D'Angelo B et al (2014) NRF2-regulation in brain health and disease: implication of cerebral inflammation. Neuropharmacology 79:298–306. https://doi.org/10.1016/j.neuropharm.2013.11.004

Sandler RH, Finegold SM, Bolte ER et al (2000) Short-term benefit from oral vancomycin treatment of regressive-onset autism. J Child Neurol 15:429–435. https://doi.org/10.1177/088307380001500701

Santangelo SL, Tsatsanis K (2005) What is known about autism: genes, brain, and behavior. Am J Pharmacogenomics 5:71–92

Sealey LA, Hughes BW, Sriskanda AN et al (2016) Environmental factors in the development of autism spectrum disorders. Environ Int 88:288–298. https://doi.org/10.1016/J.ENVINT.2015.12.021

Shultz SR, Macfabe DF, Martin S et al (2009) Intracerebroventricular injections of the enteric bacterial metabolic product propionic acid impair cognition and sensorimotor ability in the Long-Evans rat: further development of a rodent model of autism. Behav Brain Res 200:33–41. https://doi.org/10.1016/j.bbr.2008.12.023

Skladal D, Halliday J, Thorburn DR (2003) Minimum birth prevalence of mitochondrial respiratory chain disorders in children. Brain 126:1905

Slattery J, Macfabe DF, Frye RE (2016) The significance of the enteric microbiome on the development of childhood disease: a review of prebiotic and probiotic therapies in disorders of childhood. Clin Med Insights Pediatr 10:91. https://doi.org/10.4137/cmped.s38338

Sudo N (2016) The hypothalamic-pituitary-adrenal axis and gut microbiota: a target for dietary intervention? In: Gut-brain axis. Academic Press, London, pp 293–304. https://doi.org/10.1016/B978-0-12-802304-4.00013-X

Thomas RH, Meeking MM, Mepham JR et al (2012) The enteric bacterial metabolite propionic acid alters brain and plasma phospholipid molecular species: further development of a rodent model of autism spectrum disorders. J Neuroinflammation 9:153. https://doi.org/10.1186/1742-2094-9-153

Vuong HE, Hsiao EY (2017) Emerging roles for the gut microbiome in autism spectrum disorder. Biol Psychiatry 81:411–423. https://doi.org/10.1016/j.biopsych.2016.08.024

Wang L, Christophersen CT, Sorich MJ et al (2013) Increased abundance of Sutterella spp. and Ruminococcus torques in feces of children with autism spectrum disorder. Mol Autism 4:42. https://doi.org/10.1186/2040-2392-4-42

Wang Y, Wu Y, Wang Y et al (2017) Bacillus amyloliquefaciens SC06 alleviates the oxidative stress of IPEC-1 via modulating Nrf2/Keap1 signaling pathway and decreasing ROS production. Appl Microbiol Biotechnol 101:3015–3026. https://doi.org/10.1007/s00253-016-8032-4

Wang Y, Li N, Yang JJ et al (2020) Probiotics and fructo-oligosaccharide intervention modulate the microbiota-gut brain axis to improve autism spectrum reducing also the hyper-serotonergic state and the dopamine metabolism disorder. Pharmacol Res 157:104784. https://doi.org/10.1016/j.phrs.2020.104784

Wardyn JD, Ponsford AH, Sanderson CM (2015) Dissecting molecular cross-talk between Nrf2 and NF-κB response pathways. Biochem Soc Trans 43:621–626. https://doi.org/10.1042/BST20150014

West R, Roberts E, Sichel LS (2013) Improvements in gastrointestinal symptoms among children with autism spectrum disorder receiving the Delpro® Probiotic and immunomodulator formulation. J Prob Heal 1:102

Young NJ, Findling RL (2015) An update on pharmacotherapy for autism spectrum disorder in children and adolescents. Curr Opin Psychiatry 28:91–101. https://doi.org/10.1097/yco.0000000000000132

Yu M, Li H, Liu Q et al (2011) Nuclear factor p65 interacts with Keap1 to repress the Nrf2-ARE pathway. Cell Signal 23:883–892. https://doi.org/10.1016/j.cellsig.2011.01.014

Zhang DD, Hannink M (2003) Distinct cysteine residues in Keap1 are required for Keap1-dependent ubiquitination of Nrf2 and for stabilization of Nrf2 by chemopreventive agents and oxidative stress. Mol Cell Biol 23:8137–8151. https://doi.org/10.1128/mcb.23.22.8137-8151.2003

Psychobiotics for Manipulating Gut–Brain Axis in Alzheimer's Disease

Monika Kadian and Anil Kumar

Abstract

Alzheimer's disease (AD) is a major form of dementia characterized by the appearance of amyloid beta aggregation (amyloid plaque—outside the nerve terminals), formation of neurofibrillary tangles (within the neuron) hyperphosphorylated tau aggregation, reduced synaptic function, and marked neuronal death. These changes occur around and within the neurons due to the different biochemical changes in AD brain and these changes provided us the underlying cause of AD. However, a full understanding of the exact underlying mechanism involved in AD pathology remains elusive. Psychobiotics are probiotics or prebiotics or combination of these two, when ingested generate a beneficial effect on the nervous system mediated by gut–brain axis (GBA) through several ways (neuroendocrine, neuroimmune, and enteroendocrine) resulted in improved function in both gastrointestinal and CNS. We mainly focus on how gut microbiota alteration involved in AD pathogenesis and how a leading class of probiotics, i.e. psychobiotics, manipulate the gut–brain axis (bidirectional communication) and prove to be beneficial or alternative therapeutic strategy in treatment of AD. Moreover, we also summarize the current scientific findings from different researchers about the psychobiotics and their role in treatment of AD

Keywords

Inflammasome · Dysbiosis · Enterochromaffin cell signaling · Direct neural signaling · Neuroimmune signaling

M. Kadian
Department of Pharmacology, University Institute of Pharmaceutical Sciences, Panjab University, Chandigarh, India

A. Kumar (✉)
University Institute of Pharmaceutical Sciences, UGC Centre of Advanced Studies (UGC-CAS), Panjab University, Chandigarh, India

© Springer Nature Singapore Pte Ltd. 2022
P. K. Deol, S. K. Sandhu (eds.), *Probiotic Research in Therapeutics*,
https://doi.org/10.1007/978-981-16-6760-2_6

6.1 Introduction

Alzheimer's disease (AD) is a progressive, debilitating, irreversible neurodegenerative disorder, pathologically characterized by the accumulation of amyloid beta (Aβ)-associated senile plaque, neurofibrillary tangles (NFTs), hyperphosphorylated tau protein aggregation, depressing synaptic function, and neuronal death. However, a full understanding of the exact underlying mechanism involved in AD pathology remains elusive. Prevalence of AD increases in the aging population, affecting ~35 million people worldwide in 2010 (3.7 million Indian) and this number is projected to rise to 65 million (7.6 million Indian) in 2030. Almost 2/3rd of those diagnosed with AD are female and so gender is a significant factor. Several data exist for AD in terms of neuron-oriented research but now a new therapeutic approach is evolving in which researchers try to find out the link between psychobiotics or probiotics and AD pathogenesis. This chapter focuses on how gut microbiota alteration involved in pathogenesis of AD and how psychobiotics manipulate the gut–brain axis (bidirectional communication) and prove to be beneficial or alternative therapeutic strategy in treatment of AD. Moreover, we also discuss about current scientific findings of how the intestinal microbiota communicate to peripheral organs distant from the gut through different types of signaling, and how this communication affects the host physiology and disease state. Recent advances in rodents suggest that any type of changes in composition of intestinal microbiota contribute to the amyloid plaque formation but this phenomenon is yet to be defined in human subjects. Albeit, a few studies have introduced convincing proof that the gut and nervous system take part in crosstalk with one another, and that the regulation of gastrointestinal commensal microbes could be a promising treatment for neurodegenerative ailments such as AD and PD (Kocahan and Doğan 2017; Sarkar et al. 2016).

6.2 Definitions

Prebiotics (before life) is a substance usually an oligosaccharide that cannot be digested but helps to promote the growth of beneficial bacteria or probiotics termed as prebiotics.

Probiotics (for life) is a substance that contains microorganisms or bacteria that are beneficial to the host organism. Probiotics have been used as dietary supplements to target gut microbiota for prevention or therapeutic treatment of various diseases including mental disorders.

Symbiotic (plus life) is a substance containing both a prebiotic and probiotic.

6.3 Psychobiotics

Psychobiotics are probiotics or prebiotics that when ingested generate a beneficial effect on the nervous system. To generate this effect, the substances created by the microorganisms must be absorbed in the intestines and transported by the gut–brain

Table 6.1 The Gut microbes commonly used as psychobiotics

Lactobacillus species	*L. acidophilus, L. casei, L. rhamnosus, L. reuteri, L. bulgaricus, L. plantarum, L. johnsonii, L. lactis, L. helveticus, L. pentosus, L. casei Shirota, L. hilgardii,* etc.
Bifidobacterium species	*B. bifidum, B. longum, B. breve, B. infantis, B. lactis, B. adolescentis,* etc.
Other species	*Bacillus cereus, Escherichia coli, Saccharomyces cerevisiae, Enterococcus faecalis, Streptococcus thermophilus, Candid species,* etc.

axis. Elie Metchnikoff, a Nobel prize winner (1908), is the main individual who referenced the useful impact of commensal microbes on the host, without utilizing the name psychobiotics. The primary research has been documented in 1910 on the viability of probiotic microbes in the treatment of depression. After a long silence in this area of research, broad clinical and pre-clinical investigations have focussed lately on the adequacy of probiotics.

Our digestive system is brimming with microorganisms that assist us in better ingestion of food for our well-being. The grouping of microorganisms that live in the digestion tracts are called gut microbiota. Ingested substances can travel through the bloodstream and reach various body parts, and psychobiotics allow them to reach the CNS, where they will act on neurons. There are some examples of microorganisms with psychobiotic potential and are listed in Table 6.1 (Misra and Mohanty 2019).

6.4 History

- 1857—Lactic acid bacteria were first discovered by Pasteur.
- 1878—Isolation of lactic acid bacteria from rancid milk was firstly reported by Lister.
- 1889—At Pasteur Institute, Henry Tissier successfully isolated a Bifidobacterium from a breast fed infant.
- 1892—Doderlein proposed the use of micro-organism for a specific medical condition in which he proposed to treat vaginal infection with Lactobacilli.
- 1907—Introduced specific class of microbes to general public for the human health benefits by the Nobel Prize winner Elie Metchnikoff—a Russian Scientist.
- 1911—Douglas published "The Bacillus of Long Life"—supported the concept of human longevity and the consumption of fermented milk.
- 1917—Alfred Nissle isolated *E. coli* that he used to treat acute intestinal diseases such as Salmonellosis and Shigellosis.
- 1930—First stable cultures of Lactobacillus casei shirota strain were made by Dr. Minoru Shirota in Japan. He developed a fermented milk product called Yakult.
- 1935—Dr. Minoru Shirota started marketing Yakult as a probiotic yogurt-like product.

- 1935—Yale University Scientist—Retteger, proposed that Lactobacillus aci-dophilus would be an appropriate species to use for human clinical trials.
- 1965—Lilly and Stillwell coined the term Probiotics, when it was described as growth promoting factors produced by micro-organism.
- 1974—Parker described probiotics as organisms and substances, which contrib-ute to intestinal microbial balance.
- 1989—Fuller attempted to improve Parker's definition of probiotics with the further distinction- "A live microbial feed supplement which beneficially affects the host animal by improving its intestinal microbial balance.
- 2013—Dinan and colleagues defined the term Psychobiotics—as a novel class probiotics that suggest potential application in treating psychiatric diseases.

6.5   Gastrointestinal Tract Microbiota

The human gastrointestinal tract harbors an extremely diverse microbes population that plays a potential role in nutrition, metabolic activities, disinfection from pathogens, and better development of immune system. The number of microbial cells in the gut is larger than the total human cells within the same organism, and it is estimated that human microbiota comprises ten times the amount of cells in the human body. It is estimated that approx. 1200 different bacterial species cohabit into the human intestinal tract, and each individual is host to a defined set of at least 160 species in the gut. Furthermore, the intestinal microbiome contains collective microbial genome approximately 500 times as many genes as the human genome. It means bacterial microbiome genes are 500-folds than the human genome. Recently several reported studies impart that microbial signals help in modulation of pivotal functions of the healthy human system including host metabolism to brain function. Most recently, the Human Microbiome Project, using new genomic technologies, has started a catalog of specific microbiome composition and its correlation with health and specific diseases (Thursby and Juge 2017).

In humans, from the proximal to distal end of intestine, we found the increasing resides pattern of microbial density, which comprises a biological mass ranges from 1.5 to 2 kg, majority of these are strict anaerobes or facultative anaerobes. It is generally assumed that the bacterial composition of the gut is largely culturable, i.e. up to 95% of the intestinal microbiota can be cultured on other hand uncultured bacteria constitutes very low abundance. Archaeabacteria, Eukarya, and viruses also reside within the gut, but their relevance to human health is yet to be studied in details. Figure 6.1 lists and defines important terms used in the study of gut microbiota.

Moreover, the diversity of the gut microbiota can be classified into 12 different phyla and the dominant phyla accommodating about 93.5% of species belongs to 5 phyla such as *Actinobacteria, Bacteroidetes, Firmicutes, Proteobacteria,* and *Verrucomicrobia Actinobacteria* comprises several taxas used as probiotics and the key genus of this phyla is *Bifidobacterium*. The "efficient degraders of food substances" term used for the phyla *Bacteroidetes*, comprises the genera *Bacteroides, Prevotella, and Xylanibacter*. Genera like *Clostridium, Lactobacillus,*

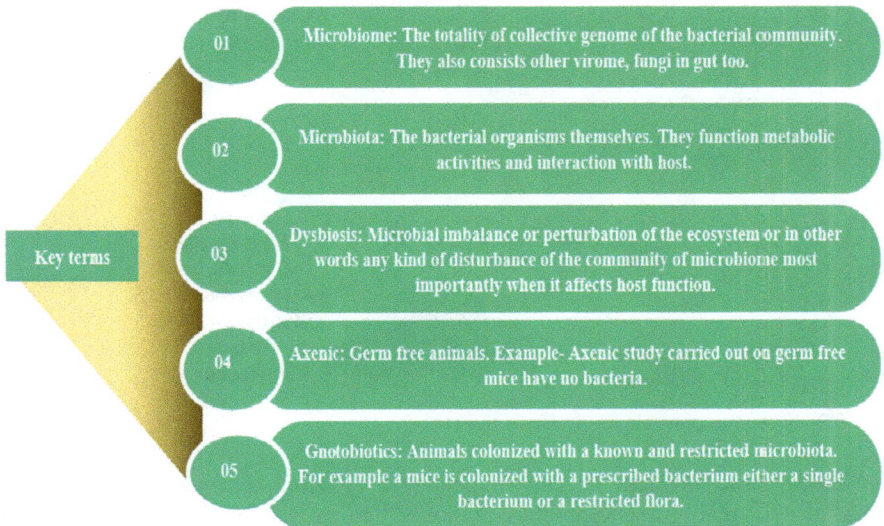

Fig. 6.1 Key terms to understand the study about gut microbiota you must be fully aware about these terms

Ruminococcus, and the butyrate producers (*Eubacterium, Fecalibacterium,* and *Roseburia*) are the fundamental members of *Firmicutes. Escherichia* and *Desulfovibrio* are the key genera of the phyla Proteobacteria, whereas mucus-degrading genus *Akkermansia* comes under the phyla *Verrucomicrobic.* The influence of the microbiota is on host physiology not restricted to the gut but involves distant organs including the brain (Schroeder and Bäckhed 2016).

6.6   Dysbiosis

A microbial flora imbalance within the host, for example, impaired intestinal microbiota due to the following factors such as congenital, dietary, chemical exposure of gut microbiota, general stress, mental stress, altitude and temperature effects, intestinal infection, environmental pollution and toxins, and noise effect provides an opportunity to bad microbes (Fig. 6.2) (Khan et al. 2020).

6.7   Gut–Brain Axis

The gut–brain axis (GBA) dwell bidirectional connectedness that monitors and integrates the central and the enteric nervous system and also establish the link between emotional and cognitive centers of the brain with peripheral intestinal functions. With time research advances explained that the gut microbiota greatly

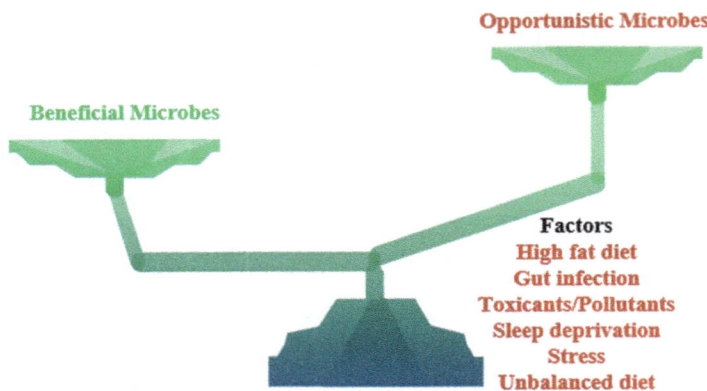

Fig. 6.2 Factors responsible for dysbiosis in gut

influenced these interactions. This bidirectional signaling between the gut and the brain also involve gut microbiota commonly known as a term gut–brain axis or microbiome gut–brain axis. Experimental studies have revealed that the microbiota uses four mechanistic pathways for connectivity, including neuroendocrine and enteroendocrine signaling, enterochromaffin cell signaling, neuroimmune signaling, and direct neural signaling, to communicate from the gut microbiota to the brain and from the brain to the gut microbiota (Fig. 6.3). Moreover, any type of disruption or miss-signaling of the gut–brain axis leads to the emergence of some kind of pathogenic diseases state, including neurodegenerative disease. Current reported studies indicate that the basis of modulation of the CNS by gut microbiota occurs mainly via neuroimmune and neuroendocrine signaling mechanisms also involving the vagus nerve. Moreover, this communication takes place by many microbes derived substances such as short-chain fatty acids (SCFAs), tryptophan metabolites, and secondary bile acids (2BAs). These microbes derived substances then transmit signals via interaction with enterochromaffin cells (ECCs), enteroendocrine cells (EECs), and mucosal immune system (Martin et al. 2018).

In addition to producing these metabolites, intestinal bacteria release various neuroactive compounds directly or indirectly regulating brain function to modulate host health and behavior through the GBA. The gut bacteria can produce several molecules with neuroactive functions such as gamma-aminobutyric acid (GABA), serotonin, catecholamines, and acetylcholine (Table 6.2). When concentration of neurotransmitters increases within the gut, it can decrease the plasma tryptophan concentrations and triggers the cells within the gut's lining to release molecules to the brain thereby improving mental illness although it is remain a mystery that, if they reach specific receptors or get appropriate levels to create a host response (Carabotti et al. 2015).

Fig. 6.3 Gut–Brain Axis—Different Signaling Mechanisms

Table 6.2 The gut microbes as a producer of neurochemical

Microbiota	Neurotransmitter
Lactobacillus and Bifidobacterium species	Gamma-aminobutyric acid (GABA)
B. infantis, Candida, Streptococcus, Escherichia, and Enterococcus species	Serotonin (5-HT)
Escherichia, Bacillus, and Saccharomyces species	Norepinephrine
Bacillus and Lactobacillus species	Dopamine
Lactobacillus species	Acetylcholine
Escherichia, Bacillus, and Saccharomyces species	Noradrenaline

6.7.1   Neuroendocrine and Enteroendocrine Signaling

EECs contributing in this pathway by establishing a bidirectional communication network between gut microbes and their metabolites with the brain. These EECs are different types in nature with several subtypes such as A, K, and L cells responsible for the key processes. These EECs are interlined between epithelial cells to entire length of the gut and these are responsible for releasing >20 different types of signaling molecule (Peptides/hormones) on the basis of chemical and mechanical stimuli. Later released signaling molecule enters into the systemic circulation and then reaches to brain centers such as nucleus tractus solitarius and hypothalamus via activating adjacent afferent fibers of Vagus nerve terminals in gut ultimately generating brain signals. Therefore, the activated vagal afferent fibers by peptides/ hormones released from EECs exert various important functions like regulation of food intake, inflammatory response, mucosal defense, motility, and secretion (Martin et al. 2018; Latorre et al. 2016).

6.7.2 Enterochromaffin Cell Signaling

ECC signaling is the best example of bidirectional connectivity between intestinal microbes, EECs, and the CNS. The major secretory substance produced from ECCs of gut is 5-HT and about 95% of the human body's 5-HT is located in the ECCs and in the enteric neurons, and just only 5% available within the CNS. Moreover, the regulation of synthesis and release of 5-HT is carried out by availability of SCFAs and 2BAs, which are primarily derived from spore-forming bacteria lying in the gut. Although, the tryptophan (Trp)- an essential amino acid is the precursor for the 5-HT and also for several other metabolites and it play a key role in GBA and also contribute to the neuroendocrine signaling within the gut–brain axis. As we know, Trp is an essential amino acid, host is unable to synthesize tryptophan, so it can be only served through dietary intake of proteins. Intestinal microbiota take part in the peripheral availability of Trp, which leads to the synthesis of 5-HT into CNS (Martin et al. 2018; Gross et al. 2016).

6.7.3 Direct Neural Signaling

Several reported evidence till date depends on vagal receptors that cognizance regulatory intestinal peptides, inflammatory substance, dietary parts, and bacterial metabolites to relay signals to the CNS, but there is also some evidence for direct activation of neurons by gut microbiota. Toll-like receptors (TLRs) 3&7, which perceive viral RNA, and TLRs 2&4, which perceive lipopolysaccharide and pepti-doglycan, are uttered in the murine and human enteric nervous systems. Moreover, L. rhamnosus, B. fragilis, and isolated polysaccharide A of B. fragilis all have appeared to involved in activation of gut afferent nerves ex vivo. Nonetheless, it stays hazy to what degree luminal microbial antigens reach neurons in vivo (Lin et al. 2019; Martin et al. 2018).

6.7.4 Neuroimmune Signaling

This signaling pathway involves the symbiotic relationship between host immune system, gut microbiota, and the CNS. Its disturbance in unique resistant microbial communication prompts significant consequences for human well-being. In this segment we talk about the interaction between inhabitant microbiota and key immunological communication, and expresses their relationship in CNS development and neurological disorders.

6.7.4.1 Inflammasome Signaling

Inflammasome, a multiprotein complexes, is able to elicit innate immune response to diverse microbial (a pathogen-associated molecular pattern—PAMP) and endogenous danger (a damage associated molecular pattern—DAMP). Moreover, these molecular pattern are perceived by pattern recognition receptors (PRRs) such as RIG-I like receptors (RLRs, in cytoplasm), TLRs (in endosomes or on cell surface). Inflammasomes are located in the cytosol of various cells such as immune cells (B-cells, T-cells, macrophages, and dendritic cells), neurons, glial cells (microglia and astrocytes), and pulmonary endothelial cells (Ma et al. 2019). Until now, different PRRs in various families, including Nod-like receptor protein 1 (NLRP1), NLRP6, NLRP7, NLR family CARD domain-containing protein 3 (NLRC3), NLRC4, and interferon-inducible protein or absent in melanoma 2 (AIM2), have been recognized to do a job of inflammasome activation (Lang et al. 2018). Prior reports demonstrate that raised amount of short-chain unsaturated fats (SCFAs), a fermented product from gut microbiota, is responsible for the activation of the NLRP3 inflammasome in gut epithelium. These activated NLRP3 inflammasome are responsible for intestinal inflammatory response. These inflammatory agents reach the brain tissue through blood circulation and can promote glial cell activation (microglia) and inflammatory response within hippocampus. This leads to worsening of the pathological condition of AD and is further responsible for cognitive deficit. Therefore, inflammasomes are responsible for the development of neurological immune, and neurodegeneration, includes AD, and Parkinson's disease (PD) (Pirzada et al. 2020).

6.7.4.2 Type I Interferon Signaling

Type I interferon (IFN-I) exert a pivotal role in both innate and adaptive immune system and thus help in maintenance of the host homeostasis. It is an ubiquitous cytokine, pleiotropic in nature and is induced by PAMPs. Its secretion depends upon the activation of different PRRs such as TLRs, RLRs, and nucleotide-binding domain and leucine rich repeat containing gene family receptors (NLRs). Moreover, in case of AD, IFN-I exert a modulatory effect on the gut microbiota to induce its responses via TLRs. These TLRs are extensively located on different cell types within gut such as dendritic cells, intestinal epithelial cells, macrophages, and T-lymphocytes. TLRs acts as a first line mediator between gut microbiota and host because of their presence on intestinal epithelial cells (Ma et al. 2019).

Under physiological conditions, in spite of steady influence to microbial-inferred TLR ligands, intestinal epithelial cell is in a condition of hypo-responsiveness with low articulation of TLRs. By the chance the digestive tract is tainted/contaminated by pathogenic microbes or when bowel infection happens, upregulation in the expression of TLRs occurs in inflammation-dependent pattern within intestinal epithelial cells and macrophages. Then accordingly, proinflammatory cytokines and chemokines are discharged into the blood. Moreover, adjusted gut microbiota profile has been found related with raised levels of plasma lipopolysaccharides, cytokines (IL-6, IL-8, IL-12, and TNF), and hyper-activated T-cells. The infected gut could likewise increase the systemic proinflammatory effect due to the release of proinflammatory cytokines and chemokine, for example, IL-6, CXCL1, CCL5, and tumor necrosis factor-alpha (TNF-alpha), which was related to the microglial activation in 3xTg-AD mouse brain.

When the cytokines are discharged by TLRs, they further move to the brain and crosses the BBB by simple diffusion and by cytokine transporters, particularly in old age subjects as the gut epithelial barrier and BBB become more porous and damaged. Furthermore, these diffused or transported cytokines act on specific receptors expressed by nerve and microglial cells and then leads to modification in their physiology and activation state. Wang and his co-workers reported buildup of amyloid beta within the brain of 5xFAD mouse model due to an alteration in gut microbiota. They also stated that the amount of activated M1 microglial cell rise in the brain resulting in the changes in pattern and number of T helper 1 cells (Th1 cells). By this study they concluded that the leaky gut results in peripheral inflammation, promote microglial activation, and amyloid plaque formation. All this contribute to cognitive deficit in longer run (Lin et al. 2019).

6.7.4.3 NF-κB Signaling

The family of NF-κB, a transcription factor responsible for innate as well as adaptive immunity of the host, also help in the maintenance of proper immune system. Any type of changes in gut microbiota arrangement adds to different inflammatory ailments by means of inborn immunity, particularly by means of NF-κB signaling pathway. In addition, the intrusion by *Campylobacter jejuni* due to dysbiosis of digestive tract microbiome likewise brought about the NF-κB activation because of release of different cytokines further activate distinctive immune cells (Ma et al. 2019).

Conversely, another strain of microbiota (Lachnospiraceae and its metabolites) provides protective function in severe inflammatory infections via NLRP12 activation, which attenuate the activation of both NF-κB/MAPK signals and high fat diet induced inflammasome. Other studies have uncovered that the association among microbiota and NF-κB signaling is additionally responsible for brain inflammation. For example, any disturbances in intestinal microbiota resulted in the reduced BDNF expression in brain regions (particularly in hippocampus) and NF-κB activation, which further leads to extreme neuroinflammation in rodent models. Gut microbiota secrete lipopolysaccharides during dysbiosis that magnify AD pathology by means of the over activation of amyloidogenic pathway. Studies have proposed that bacterial-surface lipopolysaccharides go to the brain and activate the different receptors such as TLR2, TLR4, and CD14 present on the microglia and initiate the downstream NF-κB signaling, which results in the secretion of proinflammatory cytokines that start neuroinflammatory reactions in the brain. These neuroinflammatory reactions and responsive microglia are responsible for the activation of different AD pathways, for example, over activation of BACE1 (Khan et al. 2020; Cerovic et al. 2019).

6.8 Link Between the Gut Microbiome and AD Pathogenesis

Several pre-clinical investigations supported a relationship between gut dysbiosis and neurodegeneration, while the clinical information is very limited. Most of clinical studies shows deviation in microbial composition associated to different neurodegenerative conditions. However, few pre-clinical studies are reported to investigate how gut microbiota is associated with AD pathogenesis (Cerovic et al. 2019; Bostanciklioğlu 2019).

These studies resulted in altered and reduced microbiota in AD patients while examining their feces samples. This altered and imbalanced gut microbiota, termed as dysbiosis, further leads to the peripheral inflammation, loss in integrity of intestinal, and blood–brain barriers. So that, disrupted BBB allow the entry of microbial induced proinflammatory cytokines and chemokines, activation of microglia, which further release the excess amount of proinflammatory cytokines and chemokines and ultimately results in severe neuroinflammation and neurodegeneration in AD patients (Fig. 6.4). Moreover, if we talk at the level of phyla, then there was a clear cut difference as the number of actinobacteria and fermicutes goes down and bacteroidetes levels goes up in diseased state. Therefore, the relative changes in abundance of bacterial genera are related to the levels of biomarkers in AD brain. An ongoing report inspected the connection between specific bacterial species and brain amyloid plaque production in the subjects with memory impairment.

In detailed note, this amyloid deposition was related to increase in proinflammatory cytokines producing microbes such as *Escherichia*, *Shigella* and decrease in anti-inflammatory substances producing microbe, e.g. *Eubacterium rectale*, during stool samples examination. These progressions associated well with the peripheral inflammation.

6.9 Therapeutic Role of Psychobiotics in AD

The contribution of intestinal microbiota as an alternative therapeutic approach in the AD treatment is well depicted in Tables 6.3 and 6.4. Firstly, Minter and his co-worker published the scientific report which included the antibiotic-induced imbalance in gut microbiota leads to inflammation and amyloid plaque formation in animal model of AD. Tabulated data indicates how psychobiotics manipulated the AD related symptoms in animals models of the same (Oh et al. 2020; Cheng et al. 2019; Kowalski and Mulak 2019).

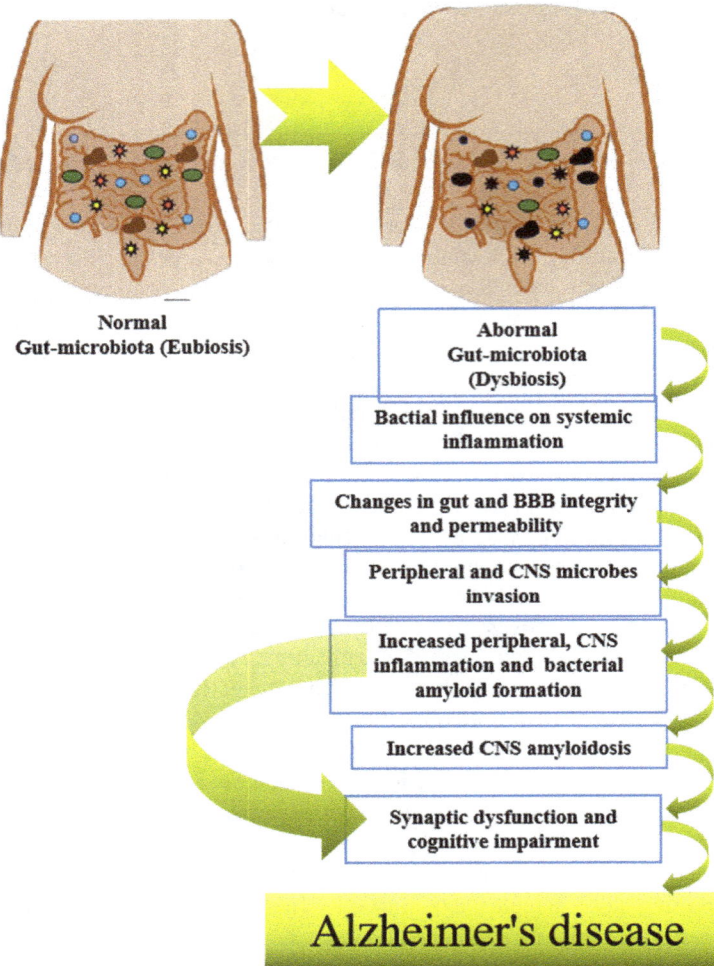

Fig. 6.4 Role of dysbiosis in pathogenesis of Alzheimer's disease

6.10   Future Prospectus and Knowledge Gap

Albeit promising proof has been acquired about psychobiotics, there are as yet numerous vulnerable areas left. For instance, it cannot be said that psychobiotics change the gut microbiota permanently. To date, there have only been a few studies on this subject and, therefore, there is a pressing need right now for long-term research (months or years) to fill up some information gaps. Moreover, this field requires more advances for better results by developing more-powerful biological

Table 6.3 Animal studies about role of gut microbiota in treatment of AD

S. no	Model used	Outcomes	Reference
1.	APP/PS1 mice	– Antibiotic-treated transgenic mice shows alterations in the gut microbiota composition and proinflammatory mediators; antibiotic-treated mice show significant decreased Aβ deposition as well as increased soluble Aβ levels and also less reactive microglia presentation around Aβ plaques – Antibiotic treatment in post-natal subjects presents long-term changes in gut microbiota, significant decreased Aβ deposition in aged transgenic mice; and also less recruitment of reactive microglia around plaque in antibiotic-exposed mice – The significant shift in the intestinal flora in transgenic mice compared to wild type mice – Pathological state of AD shift the gut microbiome toward that condition, which share features with other disorders like autism and inflammatory dysfunctions – Prebiotic supplement along with oligosaccharides (*Morinda officinalis*) responsible for the maintenance of diversity of the gut microbial community; *Morinda officinalis* decreased the neuronal apoptosis, brain swelling and also decreased the expression of amyloid beta	Minter et al. (2016, 2017), Shen et al. (2017), Bäuerl et al. (2018) and Xin et al. (2018)
2.	3 × Tg-AD mice	In transgenic mice, probiotic treatment affected the peripheral inflammatory cytokines and gut hormones, and also present a significant decrease in amyloid aggregation	Bonfili et al. (2017)
3.	5 × FAD mice	This study shows changes in microbiota composition along with age during fecal testing; reduced human amyloid precursor protein expression in the brain as well as in gut tissue	Brandscheid et al. (2017)

(continued)

Table 6.3 (continued)

S. no	Model used	Outcomes	Reference
4.	AD mouse model (ICV injection of amyloid beta)	This study includes *Bifidobacterium breve* strain A1 (administered orally) ameliorate amyloid beta-induced cognitive deficit and also suppressed the amyloid beta-induced changes in gene expression in the hippocampus; this *B.breve* did not alter the gut microbiota, but significantly raised the level of plasma acetate	Kobayashi et al. (2017)
5.	AD rat model (IP injection of D-galactose)	*Lactobacillus plantarum* significantly restored ACh level, prevent plaque formation, and ameliorated memory deficit	Nimgampalle and Kuna (2017)
6.	AD rat model (intrahippocampal injection of Aβ)	*Lactobacillus* and *Bifidobacterium* improve cognitive deficits and related oxidative stress	Athari Nik Azm et al. (2018)

techniques, for example, shotgun meta-genomics and meta-transcriptomics which further support bioinformatic and computational approaches. By these processes, we are able to do multi-omic integration between microbiota and host by using machine learning approaches. Scientist and researchers continuously doing efforts in this field to recognize new microbial community, their structures, functions and a definitive role of that individual micro-organism, but are not able to pay large attention on virome and mycobiome, which are large extent of microbiome (Martin et al. 2018; Sarkar et al. 2016).

Knowledge gap in Psychobiotics Research
- Do psychobiotics alter the architecture of the microbiome?
- Do psychobiotics exert age-specific effects, given impaired homeostatic integrity of the microbiome in aging individuals?
- What is the time-course for emergence of various psychobiotic effects, and how long do they last?
- Are psychobiotic effects are dose-sensitive?
- How long after the beginning of ingestion do psychobiotic effects emerge?
- Do psychobiotics produce long-term changes in the central nervous system?
- Are changes in one brain region broadly offset by changes in the opposing direction elsewhere?
- How do psychobiotics interact with other psychotropic substances?
- What are the dose-response functions associated with psychobiotics?
- Are there undetected psychophysiological costs alongside the observed benefits of psychobiotics?
- Does the brain adapt to long-term psychobiotic ingestion?

Table 6.4 Reported psychobiotics as a therapeutic in AD (Cheng et al. 2019)

S. no.	Model used	Psychobiotics	Duration of study	Test profile	Outcome
1.	**Subjects with severe AD** (65–90 years) F = 13 **Control gp** (M = 10; F = 13) **Probiotic gp** (M = 17; F = 18)	**1st-comb.:** *L. fermentum, L. plantarum,* and *B. lactis* **2nd-comb.:** *L. acidophilus, B. bifidum,* and *B. longum* **Dose:** 3×10^{9} CFU/day	84 days	Biochemical tests	– No effect on serum NO, GSH, TAC, and MDA – No effect on cytokines (TNF-alpha IL-6 and IL-10)
2.	**Subjects with AD** (Avg. age: 76.7 years) (M = 11; F = 9)	**Probiotic supplements preparation** (composition- *L. casei, Lactococcus lactis, L. acidophilus, B. lactis, L. paracasei, L. plantarum, B. lactis, B. bifidum,* and *L. salivarius*)	28 days	Serum and feces samples tests	– ↑Faecali bacterium – ↑Prausnitzii – ↑Serum kynurenine – No effects on serum tryptophan, phenylalanine, and tyrosine
3.	**Transgenic (3×TgAD) AD male mice model** (n = 64)	**SLAB51** (composition: *S. thermophiles, B. longum, B. breve, B. infantis, L. acidophilus, L. plantarum, L. paracasei, L. delbrueckii subsp. bulgaricus, L. brevis*) **Dose:** 200 bn bacteria/kg/day in drinking water	112 days	– Sirtuin-1 activity – Oxidative parameters	– ↑SIRT1 activity and expression – ↓Activity of GST and GPx – ↓SOD – ↓Catalase – ↓LPO
4.	**IH administration of Aβ (1–42) model** Male Wistar rats (n = 12/gp)	*L. acidophilus, L. fermentum, B. lactis,* and *B. longum* **Dose:** 10^{10} CFU/day in drinking water	56 days	– MWM task – Feces samples tests – Amyloidplaque quantification – Oxidative parameters (SOD, catalase, MDA)	– ↑Spatial memory – ↑Total Bifidobacterium and lactobacillus count – ↓Total coliform count – ↓Amyloid plaques – ↓MDA

(continued)

Table 6.4 (continued)

S. no.	Model used	Psychobiotics	Duration of study	Test profile	Outcome
					– ↓SOD profile – No effect on catalase profile
5.	ᵃICR male mice model ($n = 5$/gp)	**Fermented cow's milk** (composition- *L. fermentum* or *L. casei*) **Dose:** 10^9 CFU/0.2 mL by oral gavage	28 days	– MWM – Biochemical tests – Cytokines profile	– ↑Spatial memory – ↑SOD – ↑GSH and GPx – ↓MDA – ↓AChE activity – ↓Cytokine levels
6.	**D-galactoseinduced AD model** Wistar rats ($n = 6$/gp)	*L. plantarum* MTCC1325 **Dose:** 12×10^8 CFU/mL; 10 mL/kg body wt	60 days	– MWM – Biochemical tests (ACh and AChE) – Gross behavioral activity	– ↑Spatial memory – ↓ACh – ↓AChE activity – ↑Gross behavioral activity
7.	**Subjects with AD** (60–95 years) ($n = 30$/gp) (M = 6; F = 24)	**Probiotic milk** (composition *L. acidophilus L. casei B. bifidum, L. fermentum*) **Dose:** 2×10^9 CFU/day	84 days	– HOMA-IR – HOMA-B – Biochemical tests	– ↓Plasma MDA – Changes in insulin resistance, – Changes in beta-cell function and insulin sensitivity
8.	**Male Sprague Dawley rats** – Aged 12 months, 650–750 g – Adolescent 6 weeks, 150–180 g ($n = 4$–7/group)	*Capsosiphon fulvescens* glycoproteins **Dose:** (15 mg/kg/day) and/or *L. plantarum* **Dose:** 109 CFU/rat/day orally	28 days	– MWM – Expression of mature BDNF and phosphorylation of Nrf2 and JNK in the dorsal hippocampus	– ↑Spatial learning and memory – ↑ Nrf2 phosphorylation – ↑ BDNF expression – ↓ JNK phosphorylation

Aβ amyloid beta, *AD* Alzheimer's disease, *Avg.* average, *ACh* acetylcholine, *AChE* acetylcholinesterase, *BDNF* brain derived neurotrophic factor, *CFU* colony forming units, *Comb.* combination, *F* female, *GSH* glutathione, *GST* glutathione transferase, *GPx* glutathione peroxidase, *gp* group, *HOMA-IR* homeostasis model assessment of insulin resistance, *HOMA-B* homeostasis model assessment for beta-cell function, *IH* intrahippocampal, *IL-6* interleukin 6, *IL-10* interleukin 10, *JNK* c-Jun N-terminal kinase, *LPO* lipid peroxidation, *MDA* malondialdehyde, *M* male, *MWM* Morris water maze, *Nrf2* nuclear factor erythroid 2-related factor 2, *NO* nitric oxide, *SOD* superoxide dismutase, *SIRT 1* Sirtuin 1, *TNF-alpha* tumor necrosis factor-alpha, *TAC* total antioxidant capacity, *yrs* years
[a]ICR Strain of mice on name of initial letters of institute, i.e. Institute of Cancer Research, USA

- What are the functional implications of altered excitation–inhibition balance (due to alterations in GABA and glutamate concentrations) in specific brain regions?
- What is the direction of causality between systemic and central changes?

References

Athari Nik Azm S, Djazayeri A, Safa M et al (2018) Lactobacillus and Bifidobacterium ameliorate memory and learning deficits and oxidative stress in Aβ (1-42) injected rats. Appl Physiol Nutr Metab 43:718–726

Bäuerl C, Collado MC, Diaz Cuevas A et al (2018) Shifts in gut microbiota composition in an APP/PSS1 transgenic mouse model of Alzheimer's disease during lifespan. Lett Appl Microbiol 66:464–471

Bonfili L, Cecarini V, Berardi S et al (2017) Microbiota modulation counteracts Alzheimer's disease progression influencing neuronal proteolysis and gut hormones plasma levels. Sci Rep 7:2426

Bostanciklioğlu M (2019) The role of gut microbiota in pathogenesis of Alzheimer's disease. J Appl Microbiol 127:954–967

Brandscheid C, Schuck F, Reinhardt S et al (2017) Altered gut microbiome composition and tryptic activity of the 5xFAD Alzheimer's mouse model. J Alzheimers Dis 56:775–788

Carabotti M, Scirocco A, Maselli MA et al (2015) The gut-brain axis: interactions between enteric microbiota, central and enteric nervous systems. Ann Gastroenterol 28(2):203–209

Cerovic M, Forloni G, Balducci C et al (2019) Neuroinflammation and the gut microbiota: possible alternative therapeutic targets to counteract Alzheimer's disease? Front Aging Neurosci 11:284

Cheng LH, Liu YW, Wu CC et al (2019) Psychobiotics in mental health, neurodegenerative and neurodevelopmental disorders. J Food Drug Anal 27(3):632–648

Gross S, Garofalo DC, Balderes DA et al (2016) The novel enterochromaffin marker Lmx1a regulates serotonin biosynthesis in enteroendocrine cell lineages downstream of Nkx2.2. Development 143(14):2616–2628

Khan MS, Ikram M, Park JS et al (2020) Gut microbiota, its role in induction of Alzheimer's disease pathology, and possible therapeutic interventions: special focus on anthocyanins. Cell 9(4):1–21

Kobayashi Y, Sugahara H, Shimada K et al (2017) Therapeutic potential of Bifidobacterium breve strain A1 for preventing cognitive impairment in Alzheimer's disease. Sci Rep 7:13510

Kocahan S, Doğan Z (2017) Mechanisms of Alzheimer's disease pathogenesis and prevention: the brain, neural pathology, N-methyl-D-aspartate receptors, tau protein and other risk factors. Clin Psychopharmacol Neurosci 15(1):1–8

Kowalski K, Mulak A (2019) Brain-gut-microbiota axis in Alzheimer's disease. J Neurogastroenterol Motil 25(1):48–60

Lang Y, Chu F, Shen D et al (2018) Role of inflammasomes in neuroimmune and neurodegenerative diseases: a systematic review. Mediators Inflamm 2018:1–11

Latorre R, Sternini C, De Giorgio R et al (2016) Enteroendocrine cells: a review of their role in brain-gut communication. Neurogastroenterol Motil 28(5):620–630

Lin C, Zhao S, Zhu Y et al (2019) Microbiota-gut-brain axis and toll-like receptors in Alzheimer's disease. Comput Struct Biotechnol J 17:1309–1317

Ma Q, Xing C, Long W et al (2019) Impact of microbiota on central nervous system and neurological diseases: the gut-brain axis. J Neuroinflammation 16(1):1–14

Martin CR, Osadchiy V, Kalani A et al (2018) The brain-gut-microbiome axis. Cell Mol Gastroenterol Hepatol 6(2):133–148

Minter MR, Zhang C, Leone V et al (2016) Antibiotic-induced perturbations in gut microbial diversity influences neuro-inflammation and amyloidosis in a murine model of Alzheimer's disease. Sci Rep 6:30028

Minter MR, Hinterleitner R, Meisel M et al (2017) Antibiotic-induced perturbations in microbial diversity during post-natal development alters amyloid pathology in an aged APPSWE/PS1ΔE9 murine model of Alzheimer's disease. Sci Rep 7:10411

Misra S, Mohanty D (2019) Psychobiotics: a new approach for treating mental illness? Crit Rev Food Sci Nutr 59(8):1230–1236

Nimgampalle M, Kuna Y (2017) Anti-Alzheimer properties of probiotic, *lactobacillus plantarum* MTCC 1325 in Alzheimer's disease induced albino rats. J Clin Diagn Res 11:KC01–KC05

Oh JH, Nam TJ, Choi YH et al 2020 *Capsosiphon fulvescens* glycoproteins enhance probiotics-induced cognitive improvement in aged rats. Nutrients 12(3):1–15

Pirzada RH, Javaid N, Choi S et al (2020) The roles of the NLRP3 Inflammasome in neurodegenerative and metabolic diseases and in relevant advanced therapeutic interventions. Genes (Basel) 11(2):1–20

Sarkar A, Lehto SM, Harty S et al (2016) Psychobiotics and the manipulation of bacteria-gut-brain signals. Trends Neurosci 39(11):763–781

Schroeder BO, Bäckhed F (2016) Signals from the gut microbiota to distant organs in physiology and disease. Nat Med 22(10):1079–1089

Shen L, Liu L, Ji HF (2017) Alzheimer's disease histological and behavioral manifestations in transgenic mice correlate with specific gut microbiome state. J Alzheimers Dis 56:385–390

Thursby E, Juge N (2017) Introduction to the human gut microbiota. Biochem J 474(11): 1823–1836

Xin Y, Diling C, Jian Y et al (2018) Effects of oligosaccharides from *Morinda officinalis* on gut microbiota and metabolome of APP/PS1 transgenic mice. Front Neurol 9:412

Probiotic Supplementation in Major Depressive Disorders

7

Rahul Shukla, Mayank Handa, and Ashish Kumar

Abstract

Depressive disorders are repetitive, enervative, degenerating with ever lasting impact on socio-economic and prompt with life-threatening illness. Recent development in the area of probiotics states that in the state of chronic illness and stress there is marked decrease in levels of potentially beneficial bacteria. Bacterial presence in an intestinal tract are involved in various processes such as synthesis of vitamins, triggering of immune response, protection of defence barrier system with production of certain bioprotective molecules and neutralization of various toxins. Adaptation of western culture and modernization of lifestyle encounters with certain stressful events in daily life which leads to slow prognosis of depression and serious episodes of many disorders. Research on certain animals and humans states that stress has negative impact on an intestinal microflora, as well as affects gastrointestinal motility, imbalance in certain chemicals which will directly affect the local flora. Gastrointestinal tract is a house of around 100 million neurons, immune responsive cells and certain microorganism. Many researchers have linked certain inflammatory responses which affects the mood of humans due to intestinal disturbances. Probiotics can be explored as an adjuvant therapy along with management therapy. Some attention must be drawn towards potential benefits of already marketed probiotics in market. In this comprehensive literature review on certain factors associated with depressive disorders and probiotics as supplements for countering depressive disorders is done.

R. Shukla (✉) · M. Handa · A. Kumar
Department of Pharmaceutics, National Institute of Pharmaceutical Education and Research-Raebareli (NIPER-R), Lucknow, Uttar Pradesh, India
e-mail: rahul.shukla@niperraebareli.edu.in

Keywords

Probiotics · Depression · Microbiota · Irritable bowel · Syndrome

7.1   Introduction

Major depressive disorder (MDD) is a common, devitalizing and potentially life-threatening illness. MDD is a complex psychiatric disorder of unknown cause that will affect up to 20% of the population at some point in their lifetime (Horrobin 2002). The pathophysiology of MDD is highly multiplex and higher levels of inflammatory markers (cytokines), increased oxidative stress and decreased levels of micronutrients with dysfunction of gastrointestinal (GI) are attributes associated with MDD. The depressive disorder is characterized with loss of interest, hopelessness and lack of hunger and sleep. These symptoms may be present sub-clinically and may affect daily life. It is well recognized that subclinical levels of depressive symptoms are present in fit and healthy population. Currently, there are 350 million people racked with depressive disorder and the population of affected people is increasing briskly. MDD is the most common psychiatric disorder found in the people residing in developed country (Holzapfel et al. 1998).

During the last decade researchers showed interest in finding the relationship between the brain, gut microbiota and mood disorders. The study of changes in microbiota shows the correlation between the inflammatory conditions and mood disorders(Chang et al. 2018). In 2001, the World Health Organization (WHO) stated that probiotics, when consumed in a defined quantity could benefit the body health (World Health Organisation (WHO) 2009). The use of probiotics in the form of cheese and fermented products was well known in history. Various reports on probiotics indicate their therapeutic efficacy and efficiency in treating potential health issues (Kessler and Bromet 2013). A number of studies investigated the potential of probiotics in reduction of pathogenic microorganisms and gastric disorders and overall improvement in bowel movements. Probiotics can also boost the immune system against various microbial attacks and makes the host resistance towards allergens, pathogens as well as protective against oxidation of lipids, proteins and DNA. Probiotics could also maintain the intestinal microflora in patients undergoing the antibiotic treatment. Moreover, it has been investigated that there exists a bidirectional mechanism between gut and brain, which is associated with probiotic ingestion (Joel et al. 2005). Reported studies provide an evidence about the connection between mood and levels of microbiota in gut. Also, the gut microbiota could influence the brain functions due to their role in production of various neuropeptides such as serotonin and gamma-aminobutyric acid. This potential therapeutic profile of probiotics in mood disorders have attracted regard from the past decade (David et al. 2014).

The bidirectional signalling pathway between the gastrointestinal system and mood has been recognized and underlined that mood can impact the functioning of gut and vice versa. Many reports provided evidence that the brain–gut axis has dual communication pathway which ends up in change in mood when dysfunctional (Cepeda et al. 2017). One of the most recognized roles of brain–gut axis is to maintain digestive functions. Various alterations in brain–gut axis correspond to various gastrointestinal dysfunctions such as eating disorders, change in hunger gut inflammation, abdominal pain as well as change in mood and stress (Rhee et al. 2009). The existence of brain–gut axis link has been due to concurrence between stress and depression with gastrointestinal disorders (Dhakal et al. 2012).

Antidepressant medications are usually prescribed for various depressive disorders although 35–40% of population does not respond to these drugs. Moreover, studies on depressive symptoms mainly focussed on the behavioural, neurological and genetic aspects of disease, while the current research scenario involve the impact of immune dysregulation and environmental risk factors as important factors (Owens and Nemeroff 1994). Therefore, there is an imperative need of an effective and proper treatment for depressive symptoms. The involvement of probiotics supplementation in management of depressive symptoms could be a promising approach. Clinical research has shown that proper diet consumptions along with specific dietary components in the daily diet could aid in reducing the risk of depressive symptoms and helps in improving the mood (Dinan et al. 2013; Singh et al. 2020). Specifically, the presence of probiotics and high fibre contents in the diet improve the depressive symptoms in humans. The exact connection between the gastrointestinal microbiota and depressive disorders is being elucidated, but its involvement in dual communication between gut microbiota and brain via the gut–brain axis is a fundamental link between microbiota and brain (Akbaraly et al. 2009). Understanding the brain–gut axis could provide an impactful relationship between brain and gut microbiota which can be employed in the treatment of depression, stress and mood disorders in future.

7.2 Microbiota

During early postnatal life of mammals, bacteria colonizes in new born and the microbiota remains throughout the life of mammal without any modification. Therefore, the change of environmental conditions could reflect the microbiota of an individual. The human lower intestine consist approximately 10^{14}–10^{15} number of bacteria which sums more than the eukaryotic cells composing human body (Gill et al. 2006).

The connection between the microbiota and host is mostly favourable, however, the microbiota can be symbionts (health promoters) but could also be pathobionts (disease causing pathogens) for host. Novel high throughput and metagenomics methods helped the researchers for understanding the complexity and diversity of microbiota present in the gastrointestinal tract (Round and Mazmanian 2009). Studies carried over large centres given access to determine the microbial variations and their functions in the body along with the interindividual differences of microbiota and the stability of microbiota with the environment of gastrointestinal tract (Di Bella et al. 2013). These studies come out with the relationship between the alteration of microbiota composition and its impact on organism state.

In gut several bacteria are represented with approximately 1000 species and using metagenomics it is well established that specific microbiota is present only among a group of population (Huttenhower et al. 2012). Microbiota is highly variable and the bacterial load present in the gut also variate from one group of population to other. The amount of microbial composition increase from stomach to small intestine with change in diversity of microbiota (Morgan and Huttenhower 2012). The composition of microbiota is also variable with age, *Bacteroides* and *Firmicutes* mainly predominates in adults.

The microbial colonization of an infant starts at birth when exposed to maternal microbiota whereas babies born through caesarean section develop with different microbiota in comparison to normally delivered babies (Qin et al. 2010). Thus, the caesarean delivered babies are having abnormal immune responses and they are at high risk of development of immune disorders. Up to the age of 3, the specific microbiota is developed which is modulated by ageing and one of the impactful factors is diet (Arumugam et al. 2011). The microbiota's composition is highly dependent on individual's genetics, age and sex. The homeostatic balance of microbiota is distinct in individuals with good health while deteriorated in the elderly (Costello et al. 2009).

The existence of microbiota is of high significance for maintenance of host physiology and functioning of brain. In the gastrointestinal system microbiota is having various roles, the most important are defence against pathogenic bacteria through the competitive mechanisms and augment the production of many antimicrobial agents in the gut, strengthening of intestinal wall, activation of IgA which limits the penetration of microbes through the tissues (Yamashiro 2018). Microbiota also enhances the absorption of nutrients and facilitates the degradation of indigestible dietary compounds to further enhance their absorption. The involvement of microbiota is also established in development and maturation of host immune system. Mainly, germ-free animals are highly susceptible towards the pathogenic infection along with reduced activity of digestive enzymes and thin muscle wall (Dominguez-Bello et al. 2010). Moreover, gut microbiota tunes host towards the susceptible pathogens and has an important role in homeostasis and immune responses.

Microbiota, at a particular level is responsible for both local as well as systemic effects in the host. Also a disbalance of the microbiota or change in its composition may cause acute to chronic intestinal inflammation similar to IBS (Claesson et al. 2012). Such pathological conditions could be treated by change in dietary composition.

The relationship between the gut-brain axis was not well established and the bidirectional hypothesis of brain gut was not explored for pathophysiology of central nervous system related disorders (for example change in mood, Parkinson's disease, dementia) (Macpherson and Harris 2004). New trends in research are now more focussed towards the bidirectional communication of brain with microbiome in gut.

7.3   Gut Microbiota and Brain Functions

The dual response between the gut and brain is interconnected with severe form of mood disorders but mild forms of mood disturbances are also capable of disbalancing the gut microbiota. This may be due to the immune signalling between the brain and gut microbiota (O'Malley et al. 2010). One of the first disorders that has been suspected to be connected with gut microbiome is autism spectrum disorder (ASD). The concept of ASD is now well understood with animal models and in human subjects (Collins et al. 2009). The studies on animals suggest that an altered gut microbiota composition was found in the animals suffering with psychological disorders. The relation between the gut and brain is bidirectional, meaning link of gut microbiota with the brain and gut itself (Finegold et al. 2010). Through this pathway gut microbiome may influence the functioning of brain via endocrine, neuronal and inflammation related signals. Also the brain generated signals may impact the microbiota composition through endocrine signals (Rhee et al. 2009). The most studies investigating the link between mood disorders and gut microbiota are mainly carried out on anxiety since it has nervous, immunological and endocrinal basis (De Palma et al. 2014).

The microbiome can alter the functioning of hypothalamus–pituitary–adrenal (HPA) axis and immune system, it is also found that in a germ-free mice with gastrointestinal tract sterile, HPA axis was found to be overactive in response to the generated stress (Dinan and Cryan 2012). The stress was reversed by simply introduction of single microorganism, *Bifidobacterium infantis,* which is a common probiotic. Also the alterations in the levels of various key components of neuronal signalling (noradrenaline and 5-hydroxytryptamine) in depression was found in the germ-free mice (Sudo et al. 2004). It has been also shown that certain bacterial strains (*lactobacillus* and *Bifidobacterium)* are capable of combating the anxiety, since treatment with prebiotics (*B. longum, B. infantis, L. helveticus, or L. rhamnosus)* normalized the levels of neurotransmitters in animals. On the other hand, opportunistic pathogens are capable of causing the anxiety and depressive disorders (Clarke et al. 2013). *Campylobacter jejuni* causes anxiety like behaviour through the activation of visceral sensory system in brain and enhanced expression of c-Fos protein in hypothalamic nucleus. Similar elevation in c-Fos expression was found with *Citrobacter rodentium* (Heijtz et al. 2011).

7.3.1   Gut–Brain Axis: Focus on Gut Peptides

The brain and gut are involved in continuous dual connection. This communication may be of importance in mediation of various physiological responses ranging from gastrointestinal function to mood and behaviour (Bercik et al. 2011). These responses are responsible for the perception of satiety, nausea and pain. Moreover, stress episodes lead to altered GI motility and pattern of secretion of GI fluids. The clinical relevance of such bidirectional approach is now under investigation and a number of studies proves the connection between the gut and brain (Cryan and

Dinan 2012). This type of bidirectional signalling between the brain and microbiome could generate sensory signals from the GI tract, which is then converted to neuronal signals. These signals could lead to changes in various central nervous system processing (Hill et al. 2014).

The gut endocrine system is composed of gut peptides and many other transmitters (i.e., serotonin), which are secreted by different types of enteroendocrine cells (EECs) in the GI tract in response to food, particularly after intake of fats and carbohydrates (Tannock and Savage 1974). In the gut lumen approximately 1% of the epithelial cells are represented by EECs. These cells are further classified on the type of peptide secretion and their location. Enterochromaffins, a type of EECs have the 5-HT3 receptor expressing primary afferent nerve fibres which controls the release of serotonin and this lead to signal transmission from gut to brain (Park et al. 2013).

Peptides are formed by short chains of amino acids and consist of amide bonds. These could be released from the EECs which may act as neuropeptide on brain by binding through the GPCRs. Peptides are having the property of getting metabolized endogenously and they could reach to receptors even present away from their release site (Williams et al. 2011). Due to their metabolism endogenously, they do not get accumulated in the body. In particular to gut–brain axis there are various peptides which are involved in maintaining the physiology of brain. Many peptides secreted in the GI tract are also present in the enteric nervous system (Tillisch et al. 2013). For example, Tachykinins (i.e. substance P) and calcitonin gene related peptides are mostly found in intrinsic nervous system. The regulation of enteric motor function and gastric acid secretion is dependent on the gastrin, released on stimulation of G cells in stomach (Tan et al. 2014). There are some peptides like vasoactive intestinal peptide which functions in regulating the circadian rhythm and inhibit the secretion of GI tract which leads to decreased absorption.

Some peptides are capable of stimulating the release of enterochromaffins, a type of EECs. (Alonso et al. 2002). Due to their feature location in gastrointestinal tract, it could be possible that those peptides may be influenced by change in gut microbiome and modulate (independently or cooperatively) signalling to brain.

Gut metabolic pathway has linkage with brain disorders. Researchers have linked tryptophan-kynurenine (KYN) metabolism with major depressive disorder. Tryptophan is major dietary composition available in gut. About 1% of tryptophan is synthesized into serotonin in EECs of gut and rest in brain after crossing the BBB. In normal physiological conditions, KYN further catabolizes into compound 3-hydroxykynurenine (OHK) in the presence of kynurenine-3-monooxygenase (KMO) enzyme. Furthermore, OHK catabolizes into 3-hydroxyanthranilic acid (HAA) in the presence of kynureninase. As the catabolism proceeds further, which ends ultimately into liver with formation of ATP or erstwhile into quinolinic acid (QUIN) and ultimately into NAD. In abnormal physiological functions, when inflammation or pro-inflammatory states activate then tryptophan metabolism to serotonin decreases and kynurenine catabolism to OHK increases which promotes to neuronal apoptosis. Furthermore, researchers also reported that KYN metabolites have potential role in neuro-protection and neurodegeneration of brain. More formation of OHK enables formation of QUIN which leads to excitotoxic neurodegeneration. The cumulative effect of OHK and QUIN leads to neuronal as well as astrocyte apoptosis and inhibit the neurotrophic factors, further prognosis the depressive disorder.

Bacterial products could also stimulate the release of peptides in the GI epithelium and these peptides play important role in satiety. Various kind of peptides are released by different EECs in the gut. For example, ghrelin is secreted in high amounts at the time of meals (Banks et al. 2001). The proximal and distal small intestines contain different cell types which secretes variety of polypeptides. For example, I and K cells which are present in the proximal small intestines are responsible for production of CCK and glucose-dependent insulinotropic polypeptide, respectively (Borre et al. 2014).

Elevations in levels of pro-inflammatory cytokines such as tumour necrosis factor alpha (TNFα), interferon gamma (INFγ), interleukins have been observed in MDD. Moreover the elevations in levels of TNFα are associated with severity of depression (Raison 2019). The role of pro-inflammatory cytokines in MDD is well established. These could alter the mood through lowering the neurotransmitter levels which at extreme stages could stop the metabolism of neurotransmitters (Manach et al. 2004). Brain derived neurotrophic factor (BDNF) is also known to play important role in depressive symptoms, levels of BDNF were found low in patients with depression. Inhibition of inflammatory cytokines by use of antidepressant medication may improve the levels of BDNF in patients. It is also investigated that elevation in cytokines in periphery may not account the depression and change in sleep pattern but very low levels of cytokines might affect the complex brain functions (Kuda et al. 2000).

7.3.2 Role of Omega-3, Lipids and Microflora

The investigations carried over dietary fish consumption show reduced risk of bipolar disorder, seasonal affective disorder and MDD. The investigation is also supported by number of experiments and animal research. Clinical data shows therapeutic effects of omega-3 fatty acid in various psychological disorders. Moreover eicosapentaenoic acid (EPA) arise as strong candidate for management of MDD (Raison 2019). Increased incidence of depression are mostly associated with decreased intake of omega-3 fatty acids reported in western world over last 100 years (Appleton et al. 2015). This is largely due to the high supply of various omega-6 rich oils (corn, cottonseed and sunflower) combined already in the foods. This can probably influence the intestinal microflora, and in turn the psychiatric circuits in brain. In a preclinical study, murine was administered with 10% corn oil (first group), 9% fish oil and 1% corn oil (second group), 9% beef fat and 1% corn oil (third group). The fish oil diet led to lowest levels of bacteroides and threefold increase in bifidobacterial levels (Grosso et al. 2014b). The in vitro result shows high intake of omega-6 rich oil and pure linoleic acid could inhibit growth of *Bifidobacterium*. In arctic charr, the dietary intake containing 4% fish oil or 7%

flax oil could improve the levels of lactic acid bacteria and mainly *lactobacillus* (Simopoulos 2002).

Polyunsaturated fats could improve the adhesion of probiotics to intestinal cells. Flax seed oil causes more adhesion while arachidonic acid causes relatively less adhesion. Seal oil, which is having high quantities of EPA is known to increase adhesion of *Lactobacillus paracasei* by 12% in intestines (Sánchez-Villegas et al. 2011). The mechanism behind the increase in adhesion of bacteria by fatty acids is their potential of getting inside the bacteria thereby changing their structure and adhesion to the intestines. Studies also prove that on administration of *Bifidobacterium* to infants over 7 months increased the amount of alpha linolenic acid in plasma, it shows dual connection between probiotics and omega-3 fatty acids (Raes et al. 2002). Moreover, the number of *lactobacilli* and *Bifidobacterium* are more in healthy Japanese adults in comparison to Canada. Also, the reported cases of MDD are lower in Japan compared to Canada (Logan and Katzman 2005). The effect of omega-3 fatty acid on MDD and microbiota status is under investigation and deserves further investigation.

Moreover, a number of micronutrients are also known for reducing the risk of depressive symptoms via production and functions of monoamine neurotransmitters such as serotonin. Interventions in diet pattern are also associated with changes in depressive symptoms through microbial compositions. Adherence to diet rich in micronutrients could reduce the numbers of pathogenic bacteria such as *Escherichia coli* and increases the *bifidobacteria* (Bear et al. 2020).

Prebiotics usually promote the growth of beneficial bacteria and reduce the pathogenic forms. Fruits and whole grains are common source of prebiotics consisting of phytochemicals, polyols and polyphenols. The biological activity of different phytochemicals has been demonstrated through their antidepressant and anxiolytic effects. However, further studies are needed to build the role of phytochemicals in management of depressive symptoms (Hill et al. 2014).

7.3.3 Status of Gut Microbiota in Stress and Major Depressive Disorder

The link between gut microbiota and brain is well established. The composition and stability of gut microbiota is often changed during the depression and anxiety. Stress related disorders (i.e., anxiety, depression, etc.) alter the intestinal barrier, which could allow microbiota-driven proinflammation through the secretion of certain bacterial product from gut. Moreover studies have also proved that the gut flora secretes lipopolysaccharides (LPS) which enhances stimulation of immune responses in patients suffering from stress (Salzman et al. 2007). Patients affected with IBS with trauma in early age share a connection among the gut microbiome status and brain volume displaying a shift in the ratio of *Firmicutes/Bacteroidetes*. A correlation was found between gut microbiome status and brain volumes in patients with IBS in early age trauma, which shows a shift in *Firmicutes/Bacteroidetes* ratio. There is an higher level of Firmicutes and lower level of *Bacteroidetes* in IBS

Table 7.1 Phases of treatment of depression

	Objective	Duration of treatment	References
Acute	Remission of symptoms: – 50–75% reduction in severity – Risk of relapse – Decrease of treatment resistance – Risk of suicide Full remission: – Full response sustained for 4 weeks	– 6–8 weeks with prescribed dose of antidepressants – 12–16 weeks with psychotherapy	van Bronswijk et al. (2019)
Continuation	– Relapse prevention – Stabilize remission	– Approximately 6 months if no relapse of disease – If symptomatic breakthrough, prolongation for 9–12 months	Dunner et al. (2007)
Maintenance	– Recovery – Preventing a new episode	– Discontinue treatment after a stable recovery period of 6 months – Continuing >1 year; lifetime pharmacological maintenance for patients with three or more prior episodes	Pringsheim et al. (2016)

patients whereas in patients suffering from depression alone and not with IBS, the ratio was inversely associated with the composition of the microbiome found in IBS patients (Alonso et al. 2008). Also, it was found out that there is significant increase in levels of *Bacteroidetes* and marked downfall in the levels of *Firmicutes* in depressive-like rats. There is a decreased amount of *Bacteroides* genus in mice suffering from condition of chronic stress. Moreover, stressed mice shown to have higher populations of *clostridium* genus. Due to the alteration in the metabolites of gut there is a modification in the gut *Clostridium*, mainly tyrosine, tryptophan and phenylalanine. Such metabolites are capable of metabolism of various key neurotransmitters in mammals, including serotonin, which can further impact the enteric and central nervous system and associated with the gut–brain signalling (Taché and Yoneda 1993; Stephens and Tache 1989).

Antidepressants are widely known to have antimicrobial properties, which improve the pathophysiology of the anxiety and depression by modulating not only the brain homeostasis, but also gut microbiome (Saddiqe et al. 2010). Antibiotics such as tetracyclines and β-lactams has been found to have potential antidepressant manifestation in rodents and humans. Also, fewer groups of antibiotics like fluoroquinolones have been characterized by the development of depression and anxiety. Table 7.1 provides a brief on phases of depression. The potential interconnection between the gut microbiome and host depression is featured due to the alteration in the composition of gut microbiome and their effects on host metabolic phenotype throughout the psychiatric illness that remains unsolved. Thus, manipulation of microbiome may be useful link for understanding the physiological role of gut–brain axis in psychiatric disorders (Haast and Kiliaan 2015). Further implementation of animals with pathogenic bacteria, germ-free rodents, animals exposed to probiotics could result in beneficial results for the evolution of novel therapeutics to prevent or to treat depression and anxiety.

Fibromyalgia (FM) and chronic fatigue syndrome (CFS) are the symptoms associated with depression. Previous studies on such patients shows decreased level of *Bifidobacterium* and increased levels of *enterococcus* species. Also, it was proven that more the enterococcal count among FM/CFS patients, more severe the cognitive and neurological deficiencies, mood disbalance, nervousness (Butt et al. 2001). All of these are characteristic symptoms of MDD. Endometriosis is a condition where levels of *Lactobacilli* are low which are commonly associated with depression. Mainly in four conditions i.e., Fibromyalgia, Chronic fatigue syndrome, inflammatory bowel disease, and Endometriosis – the migration of bacterial strains from colon region to small intestine is commonly seen. This migration of bacteria results in small intestine bacterial overgrowth (SIBO). This often leads to somatic pain in patients. There is not much investigation reported about concurrence of SIBO with MDD although it is likely to occur in such patients, mainly it is caused by low stomach acid secretion (Pimentel et al. 2004). There is a decrease in the release of stomach acid and intestinal stasis in the patients suffering from the symptoms of depression. There is a role play of cytokines in the depression mainly tumour necrosis factor alpha (TNFα) and interleukin 1-beta (IL-β) which are thought to inhibit the gastric acid secretion. Depression directly influence the physical activity, it is also associated with SIBO. Various strains of probiotics are found to treat patients of SIBO (Pimentel et al. 2000).

The significance of SIBO linked to depression is not limited to abdominal malfunctions, it can affect the absorption of fat, carbohydrate, protein, vitamins and other micronutrients. Due to lack of these nutrients in the body the host defence mechanism weakens leading to various pathogenic infections in various body parts. There is decreased amounts of vitamins, zinc and folic acid in the patients suffering from depression. Decreased level of vitamins chiefly vitamin B_6 results due to reduction in the conversion of alpha linolenic acid to mood maintaining eicosapentaenoic acid (EPA) (Nowak et al. 2003). Non-digestible oligosaccharides may lead to an increase in the level of availability of nutrients due to which there is an enhancement in the condition of *Bifidobacterium*. It is interesting to know that with the treatment of SIBO various mood and behaviour disorders also improves in the patients. Cognitive disorders also have been observed in patients with Crohn's disease, and depressive symptoms are also observed with relapses (Maan et al. 2020; Ott et al. 2004). The pathophysiology of such patients shows lower levels of *lactobacilli*.

7.4 Conventional Therapies in Management of MDD

Antidepressant are prescribed particularly upon patient's medical condition and individual preferences. Studies suggest that patient suffering from severe depression is usually prescribed with antidepressant medication, while patient with acute symptoms may get benefit by other approaches (nonbiological). The effectiveness of a particular treatment is studied by clinicians and they could predict the evolution and the treatment for the depressed patient (Wirz-Justice 2006). In some cases of depression, a particular treatment is recommended. For example, bright light (BL) is usually indicated in the case of seasonal affective disorder and depressive symptoms during pregnancy. The mechanism behind the action of BL treatment may be increased in the serotonin transmission in brain and accompanying of biological rhythms (Rosenthal and Sack 1984). Optimal treatment through conventional therapy starts with nature of illness and nature of proposed mechanism of treatment. Specific psychotherapeutics are needed in case of severe form of depression while in mild to moderate symptoms of depression no specific treatment is provided. In moderate depression the prescription of antidepressant can be taken over the course of few days. In case of severe depression the use of antidepressant medication is highly recommended because the neurobiology of brain is disturbed to much extent to be responsive to psychotherapy alone (Insel et al. 1985). The choice of antidepressant is done on the basis of severity of depression with minimizing the side effects (sedation, sexual dysfunction, weight gain), presence of other psychiatric disorders, resistance to a particular antidepressant (Hay and Claudino 2012). Other factors to be considered for choice of antidepressant are contraindication, potential toxicity and its cost. Although the patient preference-after being informed about the therapeutic potential of drug-may be expected to enhance compliance (D'Sa and Duman 2002).

It has been suggested that selective serotonin reuptake inhibitors (SSRIs) are more beneficial than primarily noradrenergic antidepressants in decreasing aggression/irritability and anxious behaviour. Moreover, patients with severe depression respond more favourably to noradrenergic antidepressants than SSRIs (Åberg-Wistedt et al. 1981). Some studies suggest that monoamine oxidase inhibitors (MAOI) are more effective in patients with "atypical depression". However, given the numerous drug–drug interactions and diet restrictions, MAOIs remain a second-line treatment for depression (Eriksson et al. 1995). Table 7.2 briefs about different conventional treatment for counterfeiting depression.

Besides the different therapies available for management of depressive symptoms, still there is need of medication with least adverse effects and ease of administration. SSRIs are the most prescribed medication for the depressive symptoms but their adverse effect profile (sedation, sexual dysfunctions) and the attainment of therapeutic dose through slow dose titration makes the therapy less popular. Moreover, the association of psychotherapy combined with pharmacological intervention through drug administration is also not reported. Thus, also these factors derive the need of some new therapies for treatment of depressive symptoms.

Table 7.2 Conventional treatment of depression

S. no.	Class	Subclass	Examples	References
1	Antidepressant drugs	• Tricyclic antidepressants (TCAs)	– Imipramine – Clomipramine – Desipramine – Amoxapine – Doxepine – Maprotiline – Amitriptyline	Duval et al. (2006) and D'Sa and Duman (2002)
		• Selective serotonin reuptake inhibitors (SSRIs)	– Citalopram – Escitalopram – Fluoxetine – Paroxetine – Sertraline	
		• Selective serotonin and noradrenaline reuptake inhibitors (SNRIs)	– Venlafaxine – Milnacipran – Duloxetine	
		• Noradrenaline α_2 receptor antagonist and serotonin 5-HT$_2$ receptor antagonist	– Mianserin – Mirtazapine	
		• Serotonin (5-HT$_2$) receptor antagonists	– Trazodone – Nefazodone	
		• Noradrenaline reuptake inhibitor	– Reboxetine	
		• Dopamine reuptake inhibitors	– Bupropion – Methylphenydate	
		• Monoamine oxidase inhibitors (MAOIs)	– Iproniazide MAOI-A (Moclobemide, toloxatone) MAOI-B (Selegeline)	
		• Others	– Tianeptine – Hypericum perforatum – Tryptophan and 5 hydroxytryptophan dietary supplement	
2	Mood stabilizers	• Lithium salts		Shim et al. (2017)
		• Antiepileptics	– Carbamazepine – Oxcarbazepine – Sodium valproate – Lamotrigine	
3	Nonchemical therapies	– Sleep deprivation		Albert et al. (2009)
		– Light therapy		
		– Electroconvulsive therapy		
		– Magnetic stimulation		
		– Vagus nerve stimulation		
		– Deep brain stimulation		

(continued)

Table 7.2 (continued)

S. no.	Class	Subclass	Examples	References
4	Psychotherapy	– Cognitive therapy		van Bronswijk et al. (2019)
		– Interpersonal psychotherapy		
		– Problem-solving therapy		

7.5 Probiotics

The word probiotic is of Greek origin meaning "for life" but the actual definition of probiotic is modified timely due to its role in several therapies (Salminen et al. 2005). Usually probiotics are used to describe the substances produced by microorganisms that further enhance the growth of others. The most widely used definition of probiotics was that of Fuller: "probiotics are live microbial feed supplements which beneficially affect the host animal by improving microbial balance". The currently used definition of probiotics is given by Food and Agricultural Organization of the United Nations World Health Organization, according to which probiotics are defined as "live microorganisms which are administered in an adequate amount to pronounce health benefits on the host" (Chaucheyras-Durand and Durand 2010; Gibson and Roberfroid 1995). Taking into consideration the definition of probiotics, strain identification of microorganism is recommended in order to establish their therapeutic profile. As far the nutrition is concerned *Bifidobacterium* and *Lactobacillus* are the most popular ones.

Probiotics are known to influence gut microbiota composition in humans; therefore it may be postulated that probiotics mediates the levels of microbiome to influence the physiology of central nervous system (Ley et al. 2005). The probiotics demonstrates their anti-inflammatory and this property of probiotic help (1) treat inflammatory conditions, (2) reduce stress levels and (3) reduce intestinal permeability (Ohland et al. 2013). Table 7.3 summarizes the effect of probiotics on immune related inflammation and depression.

7.5.1 Anti-Inflammatory Properties of Probiotics

A number of investigations carried on probiotics demonstrates that probiotics reduce inflammation and improve the lifestyle of depressed patients and also has been established as an effective therapy for IBS (Pinto-Sanchez et al. 2017). For example, studies carried out in both humans and animals show lower levels of systemic pro-inflammatory cytokines after taking probiotics (Abraham and Quigley 2017). Reduction of inflammation in arthritis patients was also demonstrated upon administration of probiotics to patients. Probiotic ingestion can promote gut microbiome functions, including reduction in pathogenic species, epithelial barrier integrity and anti-inflammatory responses (Raison 2019). A probiotic strain could directly or indirectly modulate the mood by interacting with central nervous system.

Table 7.3 Effects of probiotics on depression and immune-inflammatory symptoms

Probiotic	Effects on depressive symptoms	Effect on immune / inflammation	References
Bifidobacterium infantis	Not assessed	Secretion of bioactive factors enhancing epithelial cell barrier function Increased expression of genes involved in maintaining intestinal barrier integrity	Ewaschuk et al. (2008)
Bifidobacterium longum	Reduced depression scores on hospital anxiety and depression score	No significant effect on IBS symptoms	Pinto-Sanchez et al. (2017)
Bifidobacterium longum	Not assessed	Reduction of inflammatory markers (TNFα)	Grosso et al. (2014a)
Lactobacillus casei	Not assessed	Reduced serum proinflammatory cytokines (TNFα, IL-6, IL-12) increased serum level of regulatory cytokine IL-10	Malaguarnera et al. (2012)
Lactobacillus acidophilus, Lactobacillus casei, Bifidobacterium bifidum	Reduced beck depression inventory (BDI) score	Reduction in serum insulin, HOMA-IR Beneficial effects on IBS symptoms	Akkasheh et al. (2016)
Lactobacillus helveticus and Bifidobacterium longum	No significant effect on Montgomery-Asberg depression rating scale	No significant reduction in levels of inflammatory biomarkers	Romijn et al. (2017)
Bifidobacterium bifidum, Bifidobacterium lactis, Lactobacillus acidophilus, Lactobacillus brevis, Lactobacillus casei, Lactobacillus lactis W19, Lactobacillus Lactis W58	Reduced depressive behaviour as measured by forced swim test	Reduction in hippocampal transcription of factors involved in HPA axis regulation Reduction of inflammatory cytokines	Abildgaard et al. (2017)
L. acidophilus and B. bifidum and longum	Subjects' general condition improved by 40.7%. 73% of participants rated the effect of treatment as "good" or "very good"	Reduction of levels of TNFα	Gruenwald et al. (2002)
L. helveticus and B. longum	Consumption of probiotics reduced	Reduction of measures of intestinal microbial	Messaoudi et al. (2011)

(continued)

Table 7.3 (continued)

Probiotic	Effects on depressive symptoms	Effect on immune / inflammation	References
	HADS and HSCL-90 scores	translocation and improved measures of intestinal barrier function	
L. casei	Consumption of probiotics significantly improved BAI scores. No effect on BDI scores	Reduced serum proinflammatory cytokines (TNFα, IL-6, IL-12)	Rao et al. (2009)
B. lactis and L. acidophilus, brevis, casei, lactis and salivarius	Consumption of multispecies probiotic significantly reduced overall cognitive reactivity to depression (in particular aggressive and ruminative thoughts)	Reduction in hippocampal transcription of factors involved in HPA axis regulation	Steenbergen et al. (2015)
Lactobacillus casei	Maintained salivary cortisol levels from baseline in the treatment group; a twofold increase in expression of stress response genes in placebo compared to treatment	Bacteroidaceae family was significantly lower before the exam compared to placebo	Kato-Kataoka et al. (2016)
Lactobacillus acidophilus, Lactobacillus casei, Bifidobacterium bifidum	Reduced BDI score	Not assessed	Akkasheh et al. (2016)

Gastrointestinal microbes could act directly or indirectly through the synthesis of neurotransmitters, e.g. 5-HT. Probiotics modulate the production of serotonin and its precursor, tryptophan, which may influence concentration of serotonin in brain (Slavich and Irwin 2014). Patients with depressive symptoms are likely to have lower levels of serotonin in brain. The influence of probiotics on production of neurotransmitter and further on brain is still being researched, but modulation of enteric nervous system may lead to release of gut hormones associated with mood and behaviour. Studies demonstrating effect of probiotics on mood have been reported to improve mood in humans (Miller et al. 2010).

A large number of studies evidence the association between stress and inflammation Sympathetic nervous system (SNS) and hypothalamic-pituitary adrenal (HPA) axis are major pathways whose aberrant functioning in stress leads to inflammation (Sudo 2006). For example, SNS activation due to stress induce the release of norepinephrine, which stimulates the adrenergic receptors to upregulate the transcription of pro-inflammatory immune response genes (e.g. TNFα, IL-6; Pan et al. 2017). Dysregulation of HPA axis could lead to increased levels of inflammation. Under normal conditions, the release of cortisol from the HPA attenuates the transcription of pro-inflammatory response genes. Due to aberrant functioning of HPA axis, elevation in transcription of inflammatory genes is observed which contribute to association between stress and inflammation (Parvez et al. 2006). Studies have shown impact of certain probiotic strains to normalize the functions of HPA axis in response to stress. A study by researchers found that germ-free mice subjected to stress had a very high HPA axis mediated inflammation in comparison to pathogen free mice. This aberrant functioning of HPA axis was completely normalized by treatment with probiotic *Bifidobacterium infantis* (Cohen et al. 2012). These studies in mice indicate that probiotics could be a potential therapy in reducing the stress levels by mediating the HPA axis. The study is further supported by human studies which indicates the reduction of stress by reducing the release of cortisol through HPA axis. Additional studies have been carried out in rats to demonstrate the similar effects of probiotics on stress-induced inflammation (Desbonnet et al. 2010). For example, in maternal separation stress model of rat (pups separated from their mothers), elevation in inflammatory response was there (i.e. increased levels of systemic IL-6, elevated levels of norepinephrine in the brain, increased levels of plasma corticosterone). *Bifidobacterium infantis* normalized the levels of inflammatory responses of IL-6 and other markers of inflammation (Ewaschuk et al. 2008). Similar effect was found in IBS patients where probiotic normalized the serum ratio of IL-10 to IL-12.

Studies carried out in both human and animals demonstrate that increased intestinal permeability is linked with increased permeation of inflammatory substances (Lipopolysaccharides) LPS from the gut into systemic circulation which can trigger the immune responses and increased level of circulating pro-inflammatory cytokines (IL-6, TNFα; Kawai et al. 2001). This proves the association of intestinal permeability with inflammation and may be used as a target for individuals with MDD. Probiotics have been known to reduce the intestinal permeability by various mechanisms (Owens and Nemeroff 1994). As an example, there is a secretion of bioactive factors by probiotics thus promoting the integrity of intestinal barrier and functions of epithelial barrier. Studies done on GF mouse with the probiotic *Bacteroides thetaiotamicron* increase the secretion of desmosome, protein involved in maintaining the integrity of gut epithelium (Mayer et al. 2014).

7.5.2   Probiotics for the Treatment of MDD

Investigations done in animals and humans, evidences the role of gut microbiota modulation for management of depression (Zheng et al. 2016). For example, a study demonstrated the transplantation of faecal microbiota from depressed human to GF mice leads to elevation in depressive symptoms in recipient GF mice. The results from this study provided the objective for an ongoing clinical trial with primary rationale of evaluating the faecal microbiota versus placebo on depressive symptoms in patients with MDD (Akkasheh et al. 2016). Only one clinical trial has been conducted on patients with depression to examine the effect of probiotic supplementation. The study was an 8-weeks randomized, double-blind, placebo-controlled trial that included 40 patients with MDD. *Lactobacillus acidophilus, Lactobacillus casei* and *Bifidobacterium bifidum* are the three viable freeze-dried strains of probiotic supplementation. The results indicated lower beck depression inventory scores due to probiotic supplementation in comparison to placebo (Slykerman et al. 2017). Also, the researchers also found lower levels of inflammatory markers such as homeostasis assessment of insulin resistance (HOMA-IR) and serum insulin in probiotic ingestion group compared to placebo. Similar effects of probiotic supplementation have been found in other groups of depression including females with post-partum depression and individuals with secondary depression (Zheng et al. 2016). However, it is interesting to note that not all studies reported positive results with probiotic supplementation in treating the depression. A more recent study on this topic reported the insignificance of probiotic supplementation in mood. In addition, a randomized, placebo-controlled trial of *Lactobacillus helveticus* and *Bifidobacterium longum* for the depression reports that neither of these probiotics were able to improve the low mood. Thus, the selection of specific strains and probiotic dose is of utmost importance (Ng et al. 2018). It could also be possible that probiotics may reduce the depressive symptoms in specific populations such as those with elevated inflammation, elevation in levels of cytokines (Cuello-Garcia et al. 2015).

The mechanism involved in the antidepressant effects of probiotics is multifactorial, with different probiotics exerting their action through one or more mechanisms. The effects of probiotic supplementation should be investigated in future studies to check the role of probiotics on various brain inflammatory measures. Clinical trials investigating the effects of *L. casei* shows reduced anxiety and improvement in mood (Gerritsen et al. 2011). Also, a study on students preparing for an exam shows suppression of inflammatory biomarkers on consumption of *L. casei*. It is also important to note here that probiotics are strains are highly specific in their action. For example, two different strains of *Lactobacillus rhamnosus* HN001 and *Lactobacillus rhamnosus* JB-1 showed different results upon ingestion. *Lactobacillus rhamnosus* HN001 consumption in pregnant women shows lower depression and anxiety while *Lactobacillus rhamnosus* JB-1 reported no differences in depressive symptoms (Kelly et al. 2017; Slykerman et al. 2017).

Multistrains of probiotics are also having the ability to improve the mood and behaviour of patients with depressive symptoms. The action in case of multistrain of probiotics is also dependent of specific strain and its dose. To fully understand the effect of multistrain probiotics, studies should be done in healthy as well as diseased patients to find the impact of particular strain and its dosage (Steenbergen et al. 2015).

A more recent study evaluating the effect of probiotic supplementation in major depressive disorder along with combination of gluten-free or gluten-containing diet evidenced the important link between microbiota and depressive symptoms. The approach includes the gluten-free diet which further changes the microbiota composition and alters the biochemical pathways. The study suggests the bidirectional approach using gluten-free diet and probiotics for the control of immune inflammation cascade (Karakula-Juchnowicz et al. 2019). An ongoing clinical trial of Vivomixx (a probiotic mixture of 8 strains) for major depressive disorder lights the role of probiotics in depressive disorders. The study will reveal the investigation of probiotics for improvement in standard antidepressant medication for MDD (NCT02957591 2016).

7.6   Conclusion

A majority of studies carried out on probiotic ingestion investigates the effects of probiotics on MDD. Few studies also analysed the impact of gut microbiota on brain. Also, the reports provide evidence of link between diet quality and mood. Probiotics may improve symptoms in MDD by reducing inflammation and increasing serotonin levels. They could be used as potential candidate for MDD to overcome various side effects associated with conventional tricyclic antidepressants. While there are multiple mechanisms through which the probiotics improve the lifestyle of patients with depressive symptoms. Therefore, there is immense need of continue research on probiotics as antidepressants. Use of probiotics for treatment of depressive disorders through primary or preventive treatment looks promising. With continued research, probiotics could serve as a simple dietary fraction to improve the quality of life among depressed populations.

Acknowledgments   The authors acknowledge Department of Pharmaceuticals, Ministry of Chemical and Fertilizers, Government of India. NIPER-R communication for this book chapter is NIPER-R/ Communication/093.

Conflict of Interest   The authors declare no conflict of interest.

References

Åberg-Wistedt A, Jostell KG, Ross SB, Westerlund D (1981) Effects of zimelidine and desipramine on serotonin and noradrenaline uptake mechanisms in relation to plasma concentrations and to therapeutic effects during treatment of depression. Psychopharmacology (Berl) 74(4):297–305

Abildgaard A, Elfving B, Hokland M, Wegener G, Lund S (2017) Probiotic treatment reduces depressive-like behaviour in rats independently of diet. Psychoneuroendocrinology 79:40–48

Abraham BP, Quigley EMM (2017) Probiotics in inflammatory bowel disease. Gastroenterol Clin North Am 46(4):769–782

Akbaraly TN, Brunner EJ, Ferrie JE, Marmot MG, Kivimaki M, Singh-Manoux A (2009) Dietary pattern and depressive symptoms in middle age. Br J Psychiatry 195(5):408–413

Akkasheh G, Kashani-Poor Z, Tajabadi-Ebrahimi M, Jafari P, Akbari H, Taghizadeh M et al (2016) Clinical and metabolic response to probiotic administration in patients with major depressive disorder: a randomized, double-blind, placebo-controlled trial. Nutrition 32(3):315–320

Albert GC, Cook CM, Prato FS, Thomas AW (2009) Deep brain stimulation, vagal nerve stimulation and transcranial stimulation: an overview of stimulation parameters and neurotransmitter release. Neurosci Biobehav Rev 33(7):1042–1060

Alonso M, Vianna MRM, Izquierdo I, Medina JH (2002) Signaling mechanisms mediating BDNF modulation of memory formation in vivo in the hippocampus. Cell Mol Neurobiol 22(5): 663–674

Alonso C, Guilarte M, Vicario M, Ramos L, Ramadan Z, Antolín M et al (2008) Maladaptive intestinal epithelial responses to life stress may predispose healthy women to gut mucosal inflammation. Gastroenterology 135(1):163–172

Appleton KM, Sallis HM, Perry R, Ness AR, Churchill R (2015) Omega-3 fatty acids for depression in adults. Cochrane Database Syst Rev 5(11)

Arumugam M, Raes J, Pelletier E, Le Paslier D, Yamada T, Mende DR et al (2011) Enterotypes of the human gut microbiome. Nature 473(7346):174–180

Banks WA, Farr SA, La Scola ME, Morley JE (2001) Intravenous human interleukin-1 α impairs memory processing in mice: dependence on blood-brain barrier transport into posterior division of the septum. J Pharmacol Exp Ther 299(2):536–541

Bear TLK, Dalziel JE, Coad J, Roy NC, Butts CA, Gopal PK (2020) The role of the gut microbiota in dietary interventions for depression and anxiety. Adv Nutr:1–18

Bercik P, Denou E, Collins J, Jackson W, Lu J, Jury J et al (2011) The intestinal microbiota affect central levels of brain-derived neurotropic factor and behavior in mice. Gastroenterology 141(2):599–609

Borre YE, O'Keeffe GW, Clarke G, Stanton C, Dinan TG, Cryan JF (2014) Microbiota and neurodevelopmental windows: implications for brain disorders. Trends Mol Med 20(9): 509–518

Butt H, Dunstan R, McGregor N, Roberts T (2001) Bacterial colonosis in patients with persistent fatigue. In: Proceedings of the AHMF International Clinical Sciences Conference

Cepeda MS, Katz EG, Blacketer C (2017) Microbiome-gut-brain axis: probiotics and their association with depression. J Neuropsychiatry Clin Neurosci 29(1):39–44

Chang CJ, Lu CC, Lin CS, Martel J, Ko YF, Ojcius DM et al (2018) Antrodia cinnamomea reduces obesity and modulates the gut microbiota in high-fat diet-fed mice. Int J Obes (Lond) 42(2): 231–243

Chaucheyras-Durand F, Durand H (2010) Probiotics in animal nutrition and health. Benef Microbes 1(1):3–9

Claesson MJ, Jeffery IB, Conde S, Power SE, O'connor EM, Cusack S et al (2012) Gut microbiota composition correlates with diet and health in the elderly. Nature 488(7410):173–184

Clarke G, Grenham S, Scully P, Fitzgerald P, Moloney RD, Shanahan F et al (2013) The microbiome-gut-brain axis during early life regulates the hippocampal serotonergic system in a sex-dependent manner. Mol Psychiatry 18(6):666–673

Cohen S, Janicki-Deverts D, Doyle WJ, Miller GE, Frank E, Rabin BS et al (2012) Chronic stress, glucocorticoid receptor resistance, inflammation, and disease risk. Proc Natl Acad Sci U S A 109(16):5995–5999

Collins SM, Denou E, Verdu EF, Bercik P (2009) The putative role of the intestinal microbiota in the irritable bowel syndrome. Dig Liver Dis 41(12):850–853

Costello EK, Lauber CL, Hamady M, Fierer N, Gordon JI, Knight R (2009) Bacterial community variation in human body habitats across space and time. Science 326(5960):1694–1697

Cryan JF, Dinan TG (2012) Mind-altering microorganisms: the impact of the gut microbiota on brain and behaviour. Nat Rev Neurosci 13(10):701–712

Cuello-Garcia CA, Brozek JL, Fiocchi A, Pawankar R, Yepes-Nuñez JJ, Terracciano L et al (2015) Probiotics for the prevention of allergy: a systematic review and meta-analysis of randomized controlled trials. J Allergy Clin Immunol 136(4):952–961

D'Sa C, Duman RS (2002) Antidepressants and neuroplasticity. Bipolar Disord 4(3):183–194

David LA, Maurice CF, Carmody RN, Gootenberg DB, Button JE, Wolfe BE et al (2014) Diet rapidly and reproducibly alters the human gut microbiome. Nature 505(7484):559–563

De Palma G, Collins SM, Bercik P, Verdu EF (2014) The microbiota-gut-brain axis in gastrointestinal disorders: stressed bugs, stressed brain or both? J Physiol 592(14):2989–2997

Desbonnet L, Garrett L, Clarke G, Kiely B, Cryan JF, Dinan TG (2010) Effects of the probiotic Bifidobacterium infantis in the maternal separation model of depression. Neuroscience 170(4): 1179–1188

Dhakal R, Bajpai VK, Baek KH (2012) Production of GABA (γ-aminobutyric acid) by microorganisms: a review. Braz J Microbiol 43(4):1230–1241

Di Bella JM, Bao Y, Gloor GB, Burton JP, Reid G (2013) High throughput sequencing methods and analysis for microbiome research. J Microbiol Methods 95(3):401–414

Dinan TG, Cryan JF (2012) Regulation of the stress response by the gut microbiota: implications for psychoneuroendocrinology. Psychoneuroendocrinology 37(9):1369–1378

Dinan TG, Stanton C, Cryan JF (2013) Psychobiotics: a novel class of psychotropic. Biol Psychiatry 74(10):720–726

Dominguez-Bello MG, Costello EK, Contreras M, Magris M, Hidalgo G, Fierer N et al (2010) Delivery mode shapes the acquisition and structure of the initial microbiota across multiple body habitats in newborns. Proc Natl Acad Sci U S A 107(26):11971–11975

Dunner DL, Blier P, Keller MB, Pollack MH, Thase ME, Zajecka JM (2007) Preventing recurrent depression: long-term treatment for major depressive disorder. Prim Care Companion J Clin Psychiatry 9(3):214–223

Duval F, Lebowitz BD, Macher JP (2006) Treatments in depression. Dialogues Clin Neurosci 8(2): 191–206

Eriksson E, Hedberg MA, Andersch B, Sundblad C (1995) The serotonin reuptake inhibitor paroxetin is superior to the noradrenaline reuptake inhibitor maprotiline in the treatment of premenstrual syndrome. Neuropsychopharmacology 12(2):167–176

Ewaschuk JB, Diaz H, Meddings L, Diederichs B, Dmytrash A, Backer J et al (2008) Secreted bioactive factors from Bifidobacterium infantis enhance epithelial cell barrier function. Am J Physiol Gastrointest Liver Physiol 295(5):76–85

Finegold SM, Dowd SE, Gontcharova V, Liu C, Henley KE, Wolcott RD et al (2010) Pyrosequencing study of fecal microflora of autistic and control children. Anaerobe 16(9): 444–453

Gerritsen J, Smidt H, Rijkers GT, De Vos WM (2011) Intestinal microbiota in human health and disease: the impact of probiotics. Genes Nutr 6(3):209–240

Gibson GR, Roberfroid MB (1995) Dietary modulation of the human colonic microbiota: introducing the concept of prebiotics. J Nutr 125(6):1401–1412

Gill SR, Pop M, DeBoy RT, Eckburg PB, Turnbaugh PJ, Samuel BS et al (2006) Metagenomic analysis of the human distal gut microbiome. Science 312(5778):1355–1359

Grosso G, Galvano F, Marventano S, Malaguarnera M, Bucolo C, Drago F et al (2014a) Omega-3 fatty acids and depression: scientific evidence and biological mechanisms. Oxid Med Cell Longev 7(2):86–93

Grosso G, Pajak A, Marventano S, Castellano S, Galvano F, Bucolo C et al (2014b) Role of omega-3 fatty acids in the treatment of depressive disorders: a comprehensive meta-analysis of randomized clinical trials. PLoS One 9(5):117–126

Gruenwald J, Graubaum HJ, Harde A (2002) Effect of a probiotic multivitamin compound on stress and exhaustion. Adv Ther 19(3):141–150

Haast RAM, Kiliaan AJ (2015) Impact of fatty acids on brain circulation, structure and function. Prostaglandins Leukot Essent Fatty Acids 92:3–14

Hay PJ, Claudino AM (2012) Clinical psychopharmacology of eating disorders: a research update. Int J Neuropsychopharmacol 15(2):209–222

Heijtz RD, Wang S, Anuar F, Qian Y, Björkholm B, Samuelsson A et al (2011) Normal gut microbiota modulates brain development and behavior. Proc Natl Acad Sci U S A 108(7): 3047–3052

Hill C, Guarner F, Reid G, Gibson GR, Merenstein DJ, Pot B et al (2014) Expert consensus document: the international scientific association for probiotics and prebiotics consensus statement on the scope and appropriate use of the term probiotic. Nat Rev Gastroenterol Hepatol 11(8):506–514

Holzapfel WH, Haberer P, Snel J, Schillinger U, Huis'InT veld JHJ (1998) Overview of gut flora and probiotics. Int J Food Microbiol 41(2):85–101

Horrobin DF (2002) Food, micronutrients, and psychiatry. Int Psychogeriatr 14(4):331–334

Huttenhower C, Gevers D, Knight R, Abubucker S, Badger JH, Chinwalla AT et al (2012) Structure, function and diversity of the healthy human microbiome. Nature 486(7402):207–214

Insel TR, Mueller EA, Alterman I, Linnoila M, Murphy DL (1985) Obsessive-compulsive disorder and serotonin: is there a connection? Biol Psychiatry 20(11):1174–1188

Joel D, Zohar O, Afek M, Hermesh H, Lerner L, Kuperman R et al (2005) Impaired procedural learning in obsessive-compulsive disorder and Parkinson's disease, but not in major depressive disorder. Behav Brain Res 150(2):253–263

Karakula-Juchnowicz H, Rog J, Juchnowicz D, Łoniewski I, Skonieczna-Ydecka K, Krukow P et al (2019) The study evaluating the effect of probiotic supplementation on the mental status, inflammation, and intestinal barrier in major depressive disorder patients using gluten-free or gluten-containing diet (SANGUT study): a 12-week, randomized, double-blind Nutr J 18(1): 1–13

Kato-Kataoka A, Nishida K, Takada M, Kawai M, Kikuchi-Hayakawa H, Suda K et al (2016) Fermented milk containing lactobacillus casei strain Shirota preserves the diversity of the gut microbiota and relieves abdominal dysfunction in healthy medical students exposed to academic stress. Appl Environ Microbiol 82(12):3649–3658

Kawai T, Takeuchi O, Fujita T, Inoue J, Mühlradt PF, Sato S et al (2001) Lipopolysaccharide stimulates the MyD88-independent pathway and results in activation of IFN-regulatory factor 3 and the expression of a subset of lipopolysaccharide-inducible genes. J Immunol 167(10): 5887–5894

Kelly JR, Allen AP, Temko A, Hutch W, Kennedy PJ, Farid N et al (2017) Lost in translation? The potential psychobiotic lactobacillus rhamnosus (JB-1) fails to modulate stress or cognitive performance in healthy male subjects. Brain Behav Immun 61:50–59

Kessler RC, Bromet EJ (2013) The epidemiology of depression across cultures. Annu Rev Public Health 34(1):119–138

Kuda T, Enomoto T, Yano T, Fujii T (2000) Cecal environment and TBARS level in mice fed corn oil, beef tallow and menhaden fish oil. J Nutr Sci Vitaminol (Tokyo) 46(2):65–70

Ley RE, Bäckhed F, Turnbaugh P, Lozupone CA, Knight RD, Gordon JI (2005) Obesity alters gut microbial ecology. Proc Natl Acad Sci U S A 102(31):11070–11075

Logan AC, Katzman M (2005) Major depressive disorder: probiotics may be an adjuvant therapy. Med Hypotheses 64(3):533–538

Maan G, Sikdar B, Kumar A, Shukla R, Mishra A (2020) Role of flavonoids in neurodegenerative diseases: limitations and future perspectives. Curr Top Med Chem. Apr 16 [cited 2020 Apr 19];20

Macpherson AJ, Harris NL (2004) Interactions between commensal intestinal bacteria and the immune system. Nat Rev Immunol 4(6):478–485

Malaguarnera M, Vacante M, Antic T, Giordano M, Chisari G, Acquaviva R, et al. Bifidobacterium longum with fructo-oligosaccharides in patients with nonalcoholic steatohepatitis. Dig Dis Sci 2012;57(2):545–553

Manach C, Scalbert A, Morand C, Rémésy C, Jiménez L (2004) Polyphenols: food sources and bioavailability. Am J Clin Nutr 79(5):727–747

Mayer EA, Knight R, Mazmanian SK, Cryan JF, Tillisch K (2014) Gut microbes and the brain: paradigm shift in neuroscience. J Neurosci 34(46):15490–15496

Messaoudi M, Lalonde R, Violle N, Javelot H, Desor D, Nejdi A et al (2011) Assessment of psychotropic-like properties of a probiotic formulation (lactobacillus helveticus R0052 and Bifidobacterium longum R0175) in rats and human subjects. Br J Nutr 105(5):755–764

Miller AH, Maletic V, Raison CL (2010) Inflammation and its discontents: the role of cytokines in the pathophysiology of major depression. Psiquiatr Biol 17(2):71–80

Morgan XC, Huttenhower C (2012) Chapter 12: human microbiome analysis. PLoS Comput Biol 8(12):76–88

NCT02957591 (2016) Probiotic supplementation in severe depression. https://clinicaltrials.gov/show/NCT02957591

Ng QX, Peters C, Ho CYX, Lim DY, Yeo WS (2018) A meta-analysis of the use of probiotics to alleviate depressive symptoms. J Affect Disord 228:13–19

Nowak G, Siwek M, Dudek D, Zieba A, Pilc A (2003) Effect of zinc supplementation on antidepressant therapy in unipolar depression: a preliminary placebo-controlled study. Pol J Pharmacol 55(6):1143–1147

O'Malley D, Julio-Pieper M, Gibney SM, Dinan TG, Cryan JF (2010) Distinct alterations in colonic morphology and physiology in two rat models of enhanced stress-induced anxiety and depression-like behaviour. Stress 13(2):114–122

Ohland CL, Kish L, Bell H, Thiesen A, Hotte N, Pankiv E et al (2013) Effects of lactobacillus helveticus on murine behavior are dependent on diet and genotype and correlate with alterations in the gut microbiome. Psychoneuroendocrinology 38(9):1738–1747

Ott SJ, Musfeldt M, Wenderoth DF, Hampe J, Brant O, Fölsch UR et al (2004) Reduction in diversity of the colonic mucosa associated bacterial microflora in patients with active inflammatory bowel disease. Gut 53(5):685–693

Owens MJ, Nemeroff CB (1994) Role of serotonin in the pathophysiology of depression: focus on the serotonin transporter. Clin Chem 40(2):288–295

Pan Z, Rosenblat JD, Swardfager W, McIntyre RS (2017) Role of proinflammatory cytokines in dopaminergic system disturbances, implications for anhedonic features of MDD. Curr Pharm Des 23(14):117–123

Park AJ, Collins J, Blennerhassett PA, Ghia JE, Verdu EF, Bercik P et al (2013) Altered colonic function and microbiota profile in a mouse model of chronic depression. Neurogastroenterol Motil 25(9):256–268

Parvez S, Malik KA, Ah Kang S, Kim HY (2006) Probiotics and their fermented food products are beneficial for health. J Appl Microbiol 100(6):1171–1185

Pimentel M, Hallegua D, Chow EJ, Wallace D, Bonorris G, Lin HC (2000) Eradication of small intestinal bacterial overgrowth decreases symptoms in chronic fatigue syndrome: a double blind, randomized study. Gastroenterology 118(4):414–421

Pimentel M, Wallace D, Hallegua D, Chow E, Kong Y, Park S et al (2004) A link between irritable bowel syndrome and fibromyalgia may be related to findings on lactulose breath testing. Ann Rheum Dis 63(4):450–452

Pinto-Sanchez MI, Hall GB, Ghajar K, Nardelli A, Bolino C, Lau JT et al (2017) Probiotic Bifidobacterium longum NCC3001 reduces depression scores and alters brain activity: a pilot study in patients with irritable bowel syndrome. Gastroenterology 153(2):448–459

Pringsheim T, Kelly M, Barbui C (2016) Stopping antidepressants following depression. BMJ 220 (February):4–6

Qin J, Li R, Raes J, Arumugam M, Burgdorf KS, Manichanh C et al (2010) A human gut microbial gene catalogue established by metagenomic sequencing. Nature 464(7285):59–65

Raes K, Huyghebaert G, De Smet S, Nollet L, Arnouts S, Demeyer D (2002) The deposition of conjugated linoleic acids in eggs of laying hens fed diets varying in fat level and fatty acid profile. J Nutr 132(2):182–189

Raison CL (2019) Managing inflammation while treating patients with depression. MD Edge Psychcast 35(2):393–409

Rao AV, Bested AC, Beaulne TM, Katzman MA, Iorio C, Berardi JM et al (2009) A randomized, double-blind, placebo-controlled pilot study of a probiotic in emotional symptoms of chronic fatigue syndrome. Gut Pathog 1(1):6–12

Rhee SH, Pothoulakis C, Mayer EA (2009) Principles and clinical implications of the brain-gut-enteric microbiota axis. Nat Rev Gastroenterol Hepatol 6(5):306–314

Romijn AR, Rucklidge JJ, Kuijer RG, Frampton C (2017) A double-blind, randomized, placebo-controlled trial of lactobacillus helveticus and Bifidobacterium longum for the symptoms of depression. Aust N Z J Psychiatry 51(8):810–821

Rosenthal NE, Sack DA (1984) From the Clinical Psychobiology Branch, National Institutes of Health. Arch Gen Psychiatry 41:72–80

Round JL, Mazmanian SK (2009) The gut microbiota shapes intestinal immune responses during health and disease. Nat Rev Immunol 9(5):313–323

Saddiqe Z, Naeem I, Maimoona A (2010) A review of the antibacterial activity of Hypericum perforatum L. J Ethnopharmacol 131(3):511–521

Salminen SJ, Gueimonde M, Isolauri E (2005) Probiotics that modify disease risk. J Nutr 135(5): 1294–1298

Salzman NH, Underwood MA, Bevins CL (2007) Paneth cells, defensins, and the commensal microbiota: a hypothesis on intimate interplay at the intestinal mucosa. Semin Immunol 19(2): 70–83

Sánchez-Villegas A, Verberne L, de Irala J, Ruíz-Canela M, Toledo E, Serra-Majem L et al (2011) Dietary fat intake and the risk of depression: the SUN project. PLoS One 6(1):113–121

Shim IH, Woo YS, Kim MD, Bahk WM (2017) Antidepressants and mood stabilizers: novel research avenues and clinical insights for bipolar depression. Int J Mol Sci 18(11)

Simopoulos AP (2002) Omega-3 fatty acids in inflammation and autoimmune diseases. J Am Coll Nutr 21(6):495–505

Singh A, Kumar A, Verma RK, Shukla R (2020 Apr) Silymarin encapsulated nancliquid crystals for improved activity against beta amyloid induced cytotoxicity. Int J Biol Macromol 15(149): 1198–1206

Slavich GM, Irwin MR (2014) From stress to inflammation and major depressive disorder: a social signal transduction theory of depression. Psychol Bull 236(10):3063–3079

Slykerman RF, Hood F, Wickens K, Thompson JMD, Barthow C, Murphy R et al (2017) Effect of lactobacillus rhamnosus HN001 in pregnancy on postpartum symptoms of depression and anxiety: a randomised double-blind placebo-controlled trial. EBioMedicine 24:159–165

Steenbergen L, Sellaro R, van Hemert S, Bosch JA, Colzato LS (2015) A randomized controlled trial to test the effect of multispecies probiotics on cognitive reactivity to sad mood. Brain Behav Immun 48:258–264

Stephens RL, Tache Y (1989) Intracisternal injection of a TRH analogue stimulates gastric luminal serotonin release in rats. Am J Physiol Gastrointest Liver Physiol 256(2):289–302

Sudo N (2006) Stress and gut microbiota: does postnatal microbial colonization programs the hypothalamic-pituitary-adrenal system for stress response? Int Congr Ser 1287:350–354

Sudo N, Chida Y, Aiba Y, Sonoda J, Oyama N, Yu XN et al (2004) Postnatal microbial colonization programs the hypothalamic-pituitary-adrenal system for stress response in mice. J Physiol 558(1):263–275

Taché Y, Yoneda M (1993) Central action of TRH to induce vagally mediated gastric cytoprotection and ulcer formation in rats. J Clin Gastroenterol 17:58–63

Tan J, McKenzie C, Potamitis M, Thorburn AN, Mackay CR, Macia L (2014) The role of short-chain fatty acids in health and disease. Adv Immunol 121:91–119

Tannock GW, Savage DC (1974) Influences of dietary and environmental stress on microbial populations in the murine gastrointestinal tract. Infect Immun 9(3):591–598

Tillisch K, Labus J, Kilpatrick L, Jiang Z, Stains J, Ebrat B et al (2013) Consumption of fermented milk product with probiotic modulates brain activity. Gastroenterology 144(7):52–71

van Bronswijk SC, Lemmens LHJM, Keefe JR, Huibers MJH, DeRubeis RJ, Peeters FPML (2019) A prognostic index for long-term outcome after successful acute phase cognitive therapy and interpersonal psychotherapy for major depressive disorder. Depress Anxiety 36(3):252–261

Williams BL, Hornig M, Buie T, Bauman ML, Cho Paik M, Wick I et al (2011) Impaired carbohydrate digestion and transport and mucosal dysbiosis in the intestines cf children with autism and gastrointestinal disturbances. PLoS One 6(9):321–346

Wirz-Justice A (2006) Biological rhythm disturbances in mood disorders. Int Clin Psychopharmacol 21:128–134

World Health Organisation (WHO) (2009) WHO guidelines on hand hygiene in health care: first global patient safety challenge clean care is safer care. World Health 30(1):270

Yamashiro Y (2018) Gut microbiota in health and disease. Ann Nutr Metab 71(3):242–246

Zheng P, Zeng B, Zhou C, Liu M, Fang Z, Xu X et al (2016) Gut microbiome remodeling induces depressive-like behaviors through a pathway mediated by the host's metabolism. Mol Psychiatry 21(6):786–796

Gut Dysbiosis in Insomnia and Diurnal Cycle

Rajesh Kumar

Abstract

Prolonged insomnia may shorten the life expectancy in humans, as this sleeping disorder has already been associated with miscellaneous detrimental health effects on humans such as risk of elevated blood pressure, type 2 diabetes, hyperlipidemia, obesity, coronary heart disease (CHD), and stroke. Numerous studies manifested that onset of sleeplessness and insomnia is linked to disruption of circadian and microbiome rhythms, immune response, and nutrient metabolism, however, precise mechanism is yet to be elucidated. Moreover, it is now understood that gut dysbiosis and circadian rhythm disruption can affect the quality of sleep, and are associated with other metabolic disorders. Considerable evidences have shown that gut microbiome regulates sleep and psychological states of its host through microbiome–gut–brain (MGB) axis. Research findings have indicated that diurnal variations, emotional state, and physiological stress can disrupt the composition and function of gut microbiota which in turn destabilizes host's circadian gene expression and functions. Consequently, disruptions of gut microbiome-mediated functions such as diminished bile acids conjugation or enhanced generation of H_2S, and the reduced butyrate production, in turn affect substrate oxidation and metabolic homeostasis of the host. Furthermore, gut dysbiosis and altered circadian gene expression may cause onset of insomnia, circadian misalignment, and metabolic disorders. Nevertheless, insomnia condition can be reversed by: (1) achieving daily rhythmicity of gut bacteria by prebiotics enriched dietary interventions; (2) adopting healthy lifestyle especially meal timing, selective eating patterns and sleep timing; and (3) oral administration of probiotics in order to restore the gut microbial balance. Thus, manipulation of gut microbiome structure through synbiotic diet and

R. Kumar (✉)
Central Research Institute (Ministry of Health and Family Welfare), Kasauli (Solan), Himachal Pradesh, India

© Springer Nature Singapore Pte Ltd. 2022
P. K. Deol, S. K. Sandhu (eds.), *Probiotic Research in Therapeutics*,
https://doi.org/10.1007/978-981-16-6760-2_8

chrononutrition-based approaches may therefore hold promise to consolidate host circadian rhythms.

Keywords

Gut microbiome · Dysbiosis · Microbiome–gut–brain axis · Insomnia · Circadian rhythm · Clock genes

8.1   Introduction

The diverse microbial community of human gut is termed as gut microbiome. This human gut microbiome plays an instrumental role in childhood malnutrition improvement and continued education of host's immune response. It is, therefore, clearly evident that adequate gut colonization is necessary for immunologic and metabolic health. Colonized gut bacteria communicate to host's physiology by interacting with immunocytes or through secretion of metabolic end products and in return gastrointestinal microbiome is also mutually benefitted (Carding et al. 2015; Walker 2017; Belizario and Faintuch 2018; Belizario et al. 2018). Inside human gut approximately 10^{12} microbial cell exists, which proliferates in diversity and abundance along the cephalocaudal axis with niche differentiation (Walker 2017). Human microbiome can be divided into five dominant bacterial phyla: firmicutes, bacteroidetes, proteobacteria, actinobacteria, and verrucomicrobia. Bacteriodetes and firmicutes represent more than 90% of the relative abundance of gut bacteria (Belizario and Faintuch 2018; Belizario et al. 2018). A condition in which composition of gastrointestinal microbiota gets imbalanced is called as gut dysbiosis. Alterations in the gut microbiota can be caused due to exposure to various factors, such as diet, stress, toxins, antibiotics exposure, and pathogens. Of these, expansion of enteric pathobionts harbors the greatest potential to cause gut dysbiosis (Carding et al. 2015; Walker 2017).

Sleep disorder insomnia is a condition when an individual faces difficulty falling or staying asleep. In general, insomniac people can be rarely satisfied with their sleep quality, and they often complaints about stress, lack of energy, poor concentration, mood fluctuations, and bad performance at work.

Circadian rhythm is the diurnal changes of the environment, and this term is derived from *circa diem* which is an approximate 24-h cycling period. These diurnal changes of the environment have shaped life on Earth during the course of evolution (Farre and Liu 2013; Merrow and Maas 2009; Millar 2016; Shindey et al. 2016). Circadian rhythm is governed by the brain's central circadian clock, which in turn co-ordinates rhythmic output of peripheral circadian clocks. The ability of living beings on earth to anticipate circadian and environmental disturbances has bestowed an evolutionary advantage for developing molecular mechanisms that allowed them to adapt in light and dark cycles. These molecular mechanisms are called circadian clocks and possess peculiar attributes which differentiate it from other oscillatory process. Brief definition of circadian clocks could be self-sustained, temperature compensated, and entrainable oscillators (Nobs et al. 2019).

It is now evident that gut bacteria are instrumental in regulating sleep cycle. Inadequate sleep and sleep disorders are associated with metabolic health disorders such as increased risk of hypertension, glucose intolerance, insulin sensitivity, impaired energy balance, and obesity (Nobs et al. 2019). Insomnia and its profound effect on circadian disruption and gut dysbiosis actually contribute to these metabolic health outcomes. The intestine and commensal bacteria affect energy homeostasis by controlling various physiological functions, i.e. immune function, digestion, and absorption of food, and gastric emptying which are also controlled by clock genes (Nobs et al. 2019; Parkar et al. 2019). Here we mainly discuss about circadian rhythm disruption and gut dysbiosis in insomnia and diurnal cycle. Nevertheless, approaches based on prebiotic and probiotic dietary modulation to expand gut bacterial richness and diversity in order to reverse gut dysbiosis hold the promise to consolidate host circadian rhythms.

8.2 Gut Microbiome

The human microbiome is as complex as human organ and, therefore, humans are considered as true super-organisms (The Human Microbiome Project Consortium et al. 2012; Nobs et al. 2019). Recent studies have shown that gut microbiome plays a critical role in health and diseases (Holmes et al. 2011; Shreiner et al. 2015; Buford 2017; Dinan and Cryan 2017; Nobs et al. 2019). In a symbiotic relationship human host provides a suitable environment for the colonization of gut bacteria, which in return exert several metabolic health benefits to its host. Moreover, as an auxiliary organ to the human body, gut microbiome actively contributes in the host's metabolism functions such as nutrient digestion and absorption, synthesis of vitamins, and enhancement of mucosal immune function (Nobs et al. 2019; Parkar et al. 2019). In addition, significance of the intestinal microbiota for human health is also illustrated by observations that a disrupted gut bacterial composition intensifies immune dysfunction such as auto-inflammatory colitis, while a balanced gut microbiota holds therapeutic potential to treat life-threatening infections (Elinav et al. 2011; Rohlke and Stollman 2012).

Infant meconium analysis suggested the presence of bacteria in the fetal gut prior to birth which indicates that microbe-intestinal colonization occur prenatally (Aagaard et al. 2014). During the new-born adaptation to its extra-uterine environment, appropriate initial gut colonization with commensal bacteria is particularly necessary as the colonized gut bacteria play an important role in strengthening of digestive and immune functions (Johnson and Versalovic 2012; Houghteling and Walker 2015; Walker 2017).

Gut microbial composition of an individual is shaped by aging (Palmer et al. 2007). Microbial colonization of neonatal gut starts after obtaining healthy bacterial bolus upon passing through mother's birth canal. Afterwards, microbial transfer from mother primarily shapes neonatal microbiome which further increases in heterogeneity and diversity, and leads to enhanced immunity (Schwartz et al. 2012; Scholtens et al. 2012; Nobs et al. 2019). Moreover, lactic compounds of mother's milk promote the growth of beneficial bacteria in neonatal gut. With a retention rate of approximately 60% over 5 years, gut microbial composition stabilizes in adulthood. However, gut microbial configuration preserves a pace of change in old age as distinct microbial profiles found in elderly subjects (Mariat et al. 2009; Faith et al. 2013; Nobs et al. 2019). Therefore, maintaining healthy gut microbial configuration is extremely important for short- and long-term health. Modern world's practices such as westernized diet, routine immunization, improved sanitation, and overuse of antibiotics have altered bacterial configuration of the gut, and have also been associated with the increased incidence of allergic and other immune-mediated diseases (autoimmune disease) over the last several decades (Palmer et al. 2007; Adlerberth and Wold 2009; Walker 2017). Vastest long term gut microbiome perturbations occur due to exposure to antibiotics, drugs and major dietary changes. Additionally, excess intake of high calorie food also causes a persisting and systemic change in the Firmicutes/Bacteroidetes ratio (Ley et al. 2006; Wu et al. 2011; David et al. 2014; Suez et al. 2018; Zmora et al. 2018; Nobs et al. 2019).

8.3   Dysbiosis

In simple terms gut dysbiosis is the disruption of balance of gut flora and microbiome. In brief, dysbiosis is also defined as "any change to the components of resident commensal communities relative to the community found in healthy individuals" (Maharshak et al. 2013). Digestive disturbances are the main symptoms of bacterial dysbiosis while typical dysbiosis symptoms could be indigestion, heartburn, reflux, diarrhea, abdominal pain, bloating lowered mood, food intolerances, and brain fog, etc. Several studies have demonstrated that gut microbial imbalance causes enhanced intestinal permeability and translocation of lipopolysaccharide (LPS) or endotoxins (gram^{-ve} bacterial cell components), to the systemic circulation. This further trigger low-grade chronic inflammation inside adipose tissue, liver, muscle, and the pancreas which can cause obesity and altered metabolic signaling (Carding et al. 2015).

Dysbiosis can be categorized mainly into three phases. Loss of beneficial commensal gut flora could be one dysbiosis type (Phase I) as these commensals are extremely important in the development of immune system. T regulatory cells (Tregs) play a crucial role in preventing autoimmune disease expression as Tregs are involved in preventing abnormal inflammatory responses against indigenous bacteria and self-antigens. In this context, *Bacteroides fragilis* or a mixture of *Clostridium* strains play a crucial role in the development and maintenance of Tregs (Maharshak et al. 2013). Similarly, other beneficial gut commensal bacteria, such as *Bifidobacterium breve* or *Lactobacillus acidophilus,* exhibit direct anti-inflammatory activity as they neutralize inflammatory cytokines (Backhed et al. 2005). Gut microbial equilibria gets disturbed in the absence of these probiotic bacteria which further upset the immune protection against pathogens and tolerance to self-antigens and indigenous microflora (Walker 2017).

Furthermore, Phase II dysbiosis type can occur due to an expansion of pathobionts; commensal microbes with pathogenic potential, which takes place due to the increased bacterial abundance such as *Escherichia coli, Clostridium difficile, Shigella*, vancomycin resistant *Enterococcus,* and *Klebsiella.* Moreover, genetic defects in immune function can also increase the abundance of these pathobionts which ultimately lead to various forms of colitis (Vijay-Fumar et al. 2010; Walker 2017). In case of Crohn's disease (CD), pathobionts expand due to major shift in bacterial abundances from Bacteriodetes and Enterobacteria including the concomitant depletion of Firmicutes and Bifidobacteria. Although not a single cohort study indicated toward the possible implication of a pathobiont in causing CD, yet perturbation of the host–microbe commensalism is mainly accepted as the leading factor causing bowel injury (Zechner 2017).

Finally, a Phase III dysbiosis is the loss of commensal gut bacterial diversity. More often, it has been noticed that characterization of gut microbiota in case of specific diseases was associated with less microbial diversity. Hence, gut bacterial richness offers overlapping immune protection against non-communicable diseases (NCDs). Phase III dysbiosis could be explained by the observation that single species of specific organisms could not effectively stimulate immunity in germ-free (GF) animals colonized with microbiota from a diseased animal in comparison to the multiple species of the same genera. For example, thirty different species of *Clostridium* genus could stimulate Tregs more effectively in comparison to its single species. Moreover, GF animals when colonized with multiple species of gut bacteria showed more effective IgE levels reduction than those colonized with single or a few organisms. Therefore, gut bacterial diversity has direct impact on host's health (Cahenzil et al. 2013; Atarashi et al. 2013; Walker 2017).

8.4 Causes of Dysbiosis

Infants and elderly are more vulnerable to dysbiosis than adults. During the primary stages of extra-uterine environment adaptation of young infants their immunologic and gastrointestinal metabolic development takes place. Therefore, any kind of disruption in the sequence of initial normal colonization can impose serious impact on lifelong health of an individual. Moreover, multiple environmental factors can also disrupt initial normal colonization and could cause perturbations in the maturation of immunologic and metabolic functions which further leads to disease expression throughout the lifespan (Johnson and Versalovic 2012; Walker 2017). In general, major causes of dysbiosis could be: (1) Perinatal disruption of colonization; (2) Dietary factor; (3) Underlying disease; and (4) stress.

8.4.1 Perinatal Disruption of Colonization

Intra-uterine alteration to microbial exposure could disrupt the normal colonization in the neonate. For instance, obesity during pregnancy has been known to alter mother's microbiome, which further can influence normal colonization in the neonate due to placental and birth exposure to mother's dysbiosis (Walker 2017). During their infancy, such infants are more likely to get overweight which eventually might cause obesity during their adulthood. Similarly, premature delivery of infants caused by intra-uterine infection, excessive weight, etc. also disrupt normal colonization of the neonate's gut which increases the risk of severe intestinal inflammation, i.e. necrotizing enterocolitis (Baker et al. 2004; Collado et al. 2010; Mshvidadze et al. 2010; Walker 2017). Cesarean delivery also results in dysbiotic colonization in neonates, as these newborns could not receive the healthy bacterial bolus from mother's birth canal and therefore, take substantial time to completely colonize their intestine (Jakobsson et al. 2014; Avershina and Rudi 2015). During their childhood such children are likely to acquire obesity, incidence of asthma, and celiac disease which substantiates the importance of initial normal colonization for long-term health (Vassallo and Walker 2008). During prenatal period excessive and inappropriate intake of antibiotics can alter the initial normal colonization sequence inside gut which can cause aberrations in the development of metabolic homeostasis, and later on disease expression throughout life. During pregnancy disruption of the microenvironment can alter mother's gut microbiome, which is connected with dysbiotic colonization of neonatal gut, and finally leads to neurologic diseases expression, i.e. autism and schizophrenia (Walker 2017).

8.4.2 Dietary Factors

Diet, an important environmental factor, affects gut microbiome composition. Initial colonization gets affected in those new-born infants who are not exclusively feed mother's breast milk. This gives rise to an increased incidence of non-communicable diseases (NCDs), i.e. obesity, allergy, and autoimmune disease later in life (Donnet-Hughes et al. 2015). In an interesting finding, microbiota composition was strikingly different between two adolescent populations; one (Florence, Italy) fed on the diet enriched with animal protein and fat while other (Boulpon rural African village) ate high fiber diet with complex carbohydrates, little animal protein and fat. Fecal microbial analysis of African children revealed the significant increase in Bacteroidetes and decline in Firmicutes, with unique abundance of cellulose and xylan hydrolyzing bacteria from the genera, i.e. *Prevotella* and Xylanibacter which was completely absent among European children. In addition, significantly higher short-chain fatty acids (SCFAs) were found among African children than European children. However, Enterobacteriaceae (*Escherichia* and *Shigella*) were significantly higher in European children than African children (DeFilippo et al. 2010). However, Mediterranean diet has been commonly associated with higher life expectancy and decreased incidence of heart and metabolic diseases (Azzini et al. 2011). Recent studies have indicated that food additives such as sugar substitutes and emulsifying agents are also associated with alteration in gut microbiota, and consequent disease expression (Chassaing et al. 2015; Suez et al. 2014; Walker 2017). Therefore, such observations clearly indicate that diet at any age may affect gut microbial diversity.

8.4.3 Underlying Disease

A disease can disturb gut microbial equilibrium and may lead to dysbiosis. It is often noticed among inflammatory bowel disease (IBD) patients, particularly during Crohn's disease development, that intestinal infection takes place before the onset of disease. Gut microbiome composition of sick patients differs significantly from age-matched controls. Enormous metabolic activity of such dysbiotic bacteria aggravates the phenotypic expression of disease. For example, fecal transplantation from depressive patients into GF mice could lead to the spontaneous development of depression-like behavior in animals (Zheng et al. 2016; Walker 2017).

8.4.4 Stress

Stress has now been considered as another cause of dysbiosis. Both basic and clinical evidences indicate that stress is likely to alter gut microbial composition, which ultimately leads to neurologic disease expression (Moya and Ferrer 2016). Although this particular research area is very new but stressful situations are associated with alterations in gut microbiome and brain's function among healthy individuals (Faith et al. 2013). Interestingly, the probiotic intervention could relieve neurologic changes induced with stress (Costello et al. 2012).

8.5 Insomnia

Insomnia, a prevailing sleep disorder especially among elderly, can make it hard to fall asleep or stay asleep, and to get back to sleep. It can deplete the mood, energy level, work performance, health, and life's quality of an individual. Insomnia is mainly developed due to the onset of anxiety/stress or a traumatic event. Many individuals get affected by short-term (acute) insomnia somewhere in their life, which can persist for days or weeks. However, some individuals are also affected by long-term (chronic) insomnia which could last for a month or more. Generally, insomnia could be the primary problem or it may also be associated with other underlying medical conditions, habits that disrupt sleep or the side effects of medications. Insomnia condition can be typically characterized by daytime sleepiness, low energy, irritability, depression or anxiety, and anxiety about sleep. Therefore, treating underlying cause can resolve the insomniac condition of an individual, but occasionally insomnia can last for years (Li et al. 2018; Liu et al. 2019).

8.5.1 Causes

Stress, irregular work shifts, jet lag, irregular eating timings, and inadequate sleep habits could be the main causes of chronic insomnia. In general, circadian rhythms of an individual act as an internal clock of the body which guide sleep-wake cycle, metabolism, and body temperature while its disruption can lead to insomnia (Liu et al. 2019).

8.5.2 Complications

Quality sleep is extremely important to health. Insomnia or sleep loss can affect mental and physical abilities of an individual. Insomniac patients often complaint about a lower life quality than those who are sleeping well. Prolonged insomnia may shorten the life expectancy in humans, as this sleeping disorder has already been associated with miscellaneous detrimental health effects on humans (Liu et al. 2019).

8.5.3 Gut Dysbiosis and Sleep

Lack of bacterial diversity inside gut can be toxic to the host, and might cause gut dysbiosis which mainly can be of two types; (1) small intestinal bacterial overgrowth (SIBO) and (2) small intestinal fungal overgrowth (SIFO). Persistent gut dysbiosis might lead to interrupted sleep and chronic gut inflammation. Gut dysbiotic bacterial activity can increase cortisol levels which further inhibit sleep onset. In this context, human microbiome contributes in nurturing metabolism, assisting digestion, immunomodulation, uplifting mood, and maintaining a regular sleep cycle. Vagus nerve, a cranial nerve connects brainstem with abdomen, and gut bacteria communicate with brain through vagus nerve. Equilibrium of this gut brain communication ensures the sleep quality. Human brain's pineal gland produces melatonin hormone, which predominantly governs sleep cycle. In response to changes in daylight levels, human body synthesizes melatonin. Melatonin synthesis requires tryptophan amino acid precursor which ensures sleep-wake cycle to stay in sync. In this context gut bacteria especially, probiotic bacteria expand tryptophan levels in blood, and thereby enhance melatonin supply in body (Li et al. 2018). In addition, gut microbiota effectively reduces cortisol levels. Cortisol, a stress hormone, enhances anxiety feeling and inhibits onset of sleep. Probiotic gut bacteria, i.e. *Lactobacillus* and *Bifidobacterium* can synthesize γ-aminobutyric acid (GABA), an amino acid necessary to promote deep sleep in patients suffering with depression and insomnia, such patients exhibited abnormal expression of GABA mRNA (Bravo et al. 2011; Barrett 2014). Chromaffin cells of gut mainly produce ≥90% of body's serotonin or 5-hydroxytryptamine (5-HT) which prevent both the occurrence of rapid eye movement (REM) sleep and the development of depression. Nevertheless, certain spore-forming gut bacteria can directly act on chromaffin cells to modulate the synthesis and secretion of serotonin. Therefore, it is now clear that gut bacteria execute a vital role constantly in promoting healthy sleep (Ridaura and Belkaid 2015; Li et al. 2018). Moreover, common gut inhabitants, i.e. *Escherichia coli* and *Enterococcus*, also produce 5-HT to some extent (Rieder et al. 2017; Li et al. 2018). Gut microbiota may affect central nervous system (CNS) and Hypothalamic, Pituitary, Adrenal (HPA) axis by modulating the secretion of neurotransmitters, i.e. tryptophan, cortisol, and serotonin (Powley et al. 2008).

8.6 Gut Microbiota and Depression

It is most often noticed that onset of depression and insomnia takes place mutually in an individual's life. Generally, patients having depression also complain about insomnia and poor sleep quality (Seow et al. 2016). Moreover, insomnia has been shown to intensify the severity of depression (Chung et al. 2015). Although various hypotheses have been proposed to correlate the relationship between depression and insomnia, i.e. neurotransmitter depletion, neurotrophin dysregulation, HPA axis dysregulation, and neural immune activation, but none of these could explain this complex mechanism. The HPA axis comprises stimulating inhibition loops (forward and feedback) including hypothalamus, pituitary, and adrenal glands, which controls the synthesis of glucocorticoid. The HPA axis is also implicated in the pathophysiology of cognitive functioning, anxiety, and depression. Physiological or psychological stress causes hyperactive HPA axis which can disrupt gut microbial equilibrium due to increased intestinal permeability and immune response (Li et al. 2018).

Recently quite a few studies have suggested that comorbidity of depression and insomnia can be explained on the basis of MGB axis. As described earlier, beneficial gut microbes synthesize various neurotransmitters and metabolites, i.e. serotonin, melatonin, GABA, dopamine (DA), and short-chain fatty acids (SCFAs), etc. These metabolites affect the activity of central nervous system (CNS) as well as act directly on the enteric nervous system (ENS) and the vagus nerve (Petra et al. 2015; Li et al. 2018).

Recent studies had suggested that anxiety or depression-like behavioral changes in a host could be associated with the alterations in the intestinal microbiome's composition caused by antibiotics and probiotics (Dinan and Cryan 2017; Burokas et al. 2017; Macedo et al. 2017; Guida et al. 2018; Li et al. 2018). Association between depression and gut microbiome composition can be confirmed by the fact that fecal microbiota transplantation can transmit depression-like behavior from human hosts to mice (Zheng et al. 2016; Kelly et al. 2016; Li et al. 2018). Moreover, one study also reported that oral administration of antibiotics could decrease anxious behavior, and also affected brain-derived neurotrophic factor (BDNF) expression in mice while intraperitoneal administration of antibiotics did not exhibit such effects, which suggests that antibiotic-induced alteration in gut microbiota was responsible for suppression of anxious behavior (Bercik et al. 2011; Li et al. 2018).

8.7   Intestinal Dysbacteriosis and Stress

Stress and anxiety are the most common causes of chronic insomnia. Patients under stress tend to be anxious and thereby insomnia condition often get worse. In this context, researchers now observed that gut dysbacteriosis is associated to stress and anxiety problems (Finger et al. 2011; Sun et al. 2013; Flowers and Ellingrod 2015; Kelly et al. 2016; Bravo et al. 2011; Burokas et al. 2017; Li et al. 2018). This could be easily noticed among shift workers who ordinarily suffer with short sleep time and disrupted circadian rhythms. It may trigger physiological stress reactions further causing alteration in the gut microbial composition. This could also interfere the normal function of both the immune and nervous systems, thereby making an individual vulnerable against stressful situations. Moreover, among shift workers, imbalance of glucose and lipids levels is often noticed. In addition, loss of diurnal rhythmicity perturbates physiological stress which can cause metabolic dysfunction (Kim et al. 2015; Reynolds et al. 2017; Li et al. 2018). Nevertheless, probiotic bacteria, i.e. *Lactobacillus* and *Bifidobacterium*, can improve stress and anxiety disorders by influencing ENS and the immune system (Plaza-Díaz et al. 2017).

Cortisol may act as a connecting bridge between gut microbiome and stress/anxiety. Some studies also indicated that probiotic intervention subsided stress-induced cortisol secretion in chronically stressed mice, and as a result exhibited anxiolytic and antidepressant-like effects (Desbonnet et al. 2010; Burokas et al. 2017). Moreover, it is now understood that host's clock genes control the self-sustained rhythmic production of cortisol in intestinal epithelial cells. Therefore, gut dysbiosis is clearly associated with the disruption of circadian rhythm, and it also affects the phasic production of ileal corticosterone resulting in constantly elevated levels of corticosterone (Henao-Mejia et al. 2013; Li et al. 2018). Additionally, gut bacteria also enhance body's melatonin supply (Murakami et al. 2017; Yoshikawa et al. 2017; Zhu et al. 2018; Li et al. 2018). Weaning imposes stress on mice and causes anxiety-like behavior due to weight loss and gut dysbiosis. But these effects could be reversed by melatonin supplementation as it amplified the gut microbial diversity and the relative abundance of *Lactobacillus reuteri, Lactobacillus johnsonii*, and *Lactobacillus intestinalis* while decreased the relative richness of Prevotellaceae (Henao-Mejia et al. 2013; Li et al. 2018).

8.8   The Microbiome–Gut–Brain (MGB) Axis

Quality sleep naturally clears out toxins from the body, and thus important for health. Many studies have indicated that interference in biological rhythms, nutrient metabolism, and immune function could trigger the onset of insomnia and depressive disorder, although the exact mechanism is yet to be elucidated. Moreover, considerable evidences suggested that through MGB axis the gut microbiome controls sleep and mental states of the host (Li et al. 2018).

Numerous studies have indicated that within MGB axis gut microbes communicates with brain through three pathways: (1) immunoregulatory pathway;

(2) neuroendocrine pathway, and (3) vagus nerve pathway. These three pathways mainly generate bifacial flow of information (Backhed et al. 2005; Cryan and Dinan 2012; Diaz Heijtz et al. 2011; Breit et al. 2018). The first is immunoregulatory pathway, in which the gut bacterial interaction with immune cells affects prostaglandin E2 and cytokines levels, and thus brain's function is affected. Next is the neuroendocrine pathway. Enteroendocrine cells ($\geq$20 types) of the intestine make the largest endocrine organ in human body. The gut microbiome regulates the secretion of neurotransmitters, and thus affects both CNS and HPA axis. The third is the vagus nerve pathway, in which enteric nervous system (ENS) plays an active role. The ENS makes synaptic contacts with the vagus nerve and develops a communication pathway called as the gut microbiota-ENS-vagus-brain pathway (Li et al 2018). Moreover, gut microbiota produces neurotoxic metabolites, i.e. D-lactate and NH_3, which may infiltrate into the CNS through the vagus nerve, thereby hampering stress response, sleep structure, and brain's function (Bonaz et al. 2013; Wang and Kasper 2014; Li et al. 2018). In a similar manner, gut microbiome composition may also be regulated by the CNS through these three pathways. For example, the HPA axis controls both intestinal peristalsis and gut epithelial cell functions, which in turn alters gut microbiome composition by affecting the microbial environment and permeability of intestine. However, more studies are required to test the function of these pathways. Besides, possibility of involvement of other pathways cannot be excluded (Bercik et al. 2011; Bauer et al. 2016; Rogers et al. 2016; Li et al. 2018).

8.9   Circadian Clock

Basically, circadian clocks consist of a central oscillator which generates 24-h rhythm on the basis of environmental signals transmitted by the input pathway. This is further associated to output pathways which harmonize circadian rhythm time between the periphery and central clocks, controlling different behavioral, physiological, and metabolic processes (Brandstaetter 2004; Brown et al. 2002; Buhr and Takahashi 2013; Fitzgerald et al. 2015; Nobs et al. 2019). Among mammals, circadian clocks can be characterized by a two-tier hierarchical structure: (1) master or "central" clocks which are placed inside hypothalamus's suprachiasmatic nucleus (SCN), and function in a synchronize manner and (2) "peripheral" clocks inside each body's cell. CLOCK (Circadian Locomotor Output Cycles Kaput) and ARNTL (Aryl Hydrocarbon Receptor Nuclear Translocator-Like protein) regulate a negative feedback loop with 24-h period which is placed in SCN's core. ARNTL controls ~ 15% rhythmic expression of the whole transcriptome of each mammalian cell (Panda et al. 2002; Nobs et al 2019). On the Contrary to this, prokaryotic circadian clocks are studied notably in light-responsive cyanobacteria, and are less understood. In addition, circadian clocks have also been identified in light-nonresponsive prokaryotes; not dependent on light (Paulose et al. 2016; Nobs et al. 2019). Although circadian clock consisting KaiA, KaiB, and KaiC proteins requires transcriptional feedback for stability but it can sustain even in the absence of transcription (Johnson et al. 2011; O'Neill et al. 2011; Teng et al. 2013; Johnson et al. 2017; Nobs et al. 2019). *E. aerogenes* is a commensal gut bacterium and its circadian oscillator is entrained by the human pineal gland and melatonin. Additionally, circadian oscillator of *E. aerogenes* has ~ 24-h period, identical to the human central circadian clocks, and it also controls bacterial swarming activity. Moreover, bioinformatic analyses confirmed that circadian clocks of *E. aerogenes* and cyanobacteria exhibited homogeneity to each other. Therefore, other commensal human gut microbes may also have inherent circadian clocks. Virtually diurnally changing environment of human gut with a multitude of circadian factors would constitute an adaptive circadian clock useful for few gut commensal microbes (Paulose et al. 2016; Nobs et al. 2019).

8.9.1 Mechanisms of Circadian Metaorganism Host–Microbiome Crosstalk

In general, all multicellular organisms are characterized by synergism with prokaryotes and eukaryotic species. A metaorganism is commonly referred as microbiome. Therefore, human microbiome or metaorganism constitutes a multidomain ecosystem where eukaryotes and prokaryotes interact dynamically with each other and also with the diurnal changes of the environmental conditions (Nobs et al. 2019).

8.9.1.1 The Gut Microbiota, Clock Genes, and Circadian Rhythms

Evidence suggests that gut microbiota exhibit circadian rhythms and diurnal fluctuations. *Clostridiales*, *Lactobacillales*, and Bacteroidales constitute ~60% of the gut bacteria and exhibit considerable diurnal changes that result in time-of-day-specific taxonomic forms. Bacteroidetes and firmicutes, two major phyla of mammalian gut microbiota, exhibited diurnal changes in response to dietary composition, rhythmic food ingestion, biological clock, and host's gender. Besides, relative richness of the genus *Lactobacillus* decreases during the active phase and amplifies during the resting phase in both mice and humans (Cox and Blaser 2013; Karlsson et al. 2013; Chaix et al. 2014; Thaiss et al. 2014; Voigt et al. 2014; Zarrinpar et al. 2014; Nobs et al. 2019). In addition to compositional changes, intestinal microbial communities might be doing specific tasks during resting phase; detoxification and chemotaxis, and active phase; energy harvest, DNA repair, and cell growth (Nobs et al. 2019).

Recent findings suggest that certain factors, i.e. sleep deprivation, shift experience, and circadian clock misalignment can affect intestinal microbial community diversity and expression of circadian clock genes. Therefore, gut microbiota and circadian genes are interconnected in a complex manner (Fava et al. 2013; Chaix et al. 2014; Thaiss et al. 2014; Voigt et al. 2014; Zarrinpar et al. 2014; Parekh et al. 2015; Feng et al. 2017; Chen and Devaraj 2018; Nobs et al. 2019). The gut microbiota shows rhythmic diurnal changes in its structure and function, and thereby gastrointestinal epithelium is exposed throughout the day to various microbes and their metabolites. The circadian rhythm of the microbiota responds, and turns on the transcription of host circadian clock genes, and affects oscillations at metabolite levels, which may cause epigenetic changes (Voigt et al. 2014; Nobs et al. 2019).

Both depression and insomnia are interconnected to each other, and closely related to circadian activity. Most often it has been observed that onset of these disorders is also related to biological rhythms, i.e. day length and the intensity of the light. A patient with depression disorder experiences reduced sleep time, enhanced REM sleep or shorter REM latency. Nevertheless, the exact mechanisms of comorbidity of sleep disorders and depression are yet to be elucidated (Parekh et al. 2015; Mazidi et al. 2016; Belizario and Faintuch 2018; Nobs et al. 2019).

Disturbed circadian rhythm of the host affects the balance of intestinal microbiota, and such alterations are identical to the changes take place with actual shift experience. Therefore, the body's biological and microbial clocks work with synergy. Oscillations of gut microbes mainly depend on signals received from host's clock. Circadian rhythmicity in terms of composition and functions of the intestinal microbiota, was lost in *Per1* and *Per2* clock gene knockout mice (Thaiss et al. 2014; Nobs et al. 2019). In one study, Liang et al. (2015) observed prominent changes in the daily rhythmicity of total microbial load, and diversity of fecal microbiota in *Bmal1* clock gene knockout mice. Therefore, host's clock genes, i.e. *Per1*, *Per2,* and *Bmal1* are crucial in controlling diurnal changes of intestinal microbial rhythms. However, feeding timing could be a predominant factor of the temporal orchestration of microbiome activity as the loss of daily rhythmicity of gut microbiota in $Per1/2^{-ve}$ mice could be restored by strict scheduled feeding (Nobs et al. 2019). Recently it is found that gut microbiota can amplify host's metabolism by influencing the circadian transcription factor NFIL3 (Wang et al. 2017). This can help in explaining widespread presence of metabolic syndrome among shift workers, and jet lagged people (Li et al. 2018).

Germ-free (GF) mice are better animal model for investigating the gut microbiome's effects on the brain functions of its host than conventional experimental animals. Gut microbiome can influence host's clock gene expression in the SCN. One study conducted in GF mice clearly explained that oscillations of circadian clock genes expression were decreased inside SCN (Leone et al. 2015). In addition, mutations in clock genes stimulate gut dysbiosis. Similarly, Voigt et al. (2016) observed that mutation in core clock genes lead to gut dysbacteriosis in mice, and this was further aggravated by dietary intestinal stimuli. Cumulatively it suggests that the gut microbiota performs an important role in maintaining the normal rhythmic expression of host's clock genes.

Several studies have demonstrated that polymorphisms of clock genes in depressive patients are affected by the hyperactivation of the HPA axis, that can trigger sleep dysfunction, and also performs a crucial role in the progression of depression (Wirz-Justice 2006; Kato 2007; Schulze et al. 2014). It is now clear that the circadian clock genes are associated with the progression of insomnia disorder. Furthermore, epigenetic and environmental factors can affect the expression of core clock genes (*Bmal1* and *Per1-3*) in some brain regions, i.e. cerebellum, amygdala, dorsolateral prefrontal lobe, anterior cingulate cortex, nucleus accumbens, and hippocampus (Li et al. 2018).

8.9.1.2 Circadian Misalignment in Insomnia

Recent studies indicated that misalignment of daily rhythmicity of gut bacteria and circadian rhythm is linked to sleep loss and insomnia disorder. Discovery of the electric light had made widespread shift work, and increased worldwide traveling; causing jet lag. Shift work and jet lag conditions are the major contributing factors of disrupted circadian clock which has been involved with several diseases such as sleep loss, insomnia disorder, obesity, other metabolic disorders, and neurodegenerative diseases (Bechtold et al. 2010; Brown et al. 2012; Golombek et al. 2013; Archer et al. 2014). All these conditions have also been associated with gut dysbiosis that can be characterized by a rapid decline in bacterial richness and diminished colonization resistance. The association between disturbed circadian clock and gut dysbiosis can also be recreated *in vivo* in the similar way as explained in human studies (Qin et al. 2012; Karlsson et al. 2013; Cox and Blaser 2013; Sears and Garrett 2014; Thaiss et al. 2014; Voigt et al. 2014; Nobs et al. 2019). Therefore, gut microbiome might hold the key for identifying unrecognized link between insomnia and circadian misalignment. Although circadian rhythm disorders seem to disrupt gut microbial equilibrium which is further associated with the onset of various metabolism disorders, however, this damage can be reversed through dietary adjustments and antibiotic intervention. In one such study, Poroyko et al. (2016) observed that chronic sleep fragmentation (SF) can induce enhanced food intake and reversible alterations in mice's gut microbiome such as preferential abundance of Ruminococcaceae and Lachnospiraceae, and decreased numbers of *Lactobacillaceae*. Fecal microbiota transplantation (FMT) from SF mice to GF mice lead to enhanced inflammatory reactions and disturbed metabolism, however, probiotic (*Lactobacillus* and *Bifidobacterium*) administration could subside the damaging inflammation (Vanuytsel et al. 2014; Li et al. 2018).

Unlike *in vivo* studies, related research on gut dysbacteriosis causing insomnia and other metabolic disorders associated with disrupted circadian rhythm in humans is still rare. Moreover, human subjects experiencing jet lag showed increased abundance of Firmicutes which was reversed after recovering from jet lag. Furthermore, FMT from jet lag humans into GF mice initiated similar metabolic imbalances like weight gain and reduced glucose intolerance. However, the sample size of this study was small and, therefore, no conclusive evidence could be found for many of the probable confounding factors. Hence, all these preliminary data indicate that circadian misalignment, sleep loss, and insomnia disorder in humans are clearly associated with gut dysbiosis, and altered gut bacterial configurations may thus contribute to the occurrence of metabolic imbalances. Continued sleep loss or the circadian misalignment will lead to altered host's gut microbiota and physiological stress response as a result metabolic disorders, impaired immune function, and inflammatory reactions will occur. Furthermore, it will cause nervous system dysfunction due to alteration in the metabolism of neurotransmitters. Hence, an individual will experience insomnia and depressive symptoms, which ultimately commence an impaired cycle (Li et al. 2018).

In brief, a bifacial association exists between the gut microbiome and insomnia or depression. Persistent chronic disruption of host's circadian rhythms, insomnia, and depression affect metabolic activity of gut bacteria which further cause alteration in their composition. It specifically decreases total bacterial count of the Lactobacillaceae family and increases numbers of *Bacteroides multiforme*, Enterococci, the Ruminococcaceae, and the Lachnospiraceae, which results in gut dysbacteriosis (Thaiss et al. 2014; Leone et al. 2015; Poroyko et al. 2016). Moreover, imbalanced gut bacteria can compromise integrity of intestinal epithelial cells barrier which regulates intestinal permeability (Bravo et al. 2011). Once the integrity of intestinal epithelial barrier is lost, then bacteria, LPS or endotoxins, and other harmful microbial metabolites can easily enter into the bloodstream, which can induce inflammatory reactions, and excite the vagus and spinal afferent nerves (Yarandi et al. 2016). Therefore, it is assumed that gut dysbiosis induced inflammatory reactions cause abnormal immune reactions, which further affects the CNS to cause or aggravate insomnia, depression, and metabolic disorders (Dibner et al. 2010; Bass 2012; Yi and Li 2012; Foster and McVey Neufeld 2013; Hsiao et al. 2013; Li et al. 2018).

8.10 Factors Affecting Circadian Rhythm and Gut Microbiome

Circadian rhythm and feeding behavior of host affect diurnal activity of gut bacteria and diurnally shifting microbiome also conversely affect host's circadian activity. The gut microbiome modulates host's transcriptomic oscillations in the gut and liver (Thaiss et al. 2016). In general, gut microbes are an important factor in maintaining host's circadian rhythmicity and therefore, directly entrained peripheral host's clocks (Tahara et al. 2018; Nobs et al. 2019). In this section, impact of consequent exposure to external and internal factors such as light/dark cycles, sleep and diet, on the host's circadian–microbiome interactions is described.

8.10.1 Light-Dark Cycles

Approximately 50% of the mice transcriptome had shown circadian oscillations and similarly 20–83% intestinal microbial taxa of mice also exhibit day/night rhythmicity. In mice Firmicutes peaked during dark phase restricted feeding and decreased during daytime fasting while Bacteroidetes and Verrucomicrobia peaked during daytime fasting. Both *ad libitum* and dark phase feeding in mice preserved dark phase peaks in *Bacteroidales*, and light phase peaks in *Lactobacillales* and *Clostridiales*, while light phase feeding could reverse altered microbial rhythms. Additionally, in both male and female mice diurnal oscillations remain unchanged in Bacteroidetes but disappeared in Firmicutes, while much less abundant Proteobacteria also exhibited light phase peaks in male mice (Marcobal et al.

2013; Thaiss et al. 2014; Zarrinpar et al. 2014; Zhang et al. 2014; Liang et al. 2015). Several studies in mice model indicated that dark phase peaks in Firmicutes were diet-driven, while cessation of feeding caused increased populations of Bacteroidetes, Proteobacteria, and Verrucomicrobia during the light phase (Parkar et al. 2019). Population dynamics and diet are the driving factors of the diurnal rhythmicity of the gut microbial community. Firmicutes proliferate abundantly in the presence of dietary glycans but their numbers decline when dietary source finished. While Bacteroidetes dominate during the rest phase as they feed upon host's glycans, of which intestinal mucosa is the major source. As a result of commensal microbial metabolic activities, host gets benefited in terms of energy and other growth factors, i.e. vitamins, indigenously synthesized by commensals. Moreover, light phase resting causes enhanced microbial pathways related to chemotaxis and motility, necessary for the mucus-adherent bacteria to reach closer to the intestinal epithelium (Thaiss et al. 2014, 2016; Liang et al. 2015; Parkar et al. 2019).

In a human study, Kaczmarek et al. (2017) demonstrated rhythms in gut microbiome composition and metabolic activity based on clock time, and the meal timing. Analysis of fecal samples of 28 volunteers (male: female, 1:1) confirmed that in response to the dietary intake butyrate producers, Lachnospira, Roseburia, and Eubacterium peaked initially and decreased later as the food source exhausted. Moreover, other bacterial groups emerged earlier in the day included primary feeders (survived by quick but partial breakdown of the carbohydrate into acetate and lactate such as Eggerthella), the bile-tolerant groups (such as Oscillospira and Bilophila, which survive in the post-prandial bile-rich intestinal milieu), and the Desulfovibrio (generates H_2S in response to the high lactate, and the sulfur from the bile taurocholates). As the day progress decreased food intake results in the diurnally shifting of gut microbiome and, microbial metabolic processes utilizing carbon source to generate SCFAs, also decrease in a corresponding manner. However, during overnight fasting gut microbiota re-structures its community to feed on host-derived glycans (Marcobal et al. 2013; Parkar et al. 2019).

8.10.2 Sleep Quality

Modern lifestyle associated problems, such as insufficient or poor-quality sleep, jet lag, shift work, lack of physical activity, circadian misalignment, and insomnia, are the major contributing factors of weight gain, vicious energy balance, glucose tolerance, and insulin sensitivity (Depnar et al. 2014).

Recent findings indicated that gut microbial equilibrium is associated with the regulation of sleep. In a related study, administration of antibiotics in rats exhibited reduced slow sleep wave due to the depletion of gut microbiota (Brown et al. 1990). D-lactate; produced by excessive growth of lactic acid bacteria (LAB), is hypothesized to impose neurotoxic effects once blood–brain barrier is crossed. On the contrary, neurological symptoms may be subsided by checking excessive accumulation of circulating D-lactate. In a human study, this hypothesis was tested by administering erythromycin for 6 days in 22 patients (05 male) of chronic fatigue syndrome. As a result, seven patients showed significant decrease in fecal streptococci, which could be correlated with improved sleep duration, and sleep onset latency (Jackson et al. 2015; Parkar et al. 2019).

Considering the integral role of gut microbiome in metabolic wellness, understanding precise impact of sleep quality on gut microbial composition is of considerable interest. In this context, recent human studies have indicated that due to sleep deprivation significant changes occurred in gut microbial communities which is correlated with metabolic imbalance, and obesity (Benedict et al. 2016). Overall, sleep fragmentation caused a rise in Firmicutes population by 20% and a decrease in the microbial phyla Actinobacteria and Bacteroidetes by 50% and 20%, respectively (Turnbaugh et al. 2009a; Verdam et al. 2013). Furthermore, a respective increase and decrease in the relative richness of 02 Firmicutes families, i.e. Lactobacillaceae and Lachnospiraceae, could be observed, and this bacterial configuration persisted during sleep fragmentation protocol. However, the bacterial abundance reverted to its normal composition after recovering from two weeks of sleep fragmentation. Circadian-sleep-gut microbial interactions were deduced in a metabolomic study including 24-h sleep deprived fifteen men volunteers after a normal sleep/wake cycle with 8

h of sleep, and under controlled conditions such as light, diet, sleep, and posture. As a result of acute sleep deprivation gut microbiota metabolites increased and excreted through urine. Gut microbiota metabolites are produced due to microbial breakdown of tryptophan, choline, tyrosine, and valine, i.e. acetate and 3-indoxyl sulfate, trimethylamine-N-oxide, p-cresol sulfate, and 3-hydroxyisobutyrate, respectively (Giskeodegard et al. 2015). However, in order to clearly understand the relation between gut dysbiosis and circadian disturbances, more studies are required (Parkar et al. 2019).

8.10.3 Jet Lag

Jet lag is a syndrome expressed by late physiological adaptation to frequent changes in time zone due to trans-meridian or multi-time-zone travel causes. Jet lag affects mood, focus, and bowel movement (Bobinski and Michalik 2010). Additionally, various routine activities such as meal timing, hormonal rhythms, and sleep are also changed during jet lag.

Some studies have clearly correlated jet lag condition with gut dysbiosis, and dysfunctional metabolic homeostasis. An increased abundance of Firmicutes was observed in human subjects experiencing jet lag, which could be reversed after jet lag recovery. Additionally, FMT from jet lag humans into GF mice, caused identical metabolic imbalances such as weight gain and glucose intolerance. In a similar study, four months simulated jet lagged mice did not show any significant change in the overall dietary intake, but certainly exhibited weight gain and impaired glucose tolerance. Moreover, the rhythm, functions, and abundance of the gut microbiota were suppressed especially for Ruminococcaceae, one of the major butyrate producing families. Metabolic dysfunctions caused by jet lag and gut dysbiosis were reversed by administering broad-spectrum antibiotics in mice, which clearly indicates the potential role of gut microbiota in stimulating metabolic imbalances of jet lag (Thaiss et al. 2014; Parkar et al. 2019).

8.10.4 Meal: Quality, Timing, and Frequency

Chrononutrition is a nutrition-based method of food rebalancing. It allows an individual to achieve balanced body through weight loss or gain, and by consuming food in respect of body's biological clock. Circadian and seasonal rhythms are the basic feature of all living organisms, and host's daily food intake is connected to these biological rhythms. Inside gut circadian rhythms are entrained by the central clock, meal, and eating patterns. The molecular basis for circadian timing in the intestine comprises interlocking gene expression feedback loops, which end-up in the rhythmic expression and activity of a set of clock genes, and related hormones (Konturek et al. 2011; Li et al. 2018; Nobs et al. 2019; Parkar et al. 2019).

Central clock in the brain's SCN, a light entrainable oscillator, may be entrained by meal type/timing, temperature or other factors. It may even disconnect, for instance, from SCN by restricted feeding during sleep phase. Therefore, peripheral clocks of liver and gut are entrained by the feeding regime which not only affects diurnal fluctuations in body clock but also in the gut microbiome (Bass and Takahashi 2010; Oosterman et al. 2015). Intake of a high-fat diet caused gut microbial perturbations in mice which further lead to altered oscillations in the clock genes and clock-controlled genes of brain, liver, and adipose tissue (Kohsaka et al. 2007; Verdam et al. 2013). Similarly, intake of fat- and sugar-enriched diet intensified the circadian disruption of gut microbiome with a major decrease of microbial richness, and the Firmicutes/Bacteroidetes ratio (Turnbaugh et al. 2009b; Verdam et al. 2013; Voigt et al. 2014). An individual's diet is known to cause rapid alteration in the composition and diversity of resident gut microbiota (Turnbaugh et al. 2009b; Walker et al. 2011; David et al. 2014). Plant food-derived fiber and polyphenols are the two major components which resist digestion in small intestine, and enter in the colon to regulate host's gut microbiome (Tuohy et al. 2012; Parkar et al. 2015). In general, high fiber plant-based diets encourage a butyrate-rich environment by amplifying efficient carbohydrate metabolizing bacteria of families, i.e. Ruminococcaceae (*Ruminococcus bromii*), Eubacteriaceae (*Eubacterium rectale*), and Lachnospiraceae (*Roseburia*), in comparison to low fiber animal-based diets, which are favorable to high bile-tolerant bacteria *Bacteroides, Alistipes,* and *Bilophila,* which can breakdown protein more effectively (David et al. 2014). In a study, 24-days old rats were fed diet enriched with prebiotic components, i.e. galacto-oligosaccharides, lactoferrin, polydextrose, and milk fat globule membrane. It could mitigate the impact of stress on gut bacterial diversity and composition, and also consolidated the sleep/wake cycle. Although, no major change was observed in any of the high-abundance bacterial phyla, but Deferribacteres decreased significantly, and it could be highly correlated with longer non-REM sleep (Thompson et al. 2017). Probably, dietary constituents and concentration of the bioactive polyphenols may also affect circadian rhythms in peripheral tissues of the host, by harmonizing the composition of gut commensals, which in turn produce bioactive SCFA or polyphenolic metabolites. Moreover, through microbial transformation some polyphenols convert into such metabolites, which may induce circadian entrainment (Parkar et al. 2019).

Now, meal timing and the eating window are considered equally important as the dietary composition. In a study, time-restricted feeding (TRF) could revert the circadian rhythmicity, and the relative population of *Lactobacillus* spp. in the mice fed on the high-fat diet, also reverted to the *Lactobacillus* configuration of the regular chow mice. In comparison to the *ad libitum* high-fat diet, TRF also reduced the abundance of *Lactococcus* spp. and both *Lactobacillus* and *Lactococcus* spp. are considered as obesogenic. Additionally, TRF could also restore the abundance of Clostridia and Ruminococcaceae but not their cyclical rhythms (Zarrinpar et al. 2014). Beyond doubt, TRF induced microbial *rhythms* may impose severe metabolic effects on the host in comparison to microbial abundances. Similarly, length of overnight fasting (kind of TRF) in humans was found to be proportional to the fecal propionate, which is channeled toward gluconeogenesis in the liver (Louis et al. 2007; Kaczmarek et al. 2017; Parkar et al. 2019).

8.10.5 Bile Salt Biotransformation

Host's bile acid metabolism is an example of physiological and gut microbial activities which are fixed to certain daily time points, and it also co-ordinates metabolic homeostasis. The enterohepatic circulation (EHC) of bile secreted by the liver has a clear daily rhythm. Studies demonstrated that it peaked at the light and ended at the dark phase. Some primary bile acid which had escaped EHC enters into colon and undergo microbial transformation. Microbial transformation of primary bile inside colon are deconjugation by bile salt hydrolase (bsh), dehydroxylase, and dehydrogenase activity of bacteria from Firmicutes, Bacteriodetes, and Bifidobacteria (Zhang et al. 2011; Fiorucci and Distrutti 2015; Labbe et al. 2014; Eggink et al. 2017). *Bsh* transformed *E. coli* could colonize in GF and regular experimental mice models. This could explain the contribution of bacterial bile acids—deconjugation and biotransformation activity in regulating hepatic and ileal clock genes (*Per1/2*), and clock-controlled genes which regulate metabolism of cholesterol (*Abcg5/8*) and lipid (*PPARγ, Angtpl4*) (Joyce et al. 2014). In addition, stressors like high-fat diet cause perturbations in bile metabolism, and probably affect *bsh*-activity and composition of gut microbiome, which further impact host's lipid metabolism (Yokota et al. 2012).

8.10.6 Hydrogen Sulfide

Deltaproteobacteria, such as *Desulfovibrio, Desulfobulbus, Desulfobacter*, and the taurine utilizing *Bilophila wadsworthia*, carry a low relative abundance gene encodes for dissimilatory sulfite reductase (dsrAB). The dsrAB gene performs a critical role in the cycling of the sulfated compounds, such as sulfated mucins and bile components which enter into the large intestine. These Deltaproteobacteria are mainly responsible for a net increase in H_2S inside distal colon (Carbonero et al. 2012). Sulfate-reducing *Desulfovibrio piger* prefers to metabolize microbial lactate, and competitively exclude butyrate producers, i.e. *Anaerostipes caccae* and *Eubacterium hallii*, which also need lactate (Marquet et al. 2009). Therefore, increased numbers of such Proteobacteria are prognostic of gut microbial dysbiosis. The built up H_2S inhibits microbial functions, such as butyrate production (a major source of energy in the colon) and also inhibits cytochrome c oxidase of host's mitochondrial oxidative phosphorylation, thus reducing the energy level required for cell survival (Nicholls et al. 2013). Moreover, H_2S caused phase-delay expression of circadian clock gene *Bmal1* in mice. Microbial H_2S built up was associated with inhibited substrate oxidation and high blood glucose concentration in mice fed on high-fat diet (Parkar et al. 2019).

8.10.7 Vitamins

Vitamin D_3 is important for maintaining gut microbial equilibria, while some B-complex vitamins have been linked with circadian rhythms and sleep patterns directly or indirectly (Wolf 2002; Beydoun et al. 2014). Deficiency of circulating vitamin D_3 has been linked with abdominal obesity, which further increases the risk of developing insulin resistance and inflammation. Vitamin D_3 is also associated with production of protective ileal mucus, which allows colonization of beneficial bacteria, and thus, pathobionts cannot colonize due to competitive exclusion. In a small study on 16 healthy volunteers, eight weeks dietary supplementation of vitamin D_3 improved gut microbial equilibrium by increasing gut bacterial richness. Fecal microbial analysis revealed a proportionate reduction in proteobacterial pathobionts, i.e. *Escherichia/Shigella* spp. and *Pseudomonas* spp. Thus, circulating vitamin D_3 levels can affect the gut microbiome composition and also bacterial colonization of gut (Parkar et al. 2019). Essential vitamins are diet-derived, however, about 40–65% of gut bacteria are the major producer of non-essential vitamins (Arumugam et al. 2011; Magnusdottir et al. 2015). Vitamin B_{12} (cobalamin) has been considered as an essential molecule for the longevity of both host and microbiome. Low serum vitamin B_{12} is linked with sleep disturbance and reversal of H_2S effects on substrate oxidation. Gut commensal bacteria such as *Propionibacterium freudenreichii* and *Lactobacillus reuteri* are engaged in vitamin B_{12} biosynthesis, but they mutually depend on *Bacteroides thetaiotaomicron* that pass through mucosal barrier to bind dietary corrinoids (cobalamin precursors) for making it available to bacteria for vitamin B_{12} synthesis. Similarly, vitamin B_9 (folic acid) is crucial for regulating brain clock genes, and low serum vitamin B_9 is also linked with sleep disturbance. Probiotics mainly *Bifidobacterium* spp. and *Lactobacillus* spp. are involved in the biosynthesis of folic acid (Parkar et al. 2019).

8.10.8 Biogenic Amines

Sleep deprivation impose an intensive effect on the plasma metabolome, with increased concentration of biogenic amines and their precursors (Davies et al. 2014). It could be connected to high abundance of Firmicutes which is common in sleep deprivation (Benedict et al. 2016; Poroyko et al. 2016). Brain's pineal gland produces a serotonin-derived bioamine, melatonin which predominantly governs sleep/wake functions. Inside mitochondria, melatonin regulates its bioenergetic function, and also serves as strong antioxidant. Melatonin stimulates antioxidant enzymes such as catalase, superoxide dismutase, glutathione reductase, and glutathione peroxidase. It can fix microsomal membranes, and thus, causes resistance to oxidation damage (Sharafati-Chaleshtori et al. 2017). Recent studies have confirmed bifacial associations between host's melatonin concentrations and distinct gut bacterial abundance. Moreover, gut bacteria expand tryptophan levels in the blood which is required as a precursor amino acid for melatonin synthesis. Therefore, gut microbial interactions with sleep-inducing biogenic amines, likely to affect the circadian-metabolic axis (Parkar et al. 2019).

8.11   Future Approaches: Reversal of Gut Dysbiosis and Insomnia Disorder Through Restoration of the Gut Microbiota

Probiotic and prebiotic based dietary interventions may restore intestinal microbiota disrupted in gut dysbiosis and thus can mitigate insomnia associated symptoms. Similarly, healthy lifestyle approaches such as physical activity and meal/timing may contribute significantly in balancing the gut microbiome composition. It is now known that athletes exhibit diverse richness in gut bacteria. However, it is yet to elucidate whether the effect of physical activity on gut microbiome is independent of diet effects (Carding et al. 2015). There is huge potential for restoring the microbiota in diseased individuals or those at risk. Defining "healthy microbiota composition during and throughout life" is an important pre-requisite of probiotic-based therapeutic approaches which have been used with some success for centuries in comparison to more harsh and cruder fecal microbiota transplant (FMT) approach (Gismondo et al. 1999; Wilhelm et al. 2008; Brandt and Aroniadis 2013). Restoring adequate microbial balance could be the key for any strategy to prevent gut dysbiosis associated disorders (Carding et al. 2015).

8.11.1 Probiotics

As described earlier, insomnia is often associated with gut microbial dysbiosis and is considered as a comorbid condition. In one study, oral administration of probiotic blend De Simone Formulation could mitigate sleep, and abdominal pain in irritable bowel syndrome (IBS) (Wong et al. 2015). Although gut colonization of exogenously administered probiotics is often variable, but it helps in restoring the endogenous gut microbial balance which improves gut health and sleep deficiency (Malinen et al. 2002). Impact of probiotic-based interventions in restoring gut dysbiosis and insomnia has already been described in previous sections.

8.11.2 Disciplined Diet and Eating Pattern

There is no doubt that sedentary lifestyle is the main contributor of metabolic and psychological health disorders. These lifestyle factors include intake of high calories in the form of hyper-lipidemic/or glycemic food, and poor sleep because of modern life stressors such as jet lag, shift work, and decreased exposure to sunlight. All these features not only misalign the circadian clock, but also alter gut microbiome composition. Meal quality, its type and timing of eating may also disturb circadian timing by host-derived signals and by modulating gut microbiota. Therefore, at appropriate time-of-day both quality sleep and a balanced nutritious meal may thus be necessary for regulating metabolic homeostasis. Moreover, plant-based foods are rich in fiber and bioactive polyphenols and the undigested part of such food enters into the colon to produce beneficial metabolites such as SCFAs and bioactive polyphenols. These SCFA and bioactive polyphenols not only maintain gut health, but also assist to realign circadian rhythms (Parkar et al. 2019).

8.11.3 Fecal Microbial Transplant (FMT)

In addition to probiotic and prebiotic based dietary interventions to revert gut dysbiosis, FMT could be another therapeutic approach to re-establish microbial diversity in a dysbiotic gut. Mild or moderate gut dysbiosis is treatable with synbiotic diets/products (Roberfroid 2007a, b; Hungin et al. 2013; Shanahan and Quigley 2014). However, in case of more serious dysbiotic gut conditions, these measures are inadequate. Therefore, in such cases, FMT can offer a more dynamic approach to restore microbiota in dysbiotic gut. By definition FMT procedure is not an actual transplantation, but rather a fecal microbial transfer. FMT therapy includes the physical transplantation of fecal microbiota from a healthy donor into the diseased recipient's gut. FMT therapy is an approach to re-establish diversity of gut microbiome, and consequently FMT has a regulatory drug status in the USA from the FDA. FMT therapy has been used in IBD cases (Ulcerative colitis), irritable bowel syndrome, enterocolitis, chronic constipation, and Crohn's disease (Borody et al. 2004; Angelberger et al. 2013; Wei et al. 2015; de Palma et al. 2017).

8.12   Conclusion

Recent studies have clearly demonstrated that gut microbiome also exhibit circadian rhythm in its composition, function, and colonization ability. Current evidence shows diurnal changes in the microbial metabolites such as H_2S and butyrate, which in turn influences host's metabolism and circadian rhythm. The role of gut microbiome in association with contributing pathways, in rescuing host from circadian disruption has been discussed. Gut bacteria derived bioactive SCFAs, polyphenolic metabolites, vitamins, and bioamines could be instrumental in reversing gut dysbiosis and sleep loss. Therefore, manipulating the gut microbiome could be a promising strategy to rescue the host's circadian rhythm. In this context, plant-based food holds the potential to modulate gut microbiota naturally. Plant foods are enriched with prebiotic fiber and polyphenols, and inside colon its undigestible portion is transformed by microbes into beneficial metabolites such as SCFAs and bioactive polyphenols, which are crucial in restoring gut health and circadian rhythmicity. Diets containing prebiotics and probiotics constituents may also restore the gut microbial equilibrium and insomnia. There is no doubt that sedentary lifestyle habits cannot change immediately but adopting disciplined dietary habits (quality, quantity, or timing of food) may be the easiest step of starting a healthy lifestyle. Additionally, constant use of gut-strengthening habits will certainly pave the way for long-term healthy being and peaceful sleep. Therefore, in order to align sleep cycles in sync, and to obtain a restful, rejuvenating sleep, beneficial gut bacteria need to be nurtured by adopting healthy lifestyle and eating habits. In contrary, safety concerns and narrow public acceptability of FMT approach to treat gut dysbiosis are the major issues. Thus, manipulating gut microbiome structure through synbiotic diet and chrononutrition-based approaches may therefore hold promise to improve some of the metabolic impacts of modern lifestyle-associated issues such as disturbed sleep and disrupted circadian rhythms.

References

Aagaard K, Ma J, Antony KM et al (2014) The placenta harbors a unique microbiome. Sci Transl Med 8:e66986

Adlerberth I, Wold AE (2009) Establishment of the gut microbiota in western infants. Acta Paediatr 98:229–238

Angelberger S, Reinsch W, Makristathis A et al (2013) Temporal bacterial community dynamics vary among ulcerative colitis patients after fecal microbiota transplantation. Am J Gastroenterol 108:1620–1630

Archer SN, Laing EE, Moller-Levet CS et al (2014) Mistimed sleep disrupts circadian regulation of the human transcriptome. Proc Natl Acad Sci U S A 111:E682–E691

Arumugam M, Raes J, Pelletier E et al (2011) Enterotypes of the human gut microbiome. Nature 473:174–180

Atarashi K, Tanoue T, Oshima K et al (2013) Treg induction by a rationally selected mixture of *Clostridia* strains from the human microbiota. Nature 500:232–236

Avershina E, Rudi K (2015) Confusion about the species richness of human gut microbiota. Benefic Microbes 6:657–659

Azzini E, Polita A, Fumagalli A et al (2011) Mediterranean diet effects: an Italian picture. Nutr J 10: 125

Backhed F, Ley RE, Sonnenburg JL et al (2005) Host-bacterial mutualism in the human intestine. Science 307:1915–1920

Baker JL, Michaelsen KF, Rasmussen KM et al (2004) Maternal prepregnant body mass index, duration of breast-feeding, and timing of complementary food introduction are associated with infant weight gain. Am J Clin Nutr 80:1579–1588

Barrett E (2014) gamma-Aminobutyric acid production by culturable bacteria from the human intestine. J Appl Microbiol 116:1384–1386

Bass J (2012) Circadian topology of metabolism. Nature 491:348–356

Bass J, Takahashi JS (2010) Circadian integration of metabolism and energetics. Science 330: 1349–1354

Bauer KC, Huus KE, Finlay BB (2016) Microbes and the mind: emerging hallmarks of the gut microbiota-brain axis. Cell Microbiol 18:632–644

Bechtold DA, Gibbs JE, Loudon AS (2010) Circadian dysfunction in disease. Trends Pharmacol Sci 31:191–198

Belizario JE, Faintuch J (2018) Microbiome and gut dysbiosis. Experientia Suppl 109:459–476

Belizario JE, Faintuch J, Garay-Malpartida M (2018) Gut microbiome dysbiosis and Immunometabolism: new frontiers for treatment of metabolic diseases. Mediators Inflamm 2018:2037838

Benedict C, Vogel H, Jonas W et al (2016) Gut microbiota and glucometabolic alterations in response to recurrent partial sleep deprivation in normal-weight young individuals. Mol Metab 5:1175–1186

Bercik P, Denou E, Collins J et al (2011) The intestinal microbiota affect central levels of brain-derived neurotropic factor and behaviour in mice. Gastroenterologia 141(2):599–609

Beydoun MA, Gamaldo AA, Canas JA et al (2014) Serum nutritional biomarkers and their associations with sleep among us adults in recent national surveys. PLoS One 9:e103490

Bobinski R, Michalik A (2010) Evaluation of early jet lag symptoms by passengers crossing 7 time zones. In: International Conference on Transport Systems Telematics (TST), vol 2010, pp 356–363

Bonaz B, Picq C, Sinniger V et al (2013) Vagus nerve stimulation: from epilepsy to the cholinergic anti-inflammatory pathway. Neurogastroenterol Motil 25:208–221

Borody TJ, Warren EF, Leis SM et al (2004) Bacteriotherapy using fecal flora: toying with human motions. J Clin Gastroenterol 38(6):475–483

Brandstaetter R (2004) Circadian lessons from peripheral clocks: is the time of the mammalian pacemaker up? Proc Natl Acad Sci U S A 101:5699–5700

Brandt LJ, Aroniadis OC (2013) An overview of fecal microbiota transplantation: techniques, indications, and outcomes. Gastrointest Endosc 78(2):240–249

Bravo JA, Forsythe P, Chew MV et al (2011) Ingestion of *Lactobacillus* strain regulates emotional behavior and central GABA receptor expression in a mouse via the vagus nerve. Proc Natl Acad Sci U S A 108:16050–16055

Breit S, Kupferberg A, Rogler G et al (2018) Vagus nerve as modulator of the brain–gut axis in psychiatric and inflammatory disorders. Front Psychiatry 9:44

Brown R, Price RJ, King MG et al (1990) Are antibiotic effects on sleep behavior in the rat due to modulation of gut bacteria? Physiol Behav 48:561–565

Brown SA, Zumbrunn G, Fleury-Olela F et al (2002) Rhythms of mammalian body temperature can sustain peripheral circadian clocks. Curr Biol 12:1574–1583

Brown SA, Kowalska E, Dallmann R (2012) (Re)inventing the circadian feedback loop. Dev Cell 22:477–487

Buford TW (2017) (Dis) Trust your gut: the gut microbiome in age related inflammation, health, and disease. Microbiome 5:80

Buhr ED, Takahashi JS (2013) Molecular components of the mammalian circadian clock. Handb Exp Pharmacol 217:3–27

Burokas A, Arboleya S, Moloney RD et al (2017) Targeting the microbiota-gut-brain axis: prebiotics have anxiolytic and antidepressant-like effects and reverse the impact of chronic stress in mice. Biol Psychiatry 82:472–487

Cahenzil J, Koller Y, Wyss M et al (2013) Intestinal microbial diversity during early-life colonization shapes long term IgE levels. Cell Host Microbe 14:559–570

Carbonero F, Benefiel A, Alizadeh-Ghamsari A et al (2012) Microbial pathways in colonic sulfur metabolism and links with health and disease. Front Physiol 3:448

Carding S, Verbeke K, Vipnod DT et al (2015) Dysbiosis of the gut microbiota in disease. Microb Ecol Health Dis 26:26191

Chaix A, Zarrinpar A, Miu P et al (2014) Time-restricted feeding is a preventative and therapeutic intervention against diverse nutritional challenges. Cell Metab 20:991–1005

Chassaing B, Koren O, Goodrich JK et al (2015) Dietary emulsifiers impact the mouse gut microbiota promoting colitis and metabolic syndrome. Nature 519:92–96

Chen X, Devaraj S (2018) Gut microbiome in obesity, metabolic syndrome, and diabetes. Curr Diab Rep 18:129

Chung KH, Li CY, Kuo SY et al (2015) Risk of psychiatric disorders in patients with chronic insomnia and sedative hypnotic prescription: a nationwide population-based follow-up study. J Clin Sleep Med 11:542–550

Collado MC, Isolauri E, Laitinen K et al (2010) Effect of mother's weight on infant's microbiota acquisition, composition and activity during early infancy: a prospective follow-up study initiated in early pregnancy. Am J Clin Nutr 92:1023–1030

Costello EK, Stagaman K, Dethlefsen L et al (2012) The application of ecological theory towards an understanding of the human microbiome. Science 336:1255–1262

Cox LM, Blaser MJ (2013) Pathways in microbe-induced obesity. Cell Metab 17:883–894

Cryan JF, Dinan TG (2012) Mind-altering microorganisms: the impact of the gut microbiota on brain and behaviour. Nat Rev Neurosci 13:701–712

David LA, Maurice CF, Carmody RN et al (2014) Diet rapidly and reproducibly alters the human gut microbiome. Nature 505:559

Davies SK, Ang JE, Revell VL et al (2014) Effect of sleep deprivation on the human metabolome. Proc Natl Acad Sci U S A 111:10761–10766

de Palma G, Lynch MD, Lu J et al (2017) Transplantation of fecal microbiota from patients with irritable bowel syndrome alters gut functions and behavior in recipient mice. Sci Transl Med 9(379):eaaf6397

DeFilippo C, Cavalieri D, Di Paola M et al (2010) Impact of diet in shaping gut microbiota revealed by a comparative study in children from Europe and rural Africa. Proc Natl Acad Sci U S A 107:14691–14696

Depnar CM, Stothard ER, Wright KP Jr (2014) Metabolic consequences of sleep and circadian disorders. Curr Diab Rep 14(7):507

Desbonnet L, Garrett L, Clarke G et al (2010) Effects of the probiotic Bifidobacterium infantis in the maternal separation model of depression. Neuroscience 170:1179–1188

Diaz Heijtz R, Wang S, Anuar F et al (2011) Normal gut microbiota modulates brain development and behavior. Proc Natl Acad Sci U S A 108:3047–3052

Dibner C, Schibler U, Albrecht U (2010) The mammalian circadian timing system: organization and coordination of central and peripheral clocks. Annu Rev Physiol 72:517–549

Dinan TG, Cryan JF (2017) The microbiome-gut-brain axis in health and disease. Gastroenterol Clin North Am 46:77–89

Donnet-Hughes A, Schriffin E, Walker WA (2015) Protective properties of human milk and bacterial colonization of the neonatal gut. In: Duggan C, Koletzko B, Watkins J, Walker WA (eds) Nutrition in pediatrics—basic science clinical aspects, 5th edn. Chinese Publications, New Haven, CT, pp 165–250. Chapter 30

Eggink HM, Oosterman JE, de Goede P et al (2017) Complex interaction between circadian rhythm and diet on bile acid homeostasis in male rats. Chronobiol Int 34:1339–1353

Elinav E, Strowig T, Kau AL et al (2011) NLRP6 inflammasome regulates colonic microbial ecology and risk for colitis. Cell 145:745–757

Faith JJ, Guruge JL, Charbonneau M et al (2013) The long-term stability of the human gut microbiota. Science 341:1237439

Farre EM, Liu T (2013) The PRR family of transcriptional regulators reflects the complexity and evolution of plant circadian clocks. Curr Opin Plant Biol 16:621–629

Fava F, Gitau R, Griffin BA et al (2013) The type and quantity of dietary fat and carbohydrate alter faecal microbiome and short-chain fatty acid excretion in a metabolic syndrome 'at-risk' population. Int J Obes (Lond) 37:216–223

Feng X, Uchida Y, Koch L et al (2017) Exercise prevents enhanced postoperative neuroinflammation and cognitive decline and rectifies the gut microbiome in a rat model of metabolic syndrome. Front Immunol 8:1768

Finger BC, Dinan TG, Cryan JF (2011) High-fat diet selectively protects against the effects of chronic social stress in the mouse. Neuroscience 192:351–360

Fiorucci S, Distrutti E (2015) Bile acid-activated receptors, intestinal microbiota, and the treatment of metabolic disorders. Trends Mol Med 21:702–714

Fitzgerald GA, Yang G, Paschos GK et al (2015) Molecular clocks and the human condition: approaching their characterization in human physiology and disease. Diabetes Obes Metab 17 (Suppl 1):139–142

Flowers SA, Ellingrod VL (2015) The microbiome in mental health: potential contribution of gut microbiota in disease and pharmacotherapy management. Pharmacotherapy 35:910–916

Foster JA, McVey Neufeld KA (2013) Gut-brain: how the microbiome influences anxiety and depression. Trends Neurosci 36:305–312

Giskeodegard GF, Davies SK, Revell VL et al (2015) Diurnal rhythms in the human urine metabolome during sleep and total sleep deprivation. Sci Rep 5:14843

Gismondo MR, Drago L, Lombardi A (1999) Review of probiotics available to modify gastrointestinal flora. Int J Antimicrob Agents 12(4):287–292

Golombek DA, Casiraghi LP, Agostino PV et al (2013) The times they're a-changing: effects of circadian desynchronization on physiology and disease. J Physiol Paris 107:310–322

Guida F, Turco F, Iannotta M et al (2018) Antibiotic-induced microbiota perturbation causes gut endocannabinoidome changes, hippocampal neuroglial reorganization and depression in mice. Brain Behav Immun 67:230–245

Henao-Mejia J, Strowig T, Flavell RA (2013) Microbiota keep the intestinal clock ticking. Cell 153: 741–743

Holmes E, Li JV, Athanasiou T et al (2011) Understanding the role of gut microbiome-host metabolic signal disruption in health and disease. Trends Microbiol 19:349–359

Houghteling P, Walker WA (2015) Why is initial bacterial colonization of the intestine important to the infant's and child's health. J Pediatr Gastroenterol Nutr 60:294–307

Hsiao EY, McBride SW, Hsien S et al (2013) Microbiota modulate behavioral and physiological abnormalities associated with neurodevelopmental disorders. Cell 155:1451–1463

Hungin AP, Mulligan C, Pot B et al (2013) Systematic review: probiotics in the management of lower gastrointestinal symptoms in clinical practice—an evidence-based international guide. Aliment Pharmacol Ther 38:864–886

Jackson ML, Butt H, Ball M et al (2015) Sleep quality and the treatment of intestinal microbiota imbalance in Chronic Fatigue Syndrome: a pilot study. Sleep Sci 8:124–133

Jakobsson HE, Abraharnsson TR, Jenmalum MC et al (2014) Decreased gut microbiota diversity, delayed Bacteriodetes colonization and reduced Th1 response in infants delivered by caesarean section. Gut 63:559–566

Johnson CK, Versalovic J (2012) The human microbiome and its potential importance in paediatrics. Pediatrics 129:950–960

Johnson CH, Stewart PL, Egli M (2011) The cyanobacterial circadian system: from biophysics to bioevolution. Annu Rev Biophys 40:143–167

Johnson CH, Zhao C, Xu Y et al (2017) Timing the day: what makes bacterial clocks tick? Nat Rev Microbiol 15:232–242

Joyce SA MacSharry J, Casey PG et al (2014) Regulation of host weight gain and lipid metabolism by bacterial bile acid modification in the gut. Proc Natl Acad Sci U S A 111:7421–7426

Kaczmarek JL, Musaad SM, Holscher HD (2017) Time of day and eating behaviors are associated with the composition and function of the human gastrointestinal microbiota. Am J Clin Nutr 106:1220–1231

Karlsson FH, Tremaroli V, Nookaew I et al (2013) Gut metagenome in European women with normal, impaired and diabetic glucose control. Nature 498:99–103

Kato T (2007) Molecular genetics of bipolar disorder and depression. Psychiatry Clin Neurosci 61: 3–19

Kelly JR, Clarke G, Cryan JF et al (2016) Brain-gut-microbiota axis: challenges for translation in psychiatry. Ann Epidemiol 26:366–372

Kim TW, Jeong JH, Hong SC (2015) The impact of sleep and circadian disturbance on hormones and metabolism. Int J Endocrinol 2015:591729

Kohsaka A, Laposky AD, Ramsey KM et al (2007) High-fat diet disrupts behavioral and molecular circadian rhythms in mice. Cell Metab 6:414–421

Konturek PC, Brozozowski T, Konturek SJ (2011) Gut clock: implication of circadian rhythms in the gastrointestinal tract. J Physiol Pharmacol 62(2):139–150

Labbe A, Ganopolsky JG, Martoni CJ et al (2014) Bacterial bile metabolising gene abundance in Crohn's, ulcerative colitis and type 2 diabetes metagenomes. PLoS One 9:e115175

Leone V, Gibbons SM, Martinez K et al (2015) Effects of diurnal variation of gut microbes and high-fat feeding on host circadian clock function and metabolism. Cell Host Microbe 17:681–689

Ley RE, Turnbaugh PJ, Klein S et al (2006) Microbial ecology: human gut microbes associated with obesity. Nature 444:1022–1023

Li Y, Hao Y, Fan F et al (2018) The role of microbiome in insomnia, circadian disturbance and depression. Front Psychiatry 9:669

Liang X, Bushman FD, FitzGerald GA (2015) Rhythmicity of the intestinal microbiota is regulated by gender and the host circadian clock. Proc Natl Acad Sci U S A 112:10479–10484

Liu B, Lin W, Chen S et al (2019) Gut microbiota as an objective measurement for auxiliary diagnosis of insomnia disorder. Front Microbiol 10:1770

Louis P, Scott KP, Duncan SH et al (2007) Understanding the effects of diet on bacterial metabolism in the large intestine. J Appl Microbiol 102:1197–1208

Macedo D, Filho AJMC, Soares de Sousa CN et al (2017) Antidepressants, antimicrobials or both? Gut microbiota dysbiosis in depression and possible implications of the antimicrobial effects of antidepressant drugs for antidepressant effectiveness. J Affect Disord 208:22–32

Magnusdottir S, Ravcheev D, de Crecy-Lagard V et al (2015) Systematic genome assessment of B-vitamin biosynthesis suggests co-operation among gut microbes. Front Genet 6:148

Maharshak N, Packey CD, Ellermann M et al (2013) Altered enteric microbiota ecology in interleukin 10-deficient mice during development and progression of intestinal inflammation. Gut Microbes 4(4):316–324

Malinen E, Mättö J, Salmitie M et al (2002) PCR-ELISA: II: Analysis of *Bifidobacterium* populations in human faecal samples from a consumption trial with *Bifidobacterium lactis* Bb-12 and a galacto-oligosaccharide preparation. Syst Appl Microbiol 25:249–258

Marcobal A, Southwick AM, Earle K et al (2013) A refined palate: bacterial consumption of host glycans in the gut. Glycobiology 23:1038–1046

Mariat D, Firmesse O, Levenez F et al (2009) The Firmicutes/Bacteroidetes ratio of the human microbiota changes with age. BMC Microbiol 9:123

Mazidi M, Shemshian M, Mousavi SH et al (2016) A double-blind, randomized and placebo-controlled trial of Saffron (Crocus sativus L.) in the treatment of anxiety and depression. J Complement Integr Med 13:195–199

Marquet P, Duncan SH, Chassard C et al (2009) Lactate has the potential to promote hydrogen sulphide formation in the human colon. FEMS Microbiol Lett 299:128–134

Merrow M, Maas MF (2009) Circadian clocks: evolution in the shadows. Curr Biol 19:R1042–R1045

Millar AJ (2016) The intracellular dynamics of circadian clocks reach for the light of ecology and evolution. Annu Rev Plant Biol 67:595–618

Moya A, Ferrer M (2016) Functional redundancy-induced stability of gut microbiota subjected to disturbance. Trends Microbiol 24:402–413

Mshvidadze M, Neu J, Shuster J et al (2010) Intestinal microbial ecology in premature infants assessed with non-culture-based techniques. J Pediatr 156:20–25

Murakami T, Kamada K, Mizushima K et al (2017) Changes in intestinal motility and gut microbiota composition in a rat stress model. Digestion 95:55–60

Nicholls P, Marshall DC, Cooper CE et al (2013) Sulfide inhibition of and metabolism by cytochrome c oxidase. Biochem Soc Trans 41:1312–1316

Nobs SP, Tuganbaev T, Elinav E (2019) Microbiome diurnal rhythmicity and its impact on host physiology and disease risk. EMBO Rep 20:e47129

O'Neill JS, van Ooijen G, Dixon LE et al (2011) Circadian rhythms persist without transcription in a eukaryote. Nature 469:554–558

Oosterman JE, Kalsbeek A, la Fleur SE et al (2015) Impact of nutrients on circadian rhythmicity. Am J Physiol Regul Integr Comp Physiol 308:337–350

Palmer C, Bik EM, DiGiulio DB et al (2007) Development of the human infant intestinal microbiota. PLoS Biol 5:e177

Panda S, Antoch MP, Miller BH et al (2002) Coordinated transcription of key pathways in the mouse by the circadian clock. Cell 109:307–320

Parekh PJ, Balart LA, Johnson DA (2015) The influence of the gut microbiome on obesity, metabolic syndrome and gastrointestinal disease. Clin Transl Gastroenterol 6:e91

Parkar SG, Blatchford PA, Kim CC et al (2015) New and tailored prebiotics: established applications. In: Venema K (ed) Probiotics and prebiotics: current research and future trends. Caister Academic Press, Poole, pp 289–314

Parkar SG, Kalsbeek A, Cheeseman JF (2019) Potential role for the gut microbiota in modulating host circadian rhythms and metabolic health. Microorganisms 7(2):41

Paulose JK, Wright JM, Patel AG et al (2016) Human gut bacteria are sensitive to melatonin and express endogenous circadian rhythmicity. PLoS One 11:e0146643

Petra AI, Panagiotidou S, Hatziagelaki E et al (2015) Gut-microbiota-brain axis and its effect on neuropsychiatric disorders with suspected immune dysregulation. Clin Ther 37:984–995

Plaza-Díaz J, Ruiz-Ojeda FJ, Vilchez-Padial LM et al (2017) Evidence of the anti-inflammatory effects of probiotics and synbiotics in intestinal chronic diseases. Nutrients 9:E555

Poroyko VA, Carreras A, Khalyfa A et al (2016) Chronic sleep disruption alters gut microbiota, induces systemic and adipose tissue inflammation and insulin resistance in mice. Sci Rep 6:35405

Powley TL, Wang XY, Fox EA et al (2008) Ultrastructural evidence for communication between intramuscular vagal mechanoreceptors and interstitial cells of Cajal in the rat fundus. Neurogastroenterol Motil 20:69–79

Qin J, Li Y, Cai Z et al (2012) A metagenome-wide association study of gut microbiota in type 2 diabetes. Nature 490:55–60

Reynolds AC, Paterson JL, Ferguson SA et al (2017) The shift work and health research agenda: considering changes in gut microbiota as a pathway linking shift work, sleep loss and circadian misalignment, and metabolic disease. Sleep Med Rev 34:3–9

Ridaura V, Belkaid Y (2015) Gut microbiota: the link to your second brain. Cell 161:193–194

Rieder R, Wisniewski PJ, Alderman BL et al (2017) Microbes and mental health: a review. Brain Behav Immun 66:9–17

Roberfroid MB (2007a) Inulin and oligofructose: health benefits and claims: a critical review, Inulin-type fructans: functional food ingredients. J Nutr 137:2493S–2502S

Roberfroid MB (2007b) Prebiotics: the concept revisited. J Nutr 137(3):830S–837S

Rogers GB, Keating DJ, Young RL et al (2016) From gut dysbiosis to altered brain function and mental illness: mechanisms and pathways. Mol Psychiatry 21:738–748

Rohlke F, Stollman N (2012) Fecal microbiota transplantation in relapsing Clostridium difficile infection. Therap Adv Gastroenterol 5:403–420

Scholtens PA, Oozeer R, Martin R et al (2012) The early settlers: intestinal microbiology in early life. Annu Rev Food Sci Technol 3:425–447

Schulze TG, Akula N, Breuer R et al (2014) Molecular genetic overlap in bipolar disorder, schizophrenia, and major depressive disorder. World J Biol Psychiatry 15:200–208

Schwartz S, Friedberg I, Ivanov IV et al (2012) A metagenomic study of diet-dependent interaction between gut microbiota and host in infants reveals differences in immune response. Genome Biol 13:r32

Sears CL, Garrett WS (2014) Microbes, microbiota, and colon cancer. Cell Host Microbe 15:317–328

Seow LS, Subramaniam M, Abdin E et al (2016) Sleep disturbance among people with major depressive disorders (MDD) in Singapore. J Ment Health 25:492–499

Shanahan F, Quigley EM (2014) Manipulation of the microbiota for treatment of IBS and IBD-challenges and controversies. Gastroenterology 146(6):1554–1563

Sharafati-Chaleshtori R, Shirzad H, Rafieian-Kopaei M et al (2017) Melatonin and human mitochondrial diseases. J Res Med Sci 22:2

Shindey R, Varma V, Nikhil KL et al (2016) Evolution of robust circadian clocks in Drosophila melanogaster populations reared in constant dark for over 330 generations. Naturwissenschaften 103:74

Shreiner AB, Kao JY, Young VB (2015) The gut microbiome in health and in disease. Curr Opin Gastroenterol 31:69–75

Suez J, Korem T, Zeevi D et al (2014) Artificial sweeteners induce glucose intolerance by altering the gut microbiota. Nature 514:181–186

Suez J, Zmora N, Zilberman-Schapira G et al (2018) Post-antibiotic gut mucosal microbiome reconstitution is impaired by probiotics and improved by autologous FMT. Cell 174:1406–1423 e1416

Sun Y, Zhang M, Chen CC et al (2013) Stress induced corticotropin-releasing hormone-mediated NLRP6 inflammasome inhibition and transmissible enteritis in mice. Gastroenterology 44:1478–1487

Tahara Y, Yamazaki M, Sukigara H et al (2018) Gut microbiota-derived short chain fatty acids induce circadian clock entrainment in mouse peripheral tissue. Sci Rep 8:1395

Teng SW, Mukherji S, Moffitt JR et al (2013) Robust circadian oscillations in growing cyanobacteria requires transcriptional feedback. Science 340:737–740

Thaiss CA, Zeevi D, Levy M et al (2014) Transkingdom control of microbiota diurnal oscillations promotes metabolic homeostasis. Cell 159:514–529

Thaiss CA, Levy M, Korem T et al (2016) Microbiota diurnal rhythmicity programs host transcriptome oscillations. Cell 167:1495–1510 e1412

The Human Microbiome Project Consortium, Huttenhower C, Gevers D, Knight R et al (2012) Structure, function and diversity of the healthy human microbiome. Nature 486 207–214

Thompson RS, Roller R, Mika A et al (2017) Dietary prebiotics and bioactive milk fractions improve nrem sleep, enhance rem sleep rebound and attenuate the stress-induced decrease in diurnal temperature and gut microbial alpha diversity. Front Behav Neurosci 1C:240

Tuohy KM, Conterno L, Gasperotti M (2012) Up-regulating the human intestinal microbiome using whole plant foods, polyphenols, and fiber. J Agric Food Chem 60:8776–8782

Turnbaugh PJ, Hamady M, Yatsunenko T et al (2009a) A core gut microbiome in obese and lean twins. Nature 457:480–485

Turnbaugh PJ, Ridaura VK, Faith JJ (2009b) The effect of diet on the human gut microbiome: A metagenomic analysis in humanized gnotobiotic mice. Sci Transl Med 1:6ra14

Vanuytsel T, van Wanrooy S, Vanheel H et al (2014) Psychological stress and corticotropin-releasing hormone increase intestinal permeability in humans by a mast cell-dependent mechanism. Gut 63:1293–1299

Vassallo MF, Walker WA (2008) Neonatal microbial flora and disease outcome. Nestle Nutr Workshop Ser Pediatr Program 61:211–224

Verdam FJ, Fuentes S, de Jonge C et al (2013) Human intestinal microbiota composition is associated with local and systemic inflammation in obesity. Obesity 21:E607–E615

Vijay-Fumar M, Atiken JD, Carvalho FA et al (2010) Metabolic syndrome and altered gut microbiota in mice lacking toll-like receptor 5. Science 328(5975):228–231

Voigt RM, Forsyth CB, Green SJ et al (2014) Circadian disorganization alters intestinal microbiota. PLoS One 9:e97500

Voigt RM, Summa KC, Forsyth CB et al (2016) The circadian clock mutation promotes intestinal dysbiosis. Alcohol Clin Exp Res 40:335–347

Walker WA (2017) Dysbiosis. The microbiota in gastrointestinal pathophysiology, 1st edn, pp 227–232

Walker AW, Ince J, Duncan SH et al (2011) Dominant and diet-responsive groups of bacteria within the human colonic microbiota. ISME J 5:220–230

Wang Y, Kuang Z, Yu X et al (2017) The intestinal microbiota regulates body composition through NFIL3 and the circadian clock. Science 357:913–916

Wang Y, Kasper LH (2014) The role of microbiome in central nervous system disorders. Brain Behav Immun 38:1–12

Wei Y, Zhu W, Gong J et al (2015) Fecal microbiota transplantation improves the quality of life in patients with inflammatory bowel disease. Gastroenterol Res Pract 2015:517597

Wilhelm I, Diekelmann S, Born J (2008) Sleep in children improves memory performance on declarative but not procedural tasks. Learn Mem 15(5):373–377

Wirz-Justice A (2006) Biological rhythm disturbances in mood disorders. Int Clin Psychopharmacol 21(Suppl.1):S11–S15

Wolf G (2002) Three vitamins are involved in regulation of the circadian rhythm. Nutr Rev 60:257–260

Wong RK, Yang C, Song GH et al (2015) Melatonin regulation as a possible mechanism for probiotic in irritable bowel syndrome: a randomized double-blinded placebo study. Dig Dis Sci 60:186–194

Wu GD, Chen J, Hoffmann C et al (2011) Linking long-term dietary patterns with gut microbial enterotypes. Science 334:105–108

Yarandi SS, Peterson DA, Treisman GJ et al (2016) Modulatory effects of gut microbiota on the central nervous system: how gut could play a role in neuropsychiatric health and diseases. J Neurogastroenterol Motil 22:201–212

Yi P, Li L (2012) The germfree murine animal: an important animal model for research on the relationship between gut microbiota and the host. Vet Microbiol 157:1–7

Yokota A, Fukiya S, Islam KBMS (2012) Is bile acid a determinant of the gut microbiota on a high-fat diet? Gut Microbes 3:455–459

Yoshikawa K, Kurihara C, Furuhashi H et al (2017) Psychological stress exacerbates NSAID-induced small bowel injury by inducing changes in intestinal microbiota and permeability via glucocorticoid receptor signalling. J Gastroenterol 52:61–71

Zarrinpar A, Chaix A, Yooseph S et al (2014) Diet and feeding pattern affect the diurnal dynamics of the gut microbiome. Cell Metab 20:1006–1017

Zechner LE (2017) Inflammatory disease caused by intestinal pathobionts. Curr Opin Microbiol 35:64–69

Zhang YKJ, Guo GL, Klaassen CD (2011) Diurnal variations of mouse plasma and hepatic bile acid concentrations as well as expression of biosynthetic enzymes and transporters. PLoS One 6:e16683

Zhang R, Lahens NF, Ballance HI et al (2014) A circadian gene expression atlas in mammals: Implications for biology and medicine. Proc Natl Acad Sci U S A 111:16219–16224

Zheng P, Zeng B, Zhou C et al (2016) Gut microbiome remodelling induces depressive-like behaviors through a pathway mediated by the host's metabolism. Mol Psychiatry 21:786–796

Zhu D, Ma Y, Ding S et al (2018) Effects of melatonin on intestinal microbiota and oxidative stress in colitis mice. Biomed Res Int 2018:2607679

Zmora N, Zilberman-Schapira G, Suez J et al (2018) Personalized gut mucosal colonization resistance to empiric probiotics is associated with unique host and microbiome features. Cell 174:1388–1405 e1321

Role of Gut Microbiota and Probiotic in Chronic Fatigue Syndrome

Anjali Sharma, Sharad Wakode, Supriya Sharma, and Faizana Fayaz

Abstract

Chronic fatigue syndrome (CFS) is a combination of complex illness characterized by tiredness or intense fatigue that may worsen with too much exertion. Among the wide range of neuropsychological symptoms, 97% CFS patients have been reported with neuronal disorders such as headaches and symptoms in the emotional realm. Patients with CFS also show noticeable alterations in microflora, lowering level of *Lactobacilli* and *Bifidobacterium*. Recent researches explain that probiotics in the gastrointestinal tract (GIT) can greatly influence the neuronal pathways and central nervous system (CNS) to modulate behavior. Various studies expressed the benefit of probiotic therapy in normalizing fatigue patients and also restored mitochondrial electron transport function in patients with CFS. In this chapter, we provided a historical skeleton, bidirectional communication pathophysiology, selection criteria of probiotics, CFS treatment, and clinical implications of gut–brain connections. In summary, various aspects concerning the potential and safety of probiotics in the management of chronic fatigue syndrome are discussed in this chapter.

A. Sharma · F. Fayaz
Department of Pharmaceutical Chemistry, Delhi Institute of Pharmaceutical Sciences and Research, Sector-3, New Delhi, Delhi, India

S. Wakode (✉)
Delhi Institute of Pharmaceutical Sciences and Research, Government of NCT of Delhi, Sector-3, New Delhi, Delhi, India

S. Sharma
Department of Pharmacognosy, Delhi Pharmaceutical Sciences and Research University, Sector-3, New Delhi, Delhi, India

© Springer Nature Singapore Pte Ltd. 2022
P. K. Deol, S. K. Sandhu (eds.), *Probiotic Research in Therapeutics*,
https://doi.org/10.1007/978-981-16-6760-2_9

Keywords

Chronic fatigue syndrome · Gut–brain axis · Irritable bowel syndrome ·
Bidirectional interaction · Risk factors · Microbiome

Abbreviations

5-HT	5-Hydroxytryptamine
BDNF	Brain derived neurotrophic factor
CFS	Chronic fatigue syndrome
CNS	Central Nervous System
DLMO	Delayed dim light melatonin
EECs	Enteroendocrine cells
EFSA	European Food Safety Authority
FAO/ WHO	Food and Agriculture Organization of the United Nations/World Health Organization
FDA	Food and Drug Administration
FMPP	Fermented Milk Product containing Probiotic
FMT	Fecal microbiota transplantation
GABA	Gamma-Aminobutyric Acid
GI	Gastrointestinal
GIT	Gastrointestinal tract
GLP-1	Glucagon-Like Peptide-1
GPCRs	G-Protein-Coupled Receptors
HDAC	Histone deacetylation
HIV	Human Immunodeficiency Virus
HPA	Hypothalamic-Pituitary-Adrenal
IBD	Inflammatory Bowel Disease
IBS	Irritable Bowel Syndrome
IgA	Immunoglobulin A
IL-6	Inerlukin 6
LPS	Lipopolysaccharides
MDD	Major Depressive Disorder
ME/CFS	Chronic fatigue syndrome, myalgic encephalomyelitis
MIC	Minimum Inhibitory Concentration
mRNA	Messenger RNA
NHPR	National Health Products Regulations
NMDA	N-methyl-D-aspartate receptor
PAMPs	Pathogens-Associated Molecular Patterns
PROSAFE	Biosafety assessment of probiotics used for human consumption
QPS	Qualified Presumption of Safety
SCFA	Short-Chain Fatty Acids
SIBO	Small intestinal bacterial overgrowth
USA	United States of America

9.1   Introduction

Chronic fatigue syndrome (CFS) or myalgic encephalomyelitis (ME) or ME/CFS is a medically unexplained illness characterized by severe fatigue that impairs memory, sleep and causes musculoskeletal pain (Komaroff 1996). People suffering from CFS has an impact on their routine and life quality because of the financial burden for living and in healthcare sectors (Aaron et al. 2000). In recent years, lots of development has been made to fill up the gap between the understanding of pathogenesis and mechanism of CFS along with effective ways of treatment. In addition to it, CFS can also be characterized by headaches, myalgia, arthralgia, and post-exertional malaise; cognitive difficulties; and mood swings, such as depression and anxiety (Pearce 2006). Along with this, many CFS patients complain a variety of other symptoms such as abdominal discomfort and disturbance in bowel habit (Komaroff and Buchwald 1991; Komaroff 1996). In CFS cases, patients are more often to have a diagnostic history report of irritable bowel syndrome (IBS) and have IBS related symptoms (Aaron et al. 2000).

The commensal microbiota can weigh up to 2 kg, adult gut being a host to over 100 trillion microorganisms (Bermon et al. 2015), which serves multiple host functions and play important role in human health. Currently, studies revealed the effects of gut microbiota, i.e. probiotics on the brain functions and its significant role in CFS. Species such as *Lactobacillus* and *Bifidobacterium* have been proved to be significant probiotic strains (Kailasapath and James 2000). Studies revealed that probiotic attributed some beneficial effect to human body like improvement of intestinal microflora, immune system, cancer prevention, treatment of IBS, and antihypertensive impacts. Additionally, many researchers explained bidirectional communication between gut microbiota and nervous system via different pathways. Important pathways that make connection between gut and brain functioning are vagus nerve, metabolites synthesis, cytokine activation, and alteration in neuronal circuitry (Mahgoub and Monteggia 2013). Few studies revealed that there are noticeable cross connections between the human and animal studies which describes the role of intestinal microbiota in CFS patients, including anxiety and mood disorders (Wallis et al. 2018; Newberry et al. 2018). Therefore, the concept of interconnection of microbiota-gut-brain axis can be helpful in CFS patients. In this review, we have highlighted the present scenario that how intestine microbiota affects the functions of the brain. Along with that, we have also described the potential role of probiotics and fecal microbiota transplantation FMT in treating CFS. The aim of this chapter is to put together different branches of research in order to carry out a hypothesis which reveals the potential of probiotic in the treatment of CFS.

9.2   Historical Perspectives of CFS

CFS is a global endemic, which occurred as an infectious outbreak in London, UK and Nevada, USA, in 1955 and 1984, respectively (Pose 1956; Holmes et al. 1988). It was first defined in 1988, but as its etiology of illness remains unknown several definitions and diagnostic criteria has been updated (Fukuda et al. 1994), however, many terminologies have been given to this illness, but the most frequent used are "Chronic fatigue syndrome," "myalgic encephalomyelitis," and "ME/CFS" (Sharpe 2002).

Other similar outbreaks were reported in different parts of the country between 1934 and 1990. These epidemics of unknown illness were referred by numerous names, e.g., Poliomyelitis in USA (1934), Iceland disease in Iceland (1948), ME in UK (1955), epidemic neuromyasthenia in USA (1957), Tapanui flu in New Zealand (1983), CFS in USA (1984), and Lyndonville outbreak in USA (1985) (Castro-Marrero et al. 2017).

In 2004 in Bergen, Norway CFS symptoms were observed in patients previously infected with Giardia enteritis (Naess et al. 2012) and evidences regarding the similarities of these outbreaks have been investigated in various communities (Buchwald et al. 1992; Jason et al. 1999). The follow-up studies of the patients from the outbreaks of New Zealand and USA showed that the patients, who did not recovered, exhibited the symptoms of CFS (Castro-Marrero et al. 2017).

9.3   Risk Factor of CFS

9.3.1   Gut Microbiota Composition

Evidences showed that CFS is associated with altered gut microbiota composition, with increased levels of aerobic bacteria and lower levels of *Bifidobacteria* (Logan et al. 2003). It has been found that patient suffering from CFS had greater proportions of Escherichia coli ($> 85\%$) in stool sample as compared to standard sample (49%) (Butt et al. 2001). Similarly, increased levels of *Enterococcal* were reported in CFS patients along with decreased levels of *Bifidobacteria* and severe intestinal disturbance (Butt et al. 1998). There are some evidences which revealed that changes in the microbiota composition can depend on geographical areas. For example, one study showed difference in microbiota profile based on their bacterial genera of two different countries (Belgian and Norwegian) and patients (Fremont et al. 2013). However, overall 77% of CFS cases were because of overgrowth of intestinal bacteria (Pimentel et al. 2000).

9.3.1.1   Other Risk Factors

These include family risk factor, gender, stress and injury, rest/activity, ethnicity, any recent illness, toxic, and occupational exposure. A study reported that non-genetic close contacts (partners) have the prevalence of ME/CFS of 3.2% and the risk was 7.6–13.3 as compared to the community prevalence. The prevalence was higher in non-genetic household contacts and in non-resident genetically linked relatives than in the community, thus suggesting that both household non-genetic contact and genetic relationship are risk factors of CFS (Buchwald et al. 1992). Twin studies showed that risk rate for CFS was 55% in monozygotic twins and 19% in dizygotic twins (Buchwald et al. 1992). All the age groups are prone to CFS but the peak onset age is 11–19 years in children and 30–39 in adults (Inger Johanne et al. 2014). The attack rate of the ME/CFS outbreaks varied between 1.6% and 5.1% in males and 6.4% and 10.4% in females (Acheson 1959), further, females are more prone to CFS than males due to the differing effect of immune system on male and female hormones, as female hormones favor antibody-mediated immunity and male hormones favor cell-mediated immunity (Spellberg and Edwards 2001). A community based study of CFS showed that the women particularly of minority groups and people with lower educational and occupational status were at the highest risk of CFS (Jason et al. 1999). Studies on poliomyelitis cases in Royal Free hospital outbreak showed that overexertion increased the severity of the disease (Da Silva 2018) along with the evidence that decrease in the rest period increases the chances of relapse (Compston 1957) due to reduced cell-mediated immunity (Kakanis et al. 2010). Stress, trauma, and adverse events elicit the onset of ME/CFS which are associated with the response of cortisol to stress (Dantzer et al. 1989), production of pro-inflammatory cytokines (Hoge et al. 2009), and reduced NK cell cytotoxicity (Faist et al. 1996). Toxins has shown provoking effects on ME/CFS which includes organophosphate pesticides (Tahmaz et al. 2003), Cadmium (Pacini et al. 2012), solvents (Delia et al. 2001), and mycotoxins (Joseph et al. 2013).

9.4   Etiology of CFS

The etiology of CFS is complex, unclear and remains controversial. Researchers showed the involvement of genetics, immune system, biopsychosocial model, adrenal system, sleep and nutrition to be the main factors underlying CFS (Fig. 9.1).

9.4.1   Genetics

Genetic susceptibility is also marked as a major factor in the etiology of CFS patients. A study showed that gene expression differed between CFS patients and controls after exercise (Whistler et al. 2005). Another study showed a positive interconnection between CFS, viral infections, and genetic mutations (Zhang et al. 2010).

Fig. 9.1 Factors
underlying CFS

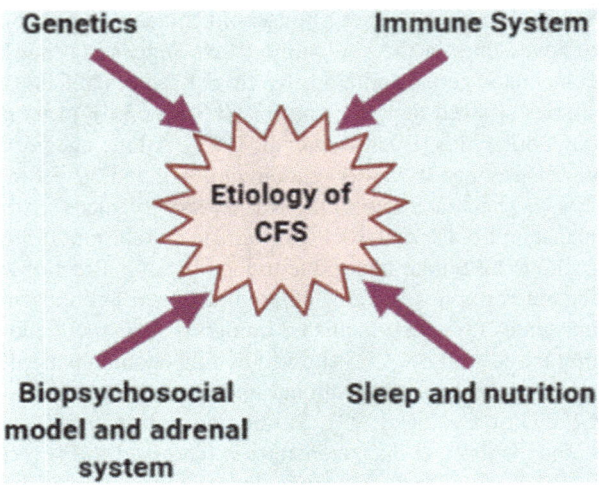

9.4.2   Immune System

As major symptoms of CFS and viral infections overlap, it has been theorized that
CFS has postinfectious etiology showing mixed evidence of an interconnection
between viral infection and CFS (Mawle et al. 1995).

9.4.2.1 Biopsychosocial Model and Adrenal System

CFS is characterized by various psychophysiological disturbances such as; arthralgia, myalgia, headache, malaise, cognition deficits, impaired memory and sleep
disorders, depression and anxiety. Which suggests it be somatic illness (Wessely
et al. 1998). A report suggested that trauma in childhood as an important risk factor
in the etiology of CFS which is associated with neuroendocrine dysfunction, a
primary hallwork of CFS (Christine et al. 2009). In another study lack of social
support to CFS patients was identified as a new factor emerging in the functional
impairments of CFS (Prins et al. 2004). This factor was further supported by a study
in which patients with poor social adjustment and pessimistic mindset showed lesser
improvement by psychological treatments in primary care of CFS (Chalder et al.
2003). Lower levels of cortisol were found in CFS patients and can contribute in the
etiology of CFS (Remzi et al. 2004; Cleare et al. 2001).

9.4.2.2 Sleep and Nutrition

A study showed an interconnection between delayed dim light melatonin (DLMO)
and CFS, giving a clue that circadian pattern might contribute to CFS. Furthermore,
it also suggested the use of melatonin as treatment therapy in CFS patients and
DLMO (Van Heukelom et al. 2006). A study also revealed that CFS patients have
lower concentration of omega-3 and omega-6 unsaturated fatty acids along with
lower zinc levels as compared to healthy patients (Maes et al. 2005).

9.5   Pathophysiological Mechanisms of Gut Microbiota in CFS

9.5.1   Host Immune Response Stimulation Leading to Diverse Patterns of Systemic Cytokine Activation

A structural component of microbial cell wall is made up of peptidoglycan that activates both the innate and acquired immune system. Innate immune response activates immune system by pro-inflammatory microbial constituent's specifically, pathogens-associated molecular patterns (PAMPs) to produce cytokines which may influence the CNS either via vagal pathway or via permeable regions of blood–brain barrier (Sherwin et al. 2016). Bacterial cell wall components may induce intestinal epithelial cells to produce molecules that can modulate neural signaling (Forsythe and Kunze 2013) thus, highlighting the potential role of gut microbiota in influencing cerebrum function in CFS.

9.5.1.1 Enteric Dysbiosis
Microbiota dysbiosis has emerged as a major factor in the pathogenesis Of CFS showing evidences of immunological deviations in CFS. Dysbiosis contributes to the growth and spread of pathobionts secreting toxic substances to the host cells due to the alteration of commensal microbes growing in the GI tract, thus triggering inflammatory immune responses (Brown 2014; Maes et al. 2007) thus can act as a major contributor toward CFS.

9.5.2   Neuroendocrine Signaling Pathway

Numerous peptides present throughout gut–brain axis pathway are responsible for various functions including gut secretion and motility. Secretory products released by bacteria stimulates enteroendocrine cells (EECs) to produce various neuropeptides such as neuropeptide Y (NPY), peptide YY, glucagon-like peptide, cholescystokinin, and substance (Cani and Knauf 2016; Cani et al. 2013) which might may play a potential role in the bidirectional gut–brain communication (Holzer and Farzi 2014). A new pathway was uncovered in a study in which neuropods (intestinal enteroendocrine cells) demonstrated direct connection between EECs and neurons of small intestine and colon (Bohórquez et al. 2015). This newly revealed signaling pathway might act as a link between gut microbiota and CFS.

9.5.3   Intestinal Metabolites

Numerous intestinal metabolites obtained from digestion and microbial fermentation of dietary food may play a crucial role in brain functions and immune responses. Alteration in the gut microbiota composition has been observed to impact the availability and regulation of fatty acids and tryptophan which further regulates immune responses. These metabolites are contributing a major role in gut–brain axis and may serve as effective targets for clinical interventions.

9.5.3.1 Fatty Acids

Fatty acids are involved in several functions of the brain such as neurotransmission, neuroinflammation, and cell survival (Bazinet and Layé 2014). Dietary fatty acids helps in eicosanoids production, a class of neurotransmitter which regulates inflammatory and immune responses via gene regulation, cytokine biosynthesis, membrane composition, and function alterations (Fritsche 2006). Short chain fatty acids (SCFA) derived as primary metabolite from gut microorganisms show potent anti-inflammatory effects. Some fatty acids such as butyrate, isobutyrate, acetate, propionate, and hexonate found in the microbe's species (Roseburia, Faecalibacterium, Eubacterium, Bifidobacterium, Enterobacter, and Lactobacillus) are produced in abundance through bacterial fermentation of indigestible carbohydrate in the normal colon. Moreover, SCFA regulate the body mainly by two systems of molecular signaling: histone deacetylation (HDAC) and G-protein-coupled receptors (GPCRs) (Tan et al. 2014). Abnormal balance in the direction of excessive HDAC was observed in Parkinson's disease (Da Silva 2018), depression and schizophrenia (Mahgoub and Monteggia 2013). Studies revealed that inhibitory activity of HDAC has appreciating effects on cancer and in animal models of CNS disorders, including dementia, brain trauma, and autoimmune encephalitis (Wallis et al. 2018; Konsoula and Barile 2012). Microbial fermentation of dietary and nutritional components also showed noteworthy effects on brain processes and immune functions. SCFA supports enteroendocrine signaling and cognate receptor (GPR44 or GPR41) binding which further stimulates the neuropeptide release of peptide YY (PYY) and glucagon-like-peptide (GLP-1). These released peptides are responsible for regulating energy homeostasis and activating the enteric and primary afferent pathway (Engel and Neurath 2010). Considering the role of SCFA in energy metabolism and in the production of secondary peptides, depicts the mechanism through which the gut microbiota influences the human health (Den Besten et al. 2013) (Fig. 9.2).

9.5.3.1.1 Tryptophan

Tryptophan, an essential amino acid is a precursor to biologically active agents, including serotonin synthesis in CNS. Serotonin, a chemical which has been implicated in depression, is found abundantly in the gut where it is synthesized in enterochromaffin cells of GIT by tryptophan. Tryptophan levels in plasma have shown strong connection with alterations in immune system which may further influence CNS functions leading to brain disorders (Song et al. 1998; Schröcksnadel et al. 2006). Tryptophan metabolism dominantly rely on Kynurenine pathway in which Kynurenine produces two different metabolites, namely, quinolinic acid (neurotoxic metabolite) and kynurenic acid (neuroprotective). The proposed neuroprotective action is indeed due to the increased metabolism of kynurenine to kynurenic acid (Agudelo et al. 2014) . Similarly, Desbonnet et al. (2008) also revealed that certain probiotics has the ability to increase tryptophan metabolism

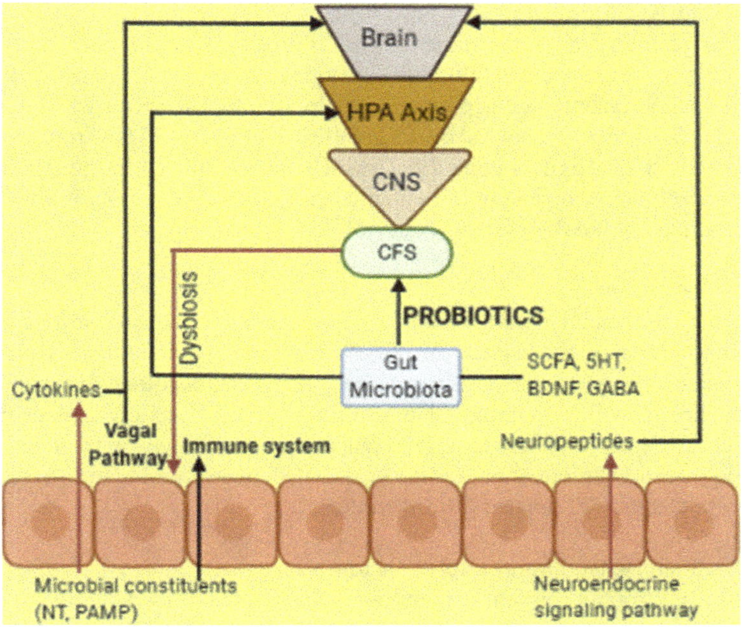

Fig. 9.2 Main signaling pathway of Probiotic, gut microbiota, and chronic fatigue syndrome. Various mechanisms involved through which microorganisms present in our gut can transmit the signals to brain. Bacteria are removed and sampled by neuroactive metabolites (fatty acids and tryptophan), stimulation of immune response (cytokines, IL-6, TNF-α), HPA axis and alteration in neuronal circuitry (GABA, 5-HT, BDNF). These routes of communication lead to regulate the brain functioning and controlled the symptoms of CFS. Dysregulation of intestinal microbiota consequently responsible for alteration of brain function which may account for chronic fatigue syndrome such as anxiety, stress and depression. The content of gut microbiota can be normalized with the use of probiotics which may be better option to treat neuronal dysfunctions

and demolishes the stress-induced depression through Kynurenine pathway (Desbonnet et al. 2008) (Fig. 9.2).

9.5.4 Alterations in Neuronal Circuitry

As we discussed above, several bacteria such as *Lactobacillus, Bifidobacterium, Escherichia,* and *Truchuris* are involved in neuroprotective mechanism. In addition to it, neurotransmitters like gamma-aminobutyric acid (GABA), serotonin (5-HT) and neuropeptides like brain derived neurotrophic factor (BDNF) produced by microorganisms transmit the signal or activate different receptors allowing neurons to communicate with each other (Cryan et al. 2012; Bercik et al. 2010). Their imbalance can influence the brain activity and mood (Barrett et al. 2012; Higuchi et al. 1997) as both are considered as major neuronal signaling messengers and their alterations can show marked effects on CFS related CNS disorders.

9.5.4.1 GABA

GABA, an inhibitory neurotransmitter of CNS is produced by glutamate metabolism and regulates neuronal excitability; its dysfunction has been related with CNS disorders such as anxiety and depression. Some microbial species such as *Lactobacillus* and *Bifidobacterium* have shown the ability to produce GABA from glutamate (Barrett et al. 2012; Higuchi et al. 1997) which can be utilized in the treatment of CNS disorders hence, revealing the potential role of gut microbiota in influencing CNS by regulating GABA system (Fig. 9.2).

9.5.4.1.1 Serotonin

Serotonin, a monoamine neurotransmitter, produced by enterochromaffin cell of GI tract is implicated in the regulation of CNS functions and influences numerous physiological processes such as sleep, mood, pain, etc. Serotonergic dysfunction has been linked with numerous CNS disorders such as anxiety and depression (Cryan et al. 2012) as gut microbiota can directly influence brain functions by controlling serotonin production (Clarke et al. 2013), hence serotonergic dysfunction might be related to CFS related CNS disorders.

9.5.4.1.1.1 Brain Derived Neurotrophic Factor

BDNF, a neurotrophin (protein) generally communicated in the CNS that regulates the endurance of existing neurons and empowers the development and differentiation of new neurons and neurotransmitters. Lower levels of BDNF have been related with CFS and their treatments (e.g., antidepressants) have been appeared to expand the expression of BDNF in the cerebrum. In particular, intestinal microbiota have been known to build levels of hippocampal BDNF in pathogen free mice on treatment with antimicrobials and fecal transplants (Bercik et al. 2011). In CFS murine model decreased expression of BDNF mRNA was found in hippocampus (Bercik et al. 2010). Several more studies conducted with mice revealed interconnection between microbiota-gut-brain (Gareau et al. 2011; Clarke et al. 2013; Heijtz et al. 2011; Sudo et al. 2004; Litleskare et al. 2018), suggesting the role of BDNF in the development of chronic fatigue syndrome. Other symptoms of chronic fatigue syndrome such as anxiety and depression decreased the expression of BDNF mRNA in hippocampus (Martinowich et al. 2007). It is considerable that the impact of the gut microbiota on BDNF is directed by sex and strain, including some additional factors. Considering the BDNF role in neuroplasticity and neurological problems, future research studies are justified to observe the interrelationship between BDNF and different neurotrophins and under what conditions these development elements are affected by the microbiota (Fig. 9.2).

9.5.5 Hypothalamic–Pituitary–Adrenal Axis (HPA Axis) and Corticosteroids

The first study revealing an interconnecting link between chronic fatigue syndrome and HPA axis came in 1981 (Poteliakhoff 1981), which hypothesized that patients with chronic fatigue syndrome showed alterations in HPA axis (Jerjes et al. 2007). However, the interrelationship of HPA axis dysfunction with chronic fatigue is still indefinable. Nevertheless, HPA axis might exacerbate the symptoms in the late course of chronic fatigue syndrome (Cleare 2004).

9.5.6   Mucosal Dysfunction

The mucosal layer of gut is made up of single layer of epithelial cells and serves as an integral part of innate immune system and shields inside of the body from outer environment. Thus, gut mucosal layer serves as a key component in acting as a defense mechanism against infection. Its working is maintained by several interconnected systems, together known as gatekeeper of the gut mucosal system. In CFS patients, increased level of serum IgA and IgM against LPS of enterobacteria has been observed, which reveals the significant role of mucosal dysfunction in the etiology of CFS. (Michael Maes et al. 2007). Due to psychological stress the mucosal layer disrupts allowing the passage of microorganisms, leading to stimulation of mucosal immune response (Chen et al. 2003). The mucosal barrier dysfunction causes modifications in mucosal secretion, gut motility, and visceral secretions, which reveals the interconnection between stress and CFS. Mucosal dysfunction and altered microbiota in IBS patients show an interconnection between IBS, CFS, and mucosal dysfunction (Zhou et al. 2009). Gut microbiome impacts motor, sensory, and immune system of gut and influences brain even at low levels (Lee and Tack 2010). As mucosal barrier dysfunction and altered intestinal gut microbiota contribute to CFS, they can be utilized as potential targets in clinical interventions against CFS (Logan et al. 2003).

9.6   Gut Microbiota and Its Effects on CFS

Gut microbiota monitors human body functions and provide numerous essential non-digestible dietary components in the body; inherent microbial family helps to digest these nutrients and energy source components for further processes in the body. In respect of it, gut microbiota maintains the gut epithelial homeostasis, immune system development and also provides protection from pathogens and several metabolic reactions. Physical exertion and mental tension is an interesting and main risk factor of CFS which alters the gut microbiota composition. For example, Bailey et al. (2004) observed dysbiosis of microbiota due to psychological stress which led to the reduction of *Bifidobacterium* and *Lactobacillus* (Bailey et al. 2004). Although, it has been noted that reduction in *Bifidobacterium* and small intestinal bacterial overgrowth (SIBO) resulted in IBS and CFS (Logan et al. 2003). Another report revealed that increased cytokine release and enhanced immune response leads to the reduction in the amount of *Bacteroides* and *Clostridium* species in the GIT of mice exposed to social stressor as compared to the controls (Bailey et al. 2011). In another study increased HPA axis reactivity in germ free

(GF) mice on exposure to mild restraint stressor was observed. This disturbance in HPA axis was then completely recovered after colonization with *Bacillus infantis* (Sudo et al. 2004; Clarke et al. 2013).

9.7 Selection Criteria of Probiotics

There are various criteria's that should be kept under consideration before selecting any bacterial strains, taking safety issues into considerations. Food and Agriculture Organization of the United Nations/World Health Organization (FAO/ WHO) was introduced in 2002, which published the "Guidelines for Evaluation of Probiotics in Food," and recognized various issues related to safety and effectiveness of probiotics (Pineiro and Embarek 2002; Taverniti and Guglielmetti 2011).

A stringent selection criterion for identification of probiotic strain is required in order to achieve consistent and positive probiotic effects. Probiotics are feasible microorganisms (bacteria or yeasts) which exert valuable effects when ingested in appropriate concentration. New microbial (e.g., yeast and *Bacillus*) strains are commonly used as probiotics; *Lactobacillus* and *Bifidobacterium* strains being considered as safest options for long-term use (Prasad et al. 1999). The most important factor to be considered before selecting any probiotic is to know how it affects our histamine levels (Sanders 2000) as histamine is a neuroregulatory molecule that performs wide array of functions in cognitive functions, digestion, and allergic reactions (Thangam et al. 2018). Certain bacteria are known to produce huge quantities of histamine, which may be an issue for those suffering from histamine intolerance, for e.g., *Lactobacillus casei*, *Lactobacillus bulgaricus,* and *Lactobacillus rheuteri*. Other bacteria such as *Lactobacillus acidophillus* is considered to be more histamine-neutral but still produce some levels of histamine.

After thorough investigations following strains are suitable for individuals:

- *Lactobacillus rhamnosus.*
- *Bifidobacterium infantis.*
- *Bifidobacterium longum.*

These above mentioned bacteria were among the ones most-commonly studied as they exhibited an overall decreased release of histamine or D-Lactic acid (Liam et al. 2011).

Selection criteria
1. While selecting any probiotic strains, it must be able to cope up with the stress that the human body imposes. When administered, it should be resistant to various enzymes present in oral cavity as amylase and lysozyme have the capability to degrade them. Although some gram-positive bacteria are usually sensitive to lysozyme, but some are more resistant and have the tendency to become the part of the resident microbiome of the oral cavity (Mikelsaar et al. 2005).

2. Another criterion for selecting probiotic is that, it should be acid and bile tolerant, in order to survive in the gut.
3. Another step is to evaluate whether probiotic has the ability to colonize the GIT epithelial cells. Adhesion of microbes to epithelial cells is a complex process, linking two membranes (i.e., microbial and human cells) which depends on the chemical and physiochemical composition of probiotic strain cell surface (Boonaert and Rouxhet 2000).

Introduction of microorganisms as probiotics should be assessed properly, as they are not considered fully safe which may cause an infection (Culligan et al. 2009). Various initiatives have been taken by European community (The European Union Novel Food regulation, QPS and PROSAFE), USA (FDA and WHO), and Canada (Health Canada: NHPR) to establish the criteria for the use of probiotic in humans. These include records of isolation history, taxonomic identification and absence of virulence, infectivity, toxicity, and transferable antibiotic resistance genes (Sanders et al. 2016).

Long-term probiotic consumption is associated with antimicrobial resistance, which was first recognized as a safety concern by EFSA in 2007 (Barlow et al. 2007) after that various multiple independent studies have been conducted to detect the presence of genes associated with antibiotic resistance in probiotics (Klein et al. 2000). In order to measure the susceptibility to antibiotics resistance various methods have been utilized such as disk diffusion and agar overlay diffusion, E-test, agar dilution or broth dilution. E-test and dilution methods are based on minimum inhibitory concentration (MIC) assay, which determines the least required concentration of an antimicrobial to inhibit microbial growth, while diffusion tests use antibiotic discs with inhibitory concentrations in agar plates. Phenotypic screening can also be considered a "true" measure, however, low reproducibility among laboratories has been observed.

Conventionally, probiotics have been selected based on stress-resistance phenotypes that guaranteed their survival through the GIT and subsequent persistence in the gut by utilizing safety assessment tools of probiotics. Safety-related genes, for e.g., antibiotic resistance and harmful metabolites production, have been reported in certain probiotic strains evaluation and are helpful in the selection of probiotics (Senan et al. 2015).

9.8 Treatment of CFS

9.8.1 Probiotics/Prebiotics

Live microbial supplement is termed as probiotic which has natural ways to resist the growth of host pathogens and support the improvement of gut microbiota functions (Fuller 1989). Bacterial families such as *Bifidobacterium, Lactobacillus,* and yeast comes under the category of probiotics used in humans (Goldin 1998). Once probiotic adheres to the gut, it produces extracellular antimicrobial components through the conversion of carbohydrates, proteins, and other minor compounds into important substances that can kill pathogenic bacteria, such as organic acids, enzymes, hydrogen peroxide, bacteriocins, and low-molecular-mass peptides. Literature justified that intake of probiotics is necessary to balance or normalize the factors of CFS. The positive outcome of probiotic established a strong association between gut microbiota and cognitive dysfunction. Furthermore, *Lactobacillus casei* showed significant effect on anxiety problem in CFS patients (Gaab et al. 2005). Similarly, administrations of other *Lactobacillus* family strains were also found to be effective in neurocognitive functions (Raber et al. 1998). Sprake et al. 2012 investigated high-dose vitamin D3 as probiotic supplement and noted a significant improvement in IBS patients which was further confirmed by other researches (Sprake et al. 2012; Hoeck and Pall 2011). In 1910, Dr. Philips invented *Lactobacillus* powder and tablets, composed of live lactic-acid bacteria and gelatin-whey which were found to be effective in melancholia and stress (Rao et al. 2009). Additionally, a study supported the use of probiotic *Bifidobacterium infantis as a* supplement in depression and or anxiety (Desbonnet et al. 2010). It has been reported that several other H_2O_2-producing species or strains of bacteria have probiotic properties, such as *Bifidobacterium bifidum* (Kawasaki et al. 2009), Lactobacillus johnsonii and commensal vaginal microbiota, such as *Lactobacillus crispatus, Lactobacillus jensenii,* and *Lactobacillus gasseri* which were found to be associated with decreased prevalence of gonorrhea, bacterial vaginosis, and HIV infection.

Messaoudi et al. (2011) observed positive effects of *Lactobacillus helveticus* R0052 and *Bifidobacterium longum* R0175 on animal and human volunteer and confirmed the improvement in CFS symptoms by checklist of Hopkins Symptom and the Hospital Anxiety and Depression Scale (Messaoudi et al. 2011). One study suggested that consumption of yogurt is helpful in bad mood and can alter psychological behavior (Benton et al. 2007). Similarly, consumption of fermented milk products also altered emotional responses of brain (Tillisch et al. 2012). Furthermore, non-digestible components of food also played similar role to probiotics and are also grouped under this category. Opie et al. (2017) reported that intake of plant based food items and fermented foods as prebiotic improved mental health as well as gut health (Opie et al. 2017). As proposed by several researchers, use of probiotics could be the potential therapeutic option in the CFS management.

9.8.1.1 Fecal Microbiota Transplantation (FMT)

Recent investigations of FMT have received a keen attention for the treatment of CFS symptoms. In this regard, Kenyon et al. reported FMT treatment in 42 CFS patient and observed promising result with FMT treatment suggesting it as a safer and promising approach in patients having CFS and IBS (Kenyon et al. 2019). Another animal study showed similar results on microbiome transplantation from NIH Swiss mice to BALB/c mice (Bercik et al. 2011). Additionally, Diaz Heijtz et al. (2011) noted reduction in stress, alteration in mood and increased motor neuron activity in GF mice as compared to pathogen free mice (Heijtz et al. 2011). One human study showed that on FMT treatment in 60 CFS patients, the response rate was 70%. After 15–20 years, data was collected from 12 patients which revealed their full recovery and 5 patients experienced CFS symptoms between 1.5 and 3 years (Borody et al. 2012). A similar study has been done on 58-year-old CFS patient which showed that on FMT treatment fecal streptococcus levels decreased from 26.39% to 0.15% and *Bifidobacterium* levels increased from 0% to 5.23% (Thurm et al. 2017). These studies showed the effectiveness of FMT in combating CFS by acting as good medical intervention.

9.9 Clinical Implications

Numerous preclinical studies evidenced the bidirectional interaction between gut microbiota and central nervous system, such interaction has been demonstrated in irritable bowel syndrome (Spiller 2008), intestinal pain (Rousseaux et al. 2007), *and* in obesity (Turnbaugh et al. 2006). A study reported that approximately 90% ME/CFS patients suffered from irritable bowel syndrome (IBS) and showed a strong association between the ME/CFS and IBS which was due to the increased levels of inflammatory mediators such as cytokines, interleukin, and TNFα (Quigley 2011a, b; Aaron et al. 2000). Likewise, numerous CFS patients reported previous diagnosis of IBS related symptoms (Aaron et al. 2000) and showed overlapping pathophysiology (Scully et al. 2010; Gomborone et al. 1996). Clinical trials on IBS showed limited proofs on improvement of IBS symptoms by the administration of probiotics and prebiotics (Spiller 2008; Moayyedi et al. 2010) which suggested possible role of commensal bacteria in the etiology of IBS but sufficient data is required to support the role of probiotics on the symptoms of IBS. As already discussed, stress is a risk factor for chronic fatigue, it increases the levels of norepinephrine in the gut which increases the pathogenicity of certain microorganisms such as *Campylobacter jejuni* (Cogan et al. 2007), *Salmonella* (Bailey et al. 1999), *and Escherichia coli* (Hughes and Sperandio 2008) which worsens the symptoms of enteritis. Various studies on germ free animals exposed to probiotics, bacteria, or antibiotics showed negative impacts on neurodevelopment and evidenced the connection between gut and central nervous system (Cryan et al. 2012) (Table 9.1). Reports suggested that administration of antibiotics, probiotics, and prebiotics can be beneficial in the treatment of IBD (Vanderpool et al. 2008) and IBS (Spiller 2008) by altering the commensal microbial flora which is supported by a placebo-controlled randomized study of IBS patients administered with preparation containing *Bifidobacterium lactis* that showed improvement in the symptoms of IBS (Scherzer 2017) (Table 9.1). Germ free mice that have imbalance of gut microbiome showed various neurological dysfunctions such as memory loss, impaired learning and recognition, and altered behavior (Gareau et al. 2011) by interfering with neurotransmitters (5-HT, BDNF and NMDA) (Bercik et al. 2011; Heijtz et al. 2011) (Table 9.1). Many studies on humans showed an interconnection between gut microbiota pathology and neuropsychiatric symptoms such as depression, anxiety, and autism (Foster et al. 2017) (Table 9.1). Moreover, gut microbiota has shown to influence central nervous system development (Tremlett et al. 2017) (Table 9.1).

Table 9.1 Clinical and preclinical implications of probiotics

Study participants	Probiotic used	Neurological symptoms evaluated	Reference
Germ free animals	Probiotics, bacteria, or antibiotics	Negative impacts on neurodevelopment	Cryan et al. (2012)
IBD and IBS patients	*Bifidobacterium lactis*	Improvement in the symptoms of IBS and IBD	Agrawal et al. (2009)
Healthy women	Fermented milk product containing probiotic (FMPP)	Altered emotional behavior	Tillisch et al. (2013)
Women with 14–16 weeks gestation	*Lactobacillus rhamnosus* HN001	Significantly reduced postpartum depression and anxiety	Slykerman et al. (2017)
IBS patients	*Bifidobacterium longum* NCC3001	Decreased depression and improved quality of life	Pinto-sanchez et al. (2017)
People with low mood	*Lactobacillus helveticus* and *Bifidobacterium longum*	No significant effects on low mood and inflammatory biomarkers	Romijn et al. (2017)
Patients suffering from major depressive disorder (MDD)	*Lactobacillus acidophilus*, *lactobacillus casei*, *Bifidobacterium bifidum*	Beneficial effects on MDD and improved insulin sensitivity	Jafari et al. (2015)
Healthy medical students having academic stress	*Lactobacillus casei* strain Shirota	Oversecretion of cortisol and reduced stress	Takada et al. (2016)
Healthy volunteers	*Bifidobacterium longum* 1714	Reduced stress and improved memory.	Allen et al. (2016)
Male volunteers	*Lactobacillus rhamnosus* JB-1	No significant effect on stress, mood, anxiety, or sleep quality	Kelly et al. (2016)
Older adults	*Lactobacillus reuteri*	No beneficial effects on stress, anxiety, and well-being	Östlund-lagerström et al. (2015)
Healthy volunteers	Prebiotic administration (fructooligosaccharides or Bimuno-galactooligosaccharides)	Showed anxiolytic effects	Schmidt et al. (2015)
Healthy volunteers	Rifaximin, an antibiotic	Stress reducing effects similar to probiotics	Wang et al. (2018)
Mice	Fructooligosaccharides and Bimuno-galactooligosaccharides	Beneficial effects on stress related behaviors	Burokas et al. (2017)
Rat	*Bifidobacterium* infantis	Reversed behavioral deficits	Desbonnet et al. (2010)
Mice	Human milk containing naturally occurring prebiotics	Reduced stress and anxiety	Tarr et al. (2015)

A clinical study by Tillisch and his team reported that ingestion of probiotic in healthy women affected brain regions and altered emotional behavior (Tillisch et al. 2013). A randomized, double-blind, placebo-controlled study on women evaluated the effects of probiotic (*Lactobacillus rhamnosus* HN001) on postnatal mood, evidenced that probiotics significantly reduced postpartum depression and anxiety (Slykerman et al. 2017). Another prospective study to evaluate the effects of probiotics on patients with IBS showed that probiotics (*Bifidobacterium longum* NCC3001) decreased depression and improved the quality of life (Pinto-sanchez et al. 2017) (Table 9.1). Another trial to investigate the effects of probiotics preparation (*Lactobacillus helveticus* and *Bifidobacterium longum*) on stress, mood, and anxiety showed that probiotics exhibited no significant effects on low mood and inflammatory biomarkers (Amy et al. 2017). A clinical trial to evaluate the effects of probiotics on patients suffering from major depressive disorder (MDD) showed that probiotics exhibited beneficial effects on MDD and improved insulin sensitivity (Jafari et al. 2015) (Table 9.1). A randomized clinical study to evaluate the effects of probiotic (*Lactobacillus casei* strain Shirota) on healthy medical students having academic stress showed that probiotic administration prevented over secretion of cortisol and reduced stress (Takada et al. 2016) (Table 9.1). A clinical study on healthy volunteers to examine the effect of probiotic (*Bifido bacterium longum* 1714) showed that probiotics reduced stress and improved memory (Allen et al. 2016) (Table 9.1). Another preclinical study on male volunteers to examine the effect of probiotic (*Lactobacillus rhamnosus* JB-1) showed that probiotic administration does not exhibit any significant effect on stress, mood, anxiety or sleep quality (Kelly et al. 2016) (Table 9.1). Another controlled clinical trial on older adults showed that probiotic (*Lactobacillus reuteri*) administration exhibited no beneficial effects on stress, anxiety, and well-being (Östlund-lagerström et al. 2015) (Table 9.1). A clinical trial on healthy volunteers showed that prebiotic administration (fructooligosaccharides or Bimuno-galactooligosaccharides) showed anxiolytic effects (Schmidt et al. 2015) (Table 9.1). A clinical study on healthy volunteers showed that rifaximin, an antibiotic exhibited stress reducing effects similar to probiotics (Wang et al. 2018) (Table 9.1). A preclinical study on mice fed with prebiotic (fructooligosaccharides and Bimuno-galactooligosaccharides) suggested the beneficial effects on stress related behaviors (Burokas et al. 2017) (Table 9.1). A preclinical study of rat showed that probiotic (*Bifidobacterium infantis*) treatment reversed behavioral deficits (Desbonnet et al. 2010) (Table 9.1). Another preclinical study of prebiotics on mice who were fed with human milk containing naturally occurring prebiotics showed reduced stress and anxiety (Tarr et al. 2015) (Table 9.1). These preclinical studies evidence the role of gut microbiome in regulating brain development in rodents, however, this relationship is debated in human beings and more controlled, well designed clinical data is required to confirm the effectiveness of probiotics and prebiotics in treatment of neuropsychiatric deficits such as stress, anxiety, and depression. The interconnection of gut microbiota and brain axis in early developmental life is unexplored and need more clinical data.

Some preclinical studies suggested that alteration in brain–gut microbiota interactions can have a role in the pathogenesis of neurodegenerative disorders such as autism (Vuong and Hsiao 2017), Alzheimer's (Vogt et al. 2017), Parkinson (Sampson et al. 2016), attention-deficit hyperactivity disorder (Aarts et al. 2017), epilepsy (Olson et al. 2018), and stroke (Wen and Wong 2017).

9.10 Concluded Remarks and Future Prospective

In this chapter, we have tried to establish the symbiotic connection between gut microbiome and CFS. Experimental studies on human volunteers showed impacts of commensal intestinal microflora on behavior and cerebrum that are contextually important and biologically noteworthy. Numerous diseased states including neurological conditions are attached to dysbiosis of human gut microbiota networks in humans. There are various gut microbiota composition risk factors that are responsible for the intestinal and mental disturbance. Modification in the normal composition of gut microbiota intercedes normal functioning of brain as suggested in animal and clinical investigations. Alteration of gut microbiota by certain agents opens another promising strategy for stress related issues, especially in gastrointestinal disorders such as inflammatory bowel syndrome. Dietary patterns additionally alter the microbial composition and function, in many ways that can differ among people, societies, and geographical areas. Probiotics, prebiotics, FMT, and fermented foods such as yogurt may impact the effect of the gut microbiome on the central nervous system and have indicated critical consequences on cerebrum functions in various preliminaries and clinical examinations. Furthermore, as the utilization of probiotics is developing exponentially, there is an urge to investigate the long-term safety of such therapeutic medications.

In ME/CFS, the immune system, metabolism, HPA axis, and mucosal dysfunction are altogether connected by the action of microbiota. A more noteworthy focus is necessary to study these interconnected frameworks which will require the expanded coordinated efforts between discrete research groups.

The primary aim of this review is to analyze the present proof for modifications in the gut microbiota by demonstrating the pathomechanism of CFS. Also, an auxiliary point looked to decide if there were any relationship between gut dysbiosis and CFS. The discoveries showed that the present proof is in confliction and we can't draw any noteworthy connection between gut dysbiosis and pathomechanism of CFS. This stresses the requirement for explicit clinical criteria to be utilized when diagnosing the condition that influence and impact CFS. Based on the current information mentioned in this review, the usefulness of modifying the gut microbiota in the treatment therapy of CFS is yet to be confirmed.

References

Aaron LA, Burke MM, Buchwald D (2000) Overlapping conditions among patients with chronic fatigue syndrome, fibromyalgia, and temporomandibular disorder. Arch Intern Med 160(2): 221–227. https://doi.org/10.1001/archinte.160.2.221

Aarts E, Ederveen THA, Naaijen J, Zwiers MP, Boekhorst J, Timmerman HM et al (2017) Gut microbiome in ADHD and its relation to neural reward anticipation. PLoS One 12(9):1–17

Acheson ED (1959) The clinical syndrome variously called benign myalgic encephalomyelitis, Iceland disease and epidemic neuromyasthenia. Am J Med 26(4):569–595. https://doi.org/10.1016/0002-9343(59)90280-3

Agrawal A, Houghton LA, Morris J, Reilly B, Guyonnet D, Goupil FN et al (2009) Clinical trial: the effects of a fermented milk product containing Bifidobacterium lactis DN-173 010 on abdominal distension and gastrointestinal transit in irritable bowel syndrome with constipation. Aliment Pharmacol Ther 29:104–114. https://doi.org/10.1111/j.1365-2036.2008.03853.x

Agudelo LZ, Femenía T, Orhan F, Porsmyr-Palmertz M, Goiny M, Martinez-Redondo V, Ruas JL et al (2014) Skeletal muscle PGC-1α1 modulates kynurenine metabolism and mediates resilience to stress-induced depression. Cell 159(1):33–45. https://doi.org/10.1016/j.cell.2014.07.051

Allen AP, Hutch W, Borre YE, Kennedy PJ, Temko A, Boylan G, Clarke G et al (2016) Bifidobacterium longum 1714 as a translational psychobiotic: modulation of stress, electrophysiology and neurocognition in healthy volunteers. Nat Publ Group 6(11):e939. https://doi.org/10.1038/tp.2016.191

Amy RR. Julia JR, Roeline GK, Frampton C (2017) A double-blind, randomized, placebocontrolled trial of lactobacillus helveticus and Bifidobacterium longum for the symptoms of depression. Aust N Z J Psychiatry 51(8):810–821. https://doi.org/10.1177/0004867416686694

Bailey MT, Karaszewski JW, Lubach GR, Coe CL, Lyte M, Al BET (1999) In vivo adaptation of attenuated salmonella typhimurium results in increased growth upon exposure to norepinephrine. Psychol Behav 67(3):359–364

Bailey MT, Lubach GR, Coe CL (2004) Prenatal stress alters bacterial colonization of the gut in infant monkeys. J Pediatr Gastroenterol Nutr 38(4):414–421. https://doi.org/10.1097/00005176-200404000-00009

Bailey MT, Dowd SE, Galley JD, Hufnagle AR, Allen RG, Lyte M (2011) Exposure to a social stressor alters the structure of the intestinal microbiota: implications for stressor-induced immunomodulation. Brain Behav Immun 25(3):397–407. https://doi.org/10.1016/j.bbi.2010.10.023

Barlow S, Chesson A, Collins JD, Dybing E, Flynn A, Fruijtier-Pölloth C, Neindre L et al (2007) Introduction of a qualified presumption of safety (QPS) approach for assessment of selected microorganisms referred to EFSA. Opinion of the Scientific Committee. EFSA J 587:1–16

Barrett E, Ross RP, O'Toole PW, Fitzgerald GF, Stanton C (2012) γ-aminobutyric acid production by culturable bacteria from the human intestine. J Appl Microbiol 113(2):411–417. https://doi.org/10.1111/j.1365-2672.2012.05344.x

Bazinet RP, Layé S (2014) Polyunsaturated fatty acids and their metabolites in brain function and disease. Nat Rev Neurosci 15(12):771–785. https://doi.org/10.1038/nrn3820

Benton D, Williams C, Brown A (2007) Impact of consuming a milk drink containing a probiotic on mood and cognition. Eur J Clin Nutr 61(3):355–361. https://doi.org/10.1038/sj.ejcn.1602546

Bercik P, Verdu EF, Foster JA, MacRi J, Potter M, Huang X, Collins SM et al (2010) Chronic gastrointestinal inflammation induces anxiety-like behavior and alters central nervous system biochemistry in mice. Gastroenterology 139(6):2102–2112. https://doi.org/10.1053/j.gastro.2010.06.063

Bercik P, Denou E, Collins J, Jackson W, Lu J, Jury J, Collins SM et al (2011) The intestinal microbiota affect central levels of brain-derived neurotropic factor and behavior in mice. Gastroenterology 141(2):599–609. https://doi.org/10.1053/j.gastro.2011.04.052

Bermon S, Petriz B, Kajeniene A, Prestes J, Castell L, Franco OL (2015) The microbiota: an exercise immunology perspective. Exerc Immunol Rev 21:70–79

Bohórquez DV, Shahid RA, Erdmann A, Kreger AM, Wang Y, Calakos N, Liddle RA et al (2015) Db_Jci78361. J Clin Invest 125(2):782–786. https://doi.org/10.1172/JCI78361DS1

Boonaert CJP, Rouxhet PG (2000) Surface of lactic acid bacteria: relationships between chemical composition and physicochemical properties surface of lactic acid bacteria : relationships between chemical composition and physicochemical properties. Appl Environ Microbiol 66(6):2548–2554. https://doi.org/10.1128/AEM.66.6.2548-2554.2000

Borody T, Nowak A, Finlayson S (2012) The GI microbiome and its role in chronic fatigue syndrome_ a summary of bacteriotherapy. J Australas Coll Nutr Env 31:3–8

Brown B (2014) Chronic fatigue syndrome: a personalized. Altern Ther Health Med 20:29–40

Buchwald D, Cheney PR, Peterson DL, Henry B, Wormsley SB, Geiger A, Komaroff AL et al (1992) A chronic illness characterized by fatigue, neurologic and immunologic disorders, and active human Herpesvirus Type 6 infection. Ann Intern Med 116(2):103–113

Burokas A, Arboleya S, Moloney RD, Peterson VL, Murphy K, Clarke G et al (2017) Archival report targeting the microbiota-gut-brain axis: prebiotics have anxiolytic and antidepressant-like effects and reverse the impact of chronic stress in mice. Biol Psychiatry 82(7):472–487. https://doi.org/10.1016/j.biopsych.2016.12.031

Butt H, Dunstan R, McGregor RN, Harrison T, Grainger TJ (1998) Faecal microbial growth inhibition in chronic fatigue/pain patients. In: Proceedings of the AHMF International Clinical and Scientific Meeting, pp 12–13

Butt H, Dunston R, McGregor N, Roberts T (2001) Bacterial colonosis' in patients with persistent fatigue. In: Proceedings of the AHMF International Clinical and Scientific Meeting, pp 1–2

Cani PD, Knauf C (2016) How gut microbes talk to organs: the role of endocrine and nervous routes. Mol Metab 5(9):743–752. https://doi.org/10.1016/j.molmet.2016.05.011

Cani PD, Everard A, Duparc T (2013) Gut microbiota, enteroendocrine functions and metabolism. Curr Opin Pharmacol 13(6):935–940. https://doi.org/10.1016/j.coph.2013.09.008

Castro-Marrero J, Faro M, Aliste L, Sáez-Francàs N, Calvo N, Martínez-Martínez A, Alegre J et al (2017) Comorbidity in chronic fatigue syndrome/Myalgic encephalomyelitis: a Nationwide population-based cohort study. Psychosomatics 58(5):533–543. https://doi.org/10.1016/j.psym.2017.04.010

Chalder T, Godfrey E, Ridsdale L, King M, Wessely S (2003) Predictors of outcome in a fatigued population in primary care following a randomized controlled trial. Psychol Med 33:283–287

Chen C, Brown DR, Xie Y, Green BT, Lyte M (2003) Catecholamines modulate Escherichia coli O157:H7 adherence to murine cecal mucosa. Shock (Augusta, GA) 20(2):183–188. https://doi.org/10.1097/01.shk.0000073867.66587.e0

Christine H, Urs MN, Elizabeth M, Roumiana B, James FJ, William CR (2009) Childhood trauma and risk for chronic fatigue syndrome. Arch Gen Psychiatry 66(1):72–80

Clarke G, Grenham S, Scully P, Fitzgerald P, Moloney RD, Shanahan F, Cryan JF et al (2013) The microbiome-gut-brain axis during early life regulates the hippocampal serotonergic system in a sex-dependent manner. Mol Psychiatry 18(6):666–673. https://doi.org/10.1038/mp.2012.77

Cleare AJ (2004) The HPA axis and the genesis of chronic fatigue syndrome. Trends Endocrinol Metab 15(2):55–59. https://doi.org/10.1016/j.tem.2003.12.002

Cleare AJ, Miell J, Heap E, Sookdeo S, Young L, Malhi GS, O'Keane V (2001) Hypothalamo-pituitary-adrenal axis dysfunction in chronic fatigue syndrome, and the effects of low-dose hydrocortisone therapy. J Clin Endocrinol Metab 86(8):3545–3554

Cogan TA, Thomas AO, Rees LEN, Taylor AH, Jepson MA, Williams PH, Humphrey TJ et al (2007) Norepinephrine increases the pathogenic potential of campylobacter jejuni. Gut 56(8):1060–1065. https://doi.org/10.1136/gut.2006.114926

Compston ND (1957) An outbreak of encephalomyelitis in the Royal Free Hospital Group, London, in 1955. Br Med J 2:895–904

Cryan JF, Dinan TG, Bernard C, Pavlov I, Beaumont W, James W, Charles E et al (2012) Mind-altering microorganisms: the impact of the gut microbiota on brain and behaviour of the nineteenth century through the pioneering work. Nat Rev Neurosci 13(10):701. https://doi.org/10.1038/nrn3346

Culligan EP, Hill C, Sleator RD (2009) Probiotics and gastrointestinal disease: successes, problems and future prospects. Gut Pathog 1(1):1–12. https://doi.org/10.1186/1757-4749-1-19

Da Silva CP (2018) Whole body vibration methods with survivors of polio. J Vis Exp 140:1–10. https://doi.org/10.3791/58449

Dantzer R, Kelley KW, Adaptatifs C (1989) Stress and immunity: an integrated view of relationships between the brain and the immune system. Life Sci 44(26):1995–2008

Delia RU, Vecchiet J, Ceccomancini A, Ricci F, Pizzigallo E (2001) Chronic fatigue syndrome following a toxic exposure. Sci Total Environ 270(1–3):27–31

Den Besten G, Van Eunen K, Groen AK, Venema K, Reijngoud DJ, Bakker BM (2013) The role of short-chain fatty acids in the interplay between diet, gut microbiota, and host energy metabolism. J Lipid Res 54(9):2325–2340. https://doi.org/10.1194/jlr.R036012

Desbonnet L, Garrett L, Clarke G, Bienenstock J, Dinan TG (2008) The probiotic Bifidobacteria infantis: an assessment of potential antidepressant properties in the rat. J Psychiatr Res 43(2): 164–174. https://doi.org/10.1016/j.jpsychires.2008.03.009

Desbonnet L, Garrett L, Clarke G, Kiely B, Cryan JF, Dinan TG (2010) Effects of the probiotic Bifidobacterium infantis in the maternal separation model of depression. Neuroscience 170(4): 1179–1188. https://doi.org/10.1016/j.neuroscience.2010.08.005

Engel MA, Neurath MF (2010) New pathophysiological insights and modern treatment of IBD. J Gastroenterol 45(6):571–583. https://doi.org/10.1007/s00535-010-0219-3

Faist E, Schinkel C, Zimmer S (1996) Update on the mechanisms of immune suppression of injury and immune modulation. World J Surg 20(4):454–459

Forsythe P, Kunze WA (2013) Voices from within: gut microbes and the CNS. Cell Mol Life Sci 70(1):55–69. https://doi.org/10.1007/s00018-012-1028-z

Foster JA, Rinaman L, Cryan JF (2017) Neurobiology of stress & the gut-brain axis: regulation by the microbiome. Neurobiol Stress 7:124–136. https://doi.org/10.1016/j.ynstr.2017.03.001

Fremont M, Coomans D, Massart S, De Meirleir K (2013) High-throughput 16s rRNA gene sequencing reveals alterations of intestinal microbiota in myalgic encephalomyelitis/chronic fatigue syndrome patients. Anaerobe 22:50–56

Fritsche K (2006) Fatty acids as modulators of the immune response. Annu Rev Nutr 26(1):45–73. https://doi.org/10.1146/annurev.nutr.25.050304.092610

Fukuda K, Straus SE, Hickie I, Sharpe MC, Psych MRC (1994) The chronic fatigue syndrome: a comprehensive approach to its definition and study. Ann Intern Med 121(12):953–959

Fuller R (1989) Probiotics in man and animals. J Appl Bacteriol 66(5):365–378. https://doi.org/10.1111/j.1365-2672.1989.tb05105.x

Gaab J, Rohleder N, Heitz V, Engert V, Schad T, Schürmeyer TH, Ehlert U (2005) Stress-induced changes in LPS-induced pro-inflammatory cytokine production in chronic fatigue syndrome. Psychoneuroendocrinology 30(2):188–198. https://doi.org/10.1016/j.psyneuen.2004.06.008

Gareau MG, Wine E, Rodrigues DM, Cho JH, Whary MT, Philpott DJ, Sherman FM et al (2011) Bacterial infection causes stress-induced memory dysfunction in mice. Gut 60(3):307–317. https://doi.org/10.1136/gut.2009.202515

Goldin BR (1998) Health benefits of probiotics. Br J Nutr 80(S2):S203–S207. https://doi.org/10.1017/s0007114500006036

Gomborone JE, Gorard DA, Dewsnap PA, Libby GW, Farthing MJG (1996) Prevalence of irritable bowel syndrome in chronic fatigue. J R Coll Physicians Lond 30(6):512–513

Heijtz RD, Wang S, Anuar F, Qian Y, Björkholm B, Samuelsson A et al (2011) Normal gut microbiota modulates brain development and behavior. Proc Natl Acad Sci U S A 108(7): 3047–3052. https://doi.org/10.1073/pnas.1010529108

Higuchi T, Hayashi H, Abe K (1997) Exchange of glutamate and γ-aminobutyrate in a lactobacillus strain. J Bacteriol 179(10):3362–3364. https://doi.org/10.1128/jb.179.10.3362-3364.1997

Hoeck AD, Pall ML (2011) Will vitamin D supplementation ameliorate diseases characterized by chronic inflammation and fatigue? Med Hypotheses 76(2):208–213. https://doi.org/10.1016/j.mehy.2010.09.032

Hoge EA, Brandstetter K, Moshier S, Pollack MH, Wong KK, Simon NM (2009) Broad spectrum of cytokine abnormalities in panic disorder and posttraumatic. Depress Anxiety 26(5):447–455. https://doi.org/10.1002/da.20564

Holzer P, Farzi A (2014) Neuropeptides and the microbiota gut-brain axis. In: Cryan JF (ed) Advances in experimental medicine and biology. Springer, New York, pp 195–219

Holmes GP, Kaplan JE, Gantz NM, Komaroff AL, Schonberger LB, Straus SE, Jones JF, Dubois RE, Cunningham-Rundles C, Pahwa S et al (1988) Chronic fatigue syndrome: a working case definition. Ann Intern Med 108(3):387–389

Hughes DT, Sperandio V (2008) Inter-kingdom signalling: communication between bacteria and their hosts. Nat Rev Microbiol 6(2):111–120. https://doi.org/10.1038/nrmicro1836

Inger Johanne B, Kari T, Nina G, Sara G, Camilla S, Lill T et al (2014) Two age peaks in the incidence of chronic fatigue syndrome/myalgic encephalomyelitis: a population-based registry study from Norway 2008–2012. BMC Med 12(1):167. https://doi.org/10.1186/s12916-014-0167-5

Jafari P, Akbari H, Taghizadeh M (2015) Clinical and metabolic response to probiotic administration in patients with major depressive disorder: a randomized, double-blind, placebo-controlled trial. Nutrition 32(3):315–320. https://doi.org/10.1016/j.nut.2015.09.003

Jason LA, Richman JA, Rademaker AW, Jordan KM (1999) A community-based study of chronic fatigue syndrome. Arch Intern Med 159(18):2129–2136

Jerjes WK, Taylor NF, Wood PJ, Cleare AJ (2007) Enhanced feedback sensitivity to prednisolone in chronic fatigue syndrome. Psychoneuroendocrinology 32(2):192–198. https://doi.org/10.1016/j.psyneuen.2006.12.005

Joseph HB, Jack DT, David CS, Roberta AM, Dennis H (2013) Detection of mycotoxins in patients with chronic fatigue syndrome. Toxins 5(4):605–617. https://doi.org/10.3390/toxins5040605

Kailasapath K, James C (2000) Survival and therapeutic potential of probiotic organisms with reference to lactobacillus acidophilus and Bifidobacterium spp. Immunol Cell Biol 78(1):80–88

Kakanis MW, Peake J, Brenu EW, Simmonds M, Gray B (2010) The open window of susceptibility to infection after acute exercise in healthy young male elite athletes. J Sci Med Sport 13:119–137

Kawasaki S, Satoh T, Todoroki M, Niimura Y (2009) b-Type dihydroorotate dehydrogenase is purified as a H_2O_2 forming NADH Oxidase from Bifidobacterium bifidum. Appl Environ Microbiol 75(3):629–636. https://doi.org/10.1128/AEM.02111-08

Kelly JR, Allen AP, Temko A, Hutch W, Paul J, Farid N et al (2016) Lost in translation? The potential psychobiotic lactobacillus rhamnosus (JB-1) fails to modulate stress or cognitive performance in healthy male subjects. Brain Behav Immun 61:50–59. https://doi.org/10.1016/j.bbi.2016.11.018

Kenyon JN, Coe S, Izadi H (2019) A retrospective outcome study of 42 patients with chronic fatigue syndrome, 30 of whom had irritable bowel syndrome. Half were treated with oral approaches, and half were treated with Faecal microbiome transplantation. Hum Microbiome J 13:100061. https://doi.org/10.1016/j.humic.2019.100061

Klein G, Hallmann C, Casas IA, Abad J, Louwers J, Reuter G (2000) Exclusion of vanA, vanB and vanC type glycopeptide resistance in strains of lactobacillus reuteri and lactobacillus rhamnosus used as probiotics by polymerase chain reaction and hybridization methods. J Appl Microbiol 89(5):815–824. https://doi.org/10.1046/j.1365-2672.2000.01187.x

Komaroff AL (1996) An examination of the working case definition of chronic fatigue syndrome. Am J Med 100(1):56–64. https://doi.org/10.1016/S0002-9343(96)90012-1

Komaroff AL, Buchwald D (1991) Symptoms and signs of chronic fatigue syndrome. Rev Infect Dis 13:S8–S11. https://doi.org/10.1093/clinids/13.Supplement_1.S8

Konsoula Z, Barile FA (2012) Epigenetic histone acetylation and deacetylation mechanisms in experimental models of neurodegenerative disorders. J Pharmacol Toxicol Methods 66(3):215–220. https://doi.org/10.1016/j.vascn.2012.08.001

Lee KJ, Tack J (2010) Altered intestinal microbiota in irritable bowel syndrome. Neurogastroenterol Motil 22(5):493–498. https://doi.org/10.1111/j.1365-2982.2010.01482.x

Liam O, Mübeccel A, Cezmi AA (2011) Probiotic lactobacillus rhamnosus downregulates FCER1 and HRH4 expression in human mast cells. J Allergy Clin Immunol 128(6):1153–1162

Litleskare S, Rortveit G, Eide GE, Hanevik K, Langeland N, Wensaas KA (2018) Prevalence of irritable bowel syndrome and chronic fatigue 10 years after giardia infection. Clin Gastroenterol Hepatol 16(7):1064–1072.e4. https://doi.org/10.1016/j.cgh.2018.01.022

Logan AC, Rao AV, Irani D (2003) Chronic fatigue syndrome: lactic acid bacteria may be of therapeutic value. Med Hypotheses 60(6):915–923. https://doi.org/10.1016/S0306-9877(03) 00096-3

Maes M, Mihaylova I, Leunis J (2005) In chronic fatigue syndrome, the decreased levels of omega-3 poly-unsaturated fatty acids are related to lowered serum zinc and defects in T cell activation. Neuro Endocrinol Lett 26(6):745–751

Maes M, Mihaylova I, Leunis JC (2007) Increased serum IgA and IgM against LPS of enterobacteria in chronic fatigue syndrome (CFS): indication for the involvement of gram-negative enterobacteria in the etiology of CFS and for the presence of an increased gut-intestinal permeability. J Affect Disord 99(1–3):237–240. https://doi.org/10.1016/j.jad.2006.08.021

Mahgoub M, Monteggia LM (2013) Epigenetics and psychiatry. Neurotherapeutics 10(4):734–741. https://doi.org/10.1007/s13311-013-0213-6

Martinowich K, Manji H, Lu B (2007) New insights into BDNF function in depression and anxiety. Nat Neurosci 10(9):1089–1093. https://doi.org/10.1038/nn1971

Mawle AC, Nisenbaum R, Dobbins JG, Gary HE, Stewart JA, Reyes M, Reeves WC et al (1995) Seroepidemiology of chronic fatigue syndrome: a case-control study. Clin Infect Dis 21(6): 1386–1389

Messaoudi M, Violle N, Bisson JF, Desor D, Javelot H, Rougeot C (2011) Beneficial psychological effects of a probiotic formulation (lactobacillus helveticus R0052 and Bifidobacterium longum R0175) in healthy human volunteers. Gut Microbes 2(4):37–41. https://doi.org/10.4161/gmic.2. 4.16108

Mikelsaar M, Leibur E, Ko P, Marcotte H, Hammarstro L (2005) Oral lactobacilli in chronic periodontitis and periodontal health : species composition and antimicrobial activity. Oral Microbiol Immunol 20:354–361

Moayyedi P, Ford AC, Talley NJ, Cremonini F, Brandt LJ, Quigley EMM (2010) The efficacy of probiotics in the treatment of irritable bowel syndrome: a systematic review. Gut 59:325–333. https://doi.org/10.1136/gut.2008.167270

Naess H, Nyland M, Hausken T, Follestad I, Nyland HI (2012) Chronic fatigue syndrome after Giardia enteritis: clinical characteristics, disability and long-term sickness absence. BMC Gastroenterol 12(1):13. https://doi.org/10.1186/1471-230X-12-13

Newberry F, Hsieh SY, Wileman T, Carding SR (2018) Does the microbiome and virome contribute to myalgic encephalomyelitis/chronic fatigue syndrome? Clin Sci 132(5):523–542. https://doi.org/10.1042/CS20171330

Olson CA, Vuong HE, Yano JM, Liang QY, Nusbaum DJ, Hsiao EY, Hsiao EY et al (2018) The gut microbiota mediates the anti-seizure effects of the ketogenic diet article the gut microbiota mediates the anti-seizure effects of the ketogenic diet. Cell 173(7):1728–1741. https://doi.org/ 10.1016/j.cell.2018.04.027

Opie RS, Itsiopoulos C, Parletta N, Sanchez-Villegas A, Akbaraly TN, Ruusunen A, Jacka FN (2017) Dietary recommendations for the prevention of depression. Nutr Neurosci 20(3): 161–171. https://doi.org/10.1179/1476830515Y.0000000043

Östlund-Lagerström L, Kihlgren A, Repsilber D, Björkstén B, Brummer RJ, Schoultz I (2015) Probiotic administration among free-living older adults: a double blinded, randomized, placebo-controlled clinical trial. Nutr J 15(1):1–10

Pacini S, Fiore MG, Magherini S, Morucci G, Branca JJV, Gulisano M, Ruggiero M (2012) Could cadmium be responsible for some of the neurological signs and symptoms of myalgic encephalomyelitis/chronic fatigue syndrome. Med Hypotheses 79(3):403–407. https://doi.org/10.1016/j.mehy.2012.06.007

Pearce JMS (2006) The enigma of chronic fatigue. Eur Neurol 56(1):31–36. https://doi.org/10.1159/000095138

Pimentel M, Hallegua D, Chow E, Wallace D, Lin HC, Program GIM, Ctr CM (2000) Eradication of small intestinal bacterial overgrowth decreases symptoms in chronic fatigue syndrome: a double blind, randomized study. Gastroenterology 118(4):A414. https://doi.org/10.1016/S0016-5085(00)83765-8

Pineiro M, Embarek PB (2002) Guidelines for the evaluation of probiotics in food. London, Ontario, Canada

Pinto-sanchez MI, Hall GB, Ghajar K, Nardelli A, Bolino C, Lau JT et al (2017) Probiotic Bifidobacterium longum NCC3001 reduces depression scores and alters brain activity: a pilot study in patients with irritable bowel syndrome. Gastroenterology 153(2):448–459. https://doi.org/10.1053/j.gastro.2017.05.003

Pose JR (1956) A new clinical entity? Lancet 268(6935):197. https://doi.org/10.1016/S0140-6736(56)91718-4

Poteliakhoff A (1981) Adrenocortical activity and some clinical findings in acute and chronic fatigue. J Psychosom Res 25(2):91–95. https://doi.org/10.1016/0022-3999(81)90095-7

Prasad J, Gill H, Smart J, Gopal PK (1999) Selection and characterisation of lactobacillus and Bifidobacterium strains for use as probiotics. Int Dairy J 8:993–1002

Prins JB, Bos E, Huibers MJH, Bleijenberg G, Servaes P, van der Werf SP, van der Meer JWM (2004) Social support and the persistence of complaints in chronic fatigue. Psychother Psychosom 73:174–182. https://doi.org/10.1159/000076455

Quigley EM (2011a) Gut microbiota and the role of probiotics in therapy. Curr Opin Pharmacol 11(6):593–603. https://doi.org/10.1016/j.coph.2011.09.010

Quigley EM (2011b) The enteric microbiota in the pathogenesis and management of constipation. Best Pract Res Clin Gastroenterol 25(1):119–126. https://doi.org/10.1016/j.bpg.2011.01.003

Raber J, Sorg O, Horn TFW, Yu N, Koob GF, Campbell IL, Bloom FE (1998) Inflammatory cytokines: putative regulators of neuronal and neuro- endocrine function. Brain Res Rev 26(2–3):320–326. https://doi.org/10.1016/S0165-0173(97)00041-6

Rao AV, Bested AC, Beaulne TM, Katzman MA, Iorio C, Berardi JM, Logan AC (2009) A randomized, double-blind, placebo-controlled pilot study of a probiotic in emotional symptoms of chronic fatigue syndrome. Gut Pathog 1(1):6. https://doi.org/10.1186/1757-4749-1-6

Remzi C, Ali G, Suat A, Kemal N, Ayşegül Jale S (2004) Hypothalamic-pituitary-gonadal axis hormones and cortisol in both menstrual phases of women with chronic fatigue syndrome and effect of depressive mood on these hormones. BMC Musculoskelet Disord 5:47

Romijn AR, Rucklidge JJ, Kuijer RG, Frampton C (2017) A double-blind, randomized, placebo-controlled trial of Lactobacillus helveticus and Bifidobacterium longum for the symptoms of depression. Aust N Z J Psychiatry 51(8):810–821. https://doi.org/10.1177/0004867416686694

Rousseaux C, Thuru X, Gelot A, Barnich N, Neut C, Dubuquoy L, Desreumaux P et al (2007) Lactobacillus acidophilus modulates intestinal pain and induces opioid and cannabinoid receptors. Nat Med 13(1):35–37. https://doi.org/10.1038/nm1521

Sampson TR, Debelius JW, Thron T, Wittung-stafshede P, Knight R, Mazmanian SK, Chesselet M et al (2016) Gut microbiota regulate motor deficits and neuroinflammation in a model of Parkinson's article gut microbiota regulate motor deficits and neuroinflammation in a model of Parkinson's disease. Cell 167(6):1469–1480. https://doi.org/10.1016/j.cell.2016.11.018

Sanders ME (2000) Considerations for use of probiotic bacteria to modulate human health. J Nutr 130(2):384S–390S. https://doi.org/10.1093/jn/130.2.384s

Sanders ME, Shane AL, Merenstein DJ (2016) Advancing probiotic research in humans in the United States: challenges and strategies. Gut Microbes 7(2):97–100. https://doi.org/10.1080/19490976.2016.1138198

Scherzer B (2017) Bowel in turmoil: when symptoms turn life into agony. Deutsche Apotheker Zeitung 157(13):237–293. https://doi.org/10.1111/j.1365-2036.2008.03853.x

Schmidt K, Cowen PJ, Harmer CJ, Tzortzis G, Errington S, Burnet PW (2015) Prebiotic intake reduces the waking cortisol response and alters emotional bias in healthy volunteers. Psychopharmacology 232(10):1793–1801

Schröcksnadel K, Wirleitner B, Winkler C, Fuchs D (2006) Monitoring tryptophan metabolism in chronic immune activation. Clin Chim Acta 364(1–2):82–90. https://doi.org/10.1016/j.cca.2005.06.013

Scully P, McKernan DP, Keohane J, Groeger D, Shanahan F, Dinan TG, Quigley EMM (2010) Plasma cytokine profiles in females with irritable bowel syndrome and extra-intestinal co-morbidity. Am J Gastroenterol 105(10):2235–2243. https://doi.org/10.1038/ajg.2010.159

Senan S, Prajapati JB, Joshi CG (2015) Feasibility of genome-wide screening for biosafety assessment of probiotics: a case study of lactobacillus helveticus MTCC 5463. Probiot Antimicrob Prot 7(4):249–258. https://doi.org/10.1007/s12602-015-9199-1

Sharpe M (2002) The report of the chief medical Officer's CFS/ME working group: what does it say and will it help? Clin Med 2(5):427–429

Sherwin E, Sandhu KV, Dinan TG, Cryan JF (2016) May the force be with you: the light and dark sides of the microbiota–gut–brain Axis in neuropsychiatry. CNS Drugs 30(11):1019–1041. https://doi.org/10.1007/s40263-016-0370-3

Slykerman RF, Hood F, Wickens K, Thompson JMD, Barthow C, Murphy R et al (2017) EBioMedicine effect of lactobacillus rhamnosus HN001 in pregnancy on postpartum symptoms of depression and anxiety: a randomised double-blind placebo-controlled trial. EBioMedicine 24:159–165. https://doi.org/10.1016/j.ebiom.2017.09.013

Song C, Lin A, Bonaccorso S, Heide C, Verkerk R, Kenis G et al (1998) The inflammatory response system and the availability of plasma tryptophan in patients with primary sleep disorders and major depression. J Affect Disord 49(3):211–219. https://doi.org/10.1016/S0165-0327(98)00025-1

Spellberg B, Edwards JE (2001) Type 1/type 2 immunity in infectious diseases. Clin Infect Dis 32(1):76–102

Spiller R (2008) Review article: probiotics and prebiotics in irritable bowel syndrome. Aliment Pharmacol Ther 28(4):385–396. https://doi.org/10.1111/j.1365-2036.2008.03750.x

Sprake EF, Grant VA, Corfe BM (2012) Vitamin D3 as a novel treatment for irritable bowel syndrome: single case leads to critical analysis of patient-centred data. BMJ Case Rep December:10–13. https://doi.org/10.1136/bcr-2012-007223

Sudo N, Chida Y, Aiba Y, Sonoda J, Oyama N, Yu XN et al (2004) Postnatal microbial colonization programs the hypothalamic-pituitary-adrenal system for stress response in mice. J Physiol 558(1):263–275. https://doi.org/10.1113/jphysiol.2004.063388

Tahmaz N, Soutar A, Cherrie JW (2003) Chronic fatigue and organophosphate pesticides in sheep farming: a retrospective study amongst people reporting to a UK pharmacovigilance scheme. Ann Occup Hyg 47(4):261–267. https://doi.org/10.1093/annhyg/meg042

Takada M, Nishida K, Kataoka-Kato A, Gondo Y, Ishikawa H, Suda K et al (2016) Probiotic lactobacillus casei strain Shirota relieves stress-associated symptoms by modulating the gut – brain interaction in human and animal models. Neurogastroenterol Motil 28:1027–1036. https://doi.org/10.1111/nmo.12804c

Tan J, McKenzie C, Potamitis M, Thorburn AN, Mackay CR, Macia L (2014) The role of short-chair fatty acids in health and disease. In: Advances in immunology, vol 121, 1st edn. Elsevier, Amsterdam, pp 91–119. https://doi.org/10.1016/B978-0-12-800100-4.00003-9

Tarr AJ. Galley JD, Fisher SE, Chichlowski M, Berg BM, Bailey MT (2015) The prebiotics 3′Sialyllactose and 6′Sialyllactose diminish stressor-induced anxiety-like behavior and colonic microbiota alterations: evidence for effects on the gut–brain axis. Brain Behav Immun 50:166–177. https://doi.org/10.1016/j.bbi.2015.06.025

Taverniti V, Guglielmetti S (2011) The immunomodulatory properties of probiotic microorganisms beyond their viability (ghost probiotics: proposal of paraprobiotic concept). Genes Nutr 6(3):261–274. https://doi.org/10.1007/s12263-011-0218-x

Thangam EB, Jemima EA, Singh H, Baig MS (2018) The role of histamine and histamine receptors in mast cell-mediated allergy and inflammation: the hunt for new therapeutic targets. Front Immunol 9:1873. https://doi.org/10.3389/fimmu.2018.01873

Thurm T, Ablin JN, Buskila D, Maharshak N (2017) Fecal microbiota transplantation for fibromyalgia: a case report and review of the literature. Open J Gastroenterol 07(04):131–139. https://doi.org/10.4236/ojgas.2017.74015

Tillisch K, Labus JS, Ebrat B, Stains J, Naliboff BD, Guyonnet D, Mayer EA et al (2012) 589 modulation of the brain-gut axis after 4-week intervention with a probiotic fermented dairy product. Gastroenterology 142(5):S-115. https://doi.org/10.1016/s0016-5085(12)60435-1

Tillisch K, Labus J, Kilpatrick L, Jiang Z, Stains J, Ebrat B et al (2013) Consumption of fermented Milk product with probiotic modulates brain activity. YGAST 144(7):1394–1401.e4. https://doi.org/10.1053/j.gastro.2013.02.043

Tremlett H, Bauer KC, Appel-cresswell S, Finlay BB, Waubant E (2017) The gut microbiome in human neurological disease : a review. Ann Neurol 81(13):369–382. https://doi.org/10.1002/ana.24901

Turnbaugh PJ, Ley RE, Mahowald MA, Magrini V, Mardis ER, Gordon JI (2006) An obesity-associated gut microbiome with increased capacity for energy harvest. Nature 444:1027–1031. https://doi.org/10.1038/nature05414

Van Heukelom RO, Prins JB, Smits MG, Bleijenberg G (2006) Influence of melatonin on fatigue severity in patients with chronic fatigue syndrome and late melatonin secretion. Eur J Neurol 4: 55–60

Vanderpool C, Yan F, Polk DB (2008) Mechanisms of probiotic action : implications for therapeutic applications in inflammatory bowel diseases. Inflamm Bowel Dis 14(11):1585–1596. https://doi.org/10.1002/ibd.20525

Vogt NM, Kerby RL, Dill-mcfarland KA, Harding SJ, Merluzzi AP, Johnson SC, Rey FE et al (2017) Gut microbiome alterations in Alzheimer's disease. Sci Rep 7(1):13537. https://doi.org/10.1038/s41598-017-13601-y

Vuong HE, Hsiao EY (2017) Emerging roles for the gut microbiome in autism spectrum disorder the microbiome in ASD emerging roles for the gut microbiome in autism spectrum disorder. Biol Psychiatry 81(5):411–423. https://doi.org/10.1016/j.biopsych.2016.08.024

Wallis A, Ball M, Butt H, Lewis DP, McKechnie S, Paull P et al (2018) Open-label pilot for treatment targeting gut dysbiosis in myalgic encephalomyelitis/chronic fatigue syndrome: neuropsychological symptoms and sex comparisons. J Transl Med 16(1):1–16. https://doi.org/10.1186/s12967-018-1392-z

Wang H, Braun C, Enck P, Braun C (2018) Effects of Rifaximin on central responses to social stress—a pilot experiment. Neurotherapeutics 15(3):807–818

Wen SW, Wong CHY (2017) An unexplored brain-gut microbiota axis in stroke. Gut Microbes 8(6):601–606. https://doi.org/10.1080/19490976.2017.1344809

Wessely S, Hotopf M, Sharpe M (1998) Chronic fatigue and its syndromes. Oxford University Press, Oxford

Whistler T, Jones JF, Unger ER, Vernon SD (2005) Exercise responsive genes measured in peripheral blood of women with chronic fatigue syndrome and matched control subjects _ BMC physiology _ full text. BMC Physiol 5(1):5

Zhang L, Gough J, Christmas D, Mattey DL, Richards SCM, Main J, Kerr JR et al (2010) Microbial infections in eight genomic subtypes of chronic fatigue syndrome / myalgic encephalomyelitis. J Clin Pathol 63:156–164. https://doi.org/10.1136/jcp.2009.072561

Zhou QQ, Zhang B, Nicholas Verne G (2009) Intestinal membrane permeability and hypersensitivity in the irritable bowel syndrome. Pain 146(1–2):41–46. https://doi.org/10.1016/j.pain.2009.06.017

Animal Models Used for Studying the Benefits of Probiotics in Neurodegeneration

G. Divyashri and S. G. Prapulla

Abstract

Neurodegenerative sicknesses, viz., Alzheimer's disease, Parkinson's disease, Huntington's disease, amyotrophic lateral sclerosis, frontotemporal dementia, and the spinocerebellar ataxias pose a significant risk to human well-being. These disorders are known to progress inexorably to severe inability and death. Presently, about 30 million people in India are known to suffer from one or the other forms of neurodegenerative disorders with an average prevalence rate of 2394 patients per 1,00,000 of the population. Probiotics are defined as "live microorganisms which, when administered in adequate amounts, confer health benefits on the host." A novel class of probiotics called "Psychobiotics," a group of probiotics that has ability to affect the central nervous system (CNS) and its related functions and behaviors are currently reviewed by scientific community for its potential application in the treatment of mental illness. An abundance of information has demonstrated that psychobiotics residing in gut play crucial roles in the prevention and treatment of various neurodegenerative disorders. Emerging evidences demonstrate that these psychobiotics protect CNS by positively modulating gut–brain axis (GBA) *via* immune, humoral, neural, and metabolic pathways. Furthermore, preclinical trials involving animal models have claimed the therapeutic benefit of several psychobiotics. However, recognizing appropriate animal model is vital for evaluating the therapeutic efficiency of psychobiotics. Thus, this chapter outlines the advantages and challenges of current animal models and discusses future research directions of neurodegenerative disorders.

G. Divyashri
Department of Biotechnology, M S Ramaiah Institute of Technology, Bengaluru, Karnataka, India

S. G. Prapulla (✉)
Department of Microbiology and Fermentation Technology, CSIR-Central Food Technological Research Institute (CSIR-CFTRI), Mysuru, Karnataka, India

© Springer Nature Singapore Pte Ltd. 2022
P. K. Deol, S. K. Sandhu (eds.), *Probiotic Research in Therapeutics*,
https://doi.org/10.1007/978-981-16-6760-2_10

Keywords

Neurodegenerative disorders · Probiotics · Psychobiotics · Central nervous
system · Animal models

10.1   Introduction

Neurodegenerative disorders pose a significant risk to human well-being. In India,
about 30 million people experience from one or the other forms of neurodegenerative disorders with a reported occurrence rate of 2394 patients per 1 lakh population
(Gourie-Devi 2014). Microbiota colonizing gastrointestinal tract (GIT) have long
been associated with both gastrointestinal and extra-gastrointestinal disorders
(Umbrello and Esposito 2016). Numerous findings have demonstrated bidirectional
communication occurring between central nervous system (CNS) and the gut
through microbiota-gut-brain axis (Umbrello and Esposito 2016). Mammalian GIT
is a metabolically active organ comprising a diversity of microbial species. This
commensal intestinal microbial species is crucial as they protect the host against
infections and maintain the body's homeostasis under normal circumstances
(Divyashri et al. 2015). Scientific evidences have demonstrated the role of dynamic
changes in the gut microbiota altering brain behavior and physiology (Rezaei Asl
et al. 2019; Ma et al. 2019; Magistrelli et al. 2019). It is now becoming more evident
that the gut bacteria residing at GIT regulate and positively influence cognitive
dysfunction and other neurodegenerative disorders (Zhu et al. 2020). And, substantial alteration in the gut microbiota composition has been implicated in CNS
disorders, viz., depression, anxiety, autism, Parkinson's disease, Huntington's disease, Alzheimer's disease, frontotemporal dementia amyotrophic lateral sclerosis,
spinocerebellar ataxias, and neurodegeneration (Finegold et al. 2012; Fung et al.
2017; Liang et al. 2018). Modification of gut microbiota using probiotics is very
well-known in human and animal disease treatment (Azad et al. 2018). Probiotics are
live microorganisms when adminstered in adequate amounts confers various health
benefits on host. A novel class of probiotics called "Psychobiotics" are currently
reviewed by the scientific community for its potential application in the treatment of
neurodegenerative disorders (Dinan et al. 2013). Dinan et al. (2013) defined them as
a group of probiotics that has the ability to affect CNS and its related functions and
behaviors. Probiotic strains that are reported to act as psychobiotics include *Lactobacillus* sp., *Bifidobacterium* sp., *Enterococcus* sp., *Saccharomyces* sp. (Sanchez
et al. 2017) through the gut–brain axis (Liu et al. 2015a, b). Psychobiotic research is
validated by employing appropriate animal models through stress induction and
conduction of behavioral tests for evaluation of anxiety, motivation, and depression
abilities (Sarkar et al. 2016).

10.2 Relationship Between Probiotics and Gut–Brain Axis (GBA)

The bidirectional cross talk between gut microbiota and CNS is known to play an essential role in maintaining human health (Giau et al. 2018). A significant number of research findings suggest that the gut microbiota influences brain and its behavior (Sudo 2019). Cross talk between gut and brain is reported to occur via neurotransmitters, immunomodulation, enteric nervous system (ENS), and short chain fatty acids (SCFAs) (Burokas et al. 2015). Metabolites produced by this gut microbiota are known to alter cognitive ability of humans diagnosed with neurodegenerative disorders (Giau et al. 2018). Furthermore, neurotransmitter related metabolites (i.e., tryptophan and gamma-amino butyric acid (GABA)) (Briguglio et al. 2018; Lyte et al. 2018) and monoamines (i.e., histamine, dopamine, and serotonin) produced by probiotics residing at GIT may affect gut-to-brain communication and are known to influence brain function involving immune, endocrine, and neural pathways (vagus nerve and enteric nervous system) (Fig. 10.1). Accumulating evidences also suggest the communication between ENS and CNS occurs through the gut bacteria via vagus nerve (Bonaz et al. 2018). Few strains of probiotics are reported to produce considerable amount of neurotransmitters, GABA, and serotonin (Divyashri and Prapulla 2015; O'Mahony et al. 2015). In addition, *Lactobacillus* sp. viz., *L. plantarum,* and *L. odontolyticus* are also reported to produce acetylcholine (Roshchina 2016). Furthermore, recently it is shown that indigenous probiotic bacteria residing in the gut regulate the host serotonin biosynthesis in the brain (Yano et al. 2015). Thus, these probiotic strains are proven worthy

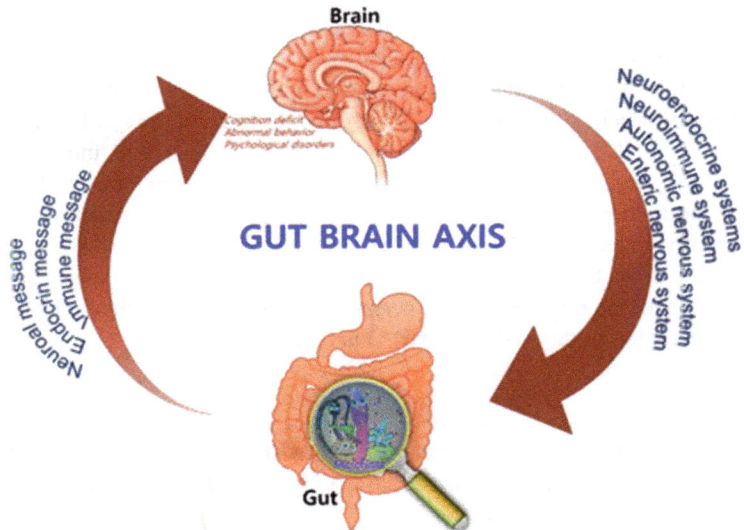

Fig. 10.1 Bidirectional cross talk between CNS and the gut microbiota regulating hormonal, neural and immunological behavior. (Image courtesy: Giau et al. 2018)

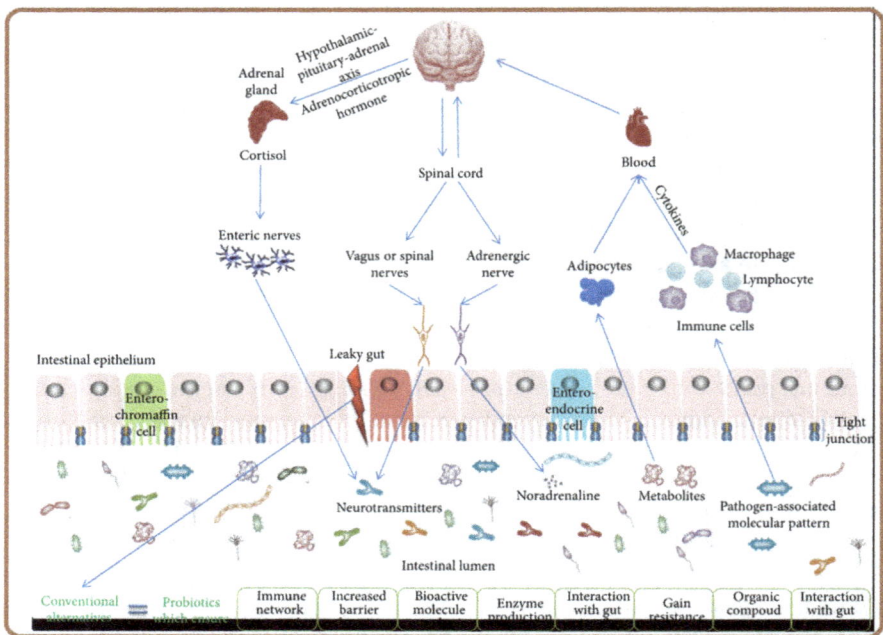

Fig. 10.2 Probable pathways in GBA communication to brain from the gut microbiota. (Image courtesy: Ilie et al. 2020)

of study to explicate their psychobiotic potential for its application to treat to neurodegenerative disorders (Cheng et al. 2019).

Quite a lot of research studies performed by many independent groups, including our own demonstrated the modulation of gut–brain axis by probiotics involving animal models of behavioral disorders (Desbonnet et al. 2014; Divyashri et al. 2015; De Palma et al. 2015). All these research findings proved that the commensal microbiota is crucial in the modulation of the host's behavioral profile (Sudo 2019). Unfortunately, the relationship between the gut microbiota and the brain is not sufficiently understood. The influence of gut microbiota on the distinctly located brain is taking place through GBA. Thus, the GBA is defined as a dense network of cells formed from the ENS, SND, and peripheral nervous system in association with the hypothalamic–pituitary–adrenal (HPA) axis (Fig. 10.2).

10.3 Experimental Approaches in Validating the Probiotic Effect on Neurodegeneration Using Animal Models

Studies using animal models have proven tremendously useful in understanding the mechanisms of neurodegenerative disorders and in assessing new therapeutic strategies. Growing evidences show that these psychobiotics have psychotropic

Table 10.1 Psychobiotics for stress, depression, and anxiety associated mental health: Experimental evidences from rodent models

Rodent model	Psychobiotics	Inference	Reference
Mice	L. paracasei Lpc-37, L. plantarum LP12407, L. plantarum LP12418, L. plantarum LP12151	Reduction in tress associated behaviors	Stenman et al. (2020)
Mice	L. plantarum PS128	Reduction in depression-like behaviors and inflammatory cytokine levels Increase in anti-inflammatory cytokine levels	Liu et al. (2016)
Rat	L. helveticus NS8	Improvement in cognitive dysfunction and enhancement of chronic restraint stress-induced behavioral symptoms	Liang et al. (2015)
Mice	E. faecium CFR 3003	Modulation of endogenous oxidative markers and redox status in brain	Divyashri et al. (2015)
Mice	B. longum1714, B. breve 1205	Reduction in stress-induced hyperthermia and anxiety	Savignac et al. (2014)
Mice	L. rhamnosus (JB-1)	Reduction of stress-induced corticosterone levels and restoration of anxiety- and depression-related behaviors	Bravo et al. (2011)
Rat	B. infantis	Normalization of the immune response and reversal of behavioral deficits	Desbonnet et al. (2010)

effects on stress, depression, and anxiety associated with neurodegenerative disorders (Table 10.1). Much of these psychobiotic researches are based on rodent models employing rodent stress inductions and rodent behavioral tests to assess overall mental health (Sarkar et al. 2016).

10.3.1 Psychobiotics in the Treatment of Anxiety and Depression

Over the decade, in animal models, psychobiotics have shown to inhibit inflammation and reduction of cortisol levels, resulting in an amelioration of anxiety and depression-related symptoms, thereby proving efficacious in restoring neurodegenerative and neurodevelopmental disorders, including Parkinson's disease and Alzheimer's disease (Cheng et al. 2019). More specifically, the administration of *L. plantarum* PS128 could reduce depression and anxiety-like behaviors in mice by decreasing corticosterone levels and inflammation, thereby increasing levels of serotonin and dopamine in the prefrontal cortex and striatum regions compared with control mice (Liu et al. 2015a, b, 2016). In addition, the administration of *L. helveticus* NS8 could increase norepinephrine, serotonin, and brain-derived neurotrophic factor (BDNF) levels in hippocampus region, thereby contributing better mental health (Liang et al. 2015). Administration of *L. rhamnosus* (JB-1) caused region dependent alterations in GABA receptor expression and reduced plasma corticosterone levels, thereby decreasing the symptoms of anxiety and depression (Bravo et al. 2011). On similar lines, *Bifidobacterium* sp. is also proven effective in treating anxiety and depression-like behaviors (Bercik et al. 2010; Desbonnet et al. 2010). Takahashi et al. (2019) evaluated the anti-depression ability of *Enterococcus faecalis* 2001 in dextran sulfate sodium treated mice showing the involvement of the brain–gut axis. *E. faecalis* 2001 demonstrated anti-depression ability by decreasing the levels of inflammatory cytokines in the rectal and hippocampal regions, facilitating the NFκB p65/XIAP pathway in the hippocampus. The propensity of a probiotic strain, *E. faecium* CFR 3003 in the modulation of endogenous oxidative markers and redox status in specific brain regions was assessed (Divyashri et al. 2015). Young mice supplemented with *E. faecium* CFR 3003 showed reduction in oxidative markers level, improved antioxidant enzymes activities with the concomitant increase in the levels of dopamine and GABA. Taken together, these findings evidently indicate the propensity of this bacterium to protect against tissue damage mediated through free radicals and inflammatory cytokines (Divyashri et al. 2015). In addition, a treatment with fermented cow's milk using *L. fermentum* or *L. casei* increased learning and memory behavior with the significant increase in levels of antioxidant enzymes viz., superoxide dismutase and glutathione peroxidase compared to the control group (Musa et al. 2017).

10.3.2 Psychobiotics in the Treatment of Alzheimer's Disease

Alzheimer's disease is a chronic neurodegenerative disorder characterized by cognitive and memory impairments (Kumar and Singh 2015). It also results in the formation of neurofibrillary tangles from abnormally phosphorylated tau and abnormal accumulation of amyloid plaques (Drummond and Wisniewski 2017). Animal models play a crucial role in defining disease associated mechanisms and have been of prime importance in evaluating the effectiveness of novel therapeutic agents, owing their initial studies performed in rodent models (Drummond and Wisniewski 2017). Animal models employed for Alzheimer's disease research are listed in (Table 10.2). Spontaneous (normal, age related, and senescence accelerated mouse (SAM)) and chemically induced animal models are available to evaluate cognitive dysfunction as a surrogate for human dementia and Alzheimer's disease. Spontaneous models develop the conditions of Alzheimer's disease without artificial manipulation. However, induced models require artificial manipulation, viz., the administration of drug or surgical procedures for the onset of the disease conditions (Kaushal et al. 2013). Whereas, transgenic models are constructed with either bearing an exogenous gene or knockout of a particular gene and rodent transgenic models (more than 150) are made available for research community. The underlying principle employed in generating transgenic model of Alzheimer's disease is the over expression of the transgene carrying familial Alzheimer's disease mutations using different promoters. Genes, viz., mutant APP, PS-1 and PS-2, apolipoprotein E

Table 10.2 Animal models are employed for Alzheimer's disease research

Model type	Animal paradigm	Phenotype
Spontaneous	Normal and aged rodents	Normal aging process
	SAM	Accelerated senescence, increased accumulation of hippocampal Aβ with noted behavioral impairment
Chemically induced	Scopolamine	Muscarinic learning impairment
	Mecamylamine	Nicotinic learning impairment
	STZ (streptozotocin)	Progressive memory loss
	Aβ infusion	Learning and cognitive deficits
Transgenic	APP mice and rats	Cognitive deficits and behavioral impairment
	Tau mice and rats	Behavioral and motor deficits
	APP/PSEN1 double transgenic mice	Aβ plaques accumulation
	APP/tau/PSN1 triple transgenic mice	Accumulation of Aβ plaques and intra-neuronal tangles of tau protein

(ApoE), and tau are utilized to construct transgenic mouse models of Alzheimer's disease (Kaushal et al. 2013). In addition, Tau transgenic mice have also been developed by expressing human tau protein using specific promoters.

Scientific evidences demonstrating the effect of psychobiotics in ameliorating cognitive disorders are very short and the same are displayed in Table 10.3. Based on the research findings from various animal studies, psychobiotics have shown the ability toameliorate cognitive and memory deficits compared with the control groups (Bonfili et al. 2018; Athari Nik Azm et al. 2018; Athari Nik Azm et al. 2017; Nimgampalle and Kuna 2017). The administration of probiotic formulation (SLAB51) on transgenic Alzheimer's disease mice could significantly reduce oxidative stress by inducing SIRT-1-dependent mechanisms (Bonfili et al. 2018). The effect of multiple strains, *L. acidophilus, L. fermentum, B. lactis,* and *B. longum* on animal model of Alzheimer's disease was evaluated. It was shown, after psychobiotic intervention, total counts of *Bifidobacterium* sp. and *Lactobacillus* sp. were increased and the levels of coliform was decreased in the stool. Furthermore, improvement in learning and memory deficits in Alzheimer's disease rats compared to control was observed with psychobiotic supplementation. In addition, significant reductions in the number of amyloid plaques, inflammation, and oxidative stress were observed in the Alzheimer-probiotics group (Athari Nik Azm et al. 2018). Athari Nik Azm et al. (2017) showed that the administration of psychobiotics could decrease the levels of insulin and improve insulin resistance in comparison to control group of Alzheimer's disease rats. However, no reduction in serum triglyceride level was observed between control and psychobiotic treated Alzheimer's disease rats. Administration of *L. plantarum* MTCC1325 to D-galactose-induced Alzheimer's disease rats not only improved cognitive behaviors but also restored acetylcholine levels (Nimgampalle and Kuna 2017). Supplementation of De Simone Formulation (a probiotic mixture comprising of eight different bacterial species) demonstrated significant modification in gut microbiota with concomitant increase

Table 10.3 Psychobiotics in the treatment of Alzheimer's disease

Study model and sample size (N)	Psychobiotic name	Dosage and route of administration	Study duration	Inference	Reference
Male transgenic Alzheimer's disease mice $N = 64$	S. thermophilus, B. longum, B. breve, B. infantis, L. acidophilus, L. plantarum, L. paracasei, L. delbrueckii subsp. bulgaricus, L. brevis	200 billion bacteria/ kg/day in drinking water	16 weeks	Reduced oxidative stress in Alzheimer's disease mice brain by activating SIRT1- dependent mechanisms	Bonfili et al. (2018)
Male Wistar rats $N = 60$	L. acidophilus, L. fermentum, B. lactis, and B. longum	10^{10} CFU/day in the drinking water	8 weeks	Improvement in oxidative stress biomarkers (malondialdehyde levels and superoxide dismutase activity)	Athari Nik Azm et al. (2018)
Male Wistar rats $N = 60$	L. acidophilus, L. fermentum, B. lactis, and B. longum	10^{10} CFU/day in drinking water	8 weeks	Effective role in controlling glycemic status of Alzheimer's disease by decreasing only insulin level and HOMA-IR index	Athari Nik Azm et al. (2017)
Wistar rats $N = 48$	L. plantarum MTCC1325	12×10^{8} CFU/mL 10 mL/kg body weight	60 days	Probiotic treatment ameliorated cognition deficits and restored acetylcholine levels	Nimgampalle and Kuna (2017)

in Actinobacteria and Bacteroidetes. Conversely, both of them were significantly reduced in vehicle-treated animals with impact on long-term potentiation, inflammation, and neural plasticity (Distrutti et al. 2014). Moreover, the administration of *B. breve* A1 to Alzheimer's disease mice indicated its potential to prevent cognitive dysfunction associated with Alzheimer's disease by diminishing the impairment of alternation behavior (Kobayashi et al. 2017). Together, these research evidences symbolizes the proof of principle that modulation of gut microbiota using probiotics will impact positively on cognitive functions. This also suggests a probable role of memory enhancing ability of probiotic strains in the prevention of cognitive impairments associated with Alzheimer's disease (Wong et al. 2018).

10.3.3 Psychobiotics in the Treatment of Parkinson's Disease

Parkinson's disease is the second most common age-related human neurodegenerative disorder (Ribeiro et al. 2013). It is characterized by the progressive loss of dopamine neurons in the region of midbrain (substantia nigra pars compacta) and accumulation of the misfolded presynaptic neuronal protein, viz., α-synuclein throughout the nervous system (Dawson et al. 2018). In spite of the fact that many individuals with Parkinson's disease have no clear genetic cause, research has been guided by the discovery of mutated genes that deterministically drive this disease for risk. These genetic guideposts, along with the identification of protein (α-synuclein accumulation), underline the pathophysiology of Parkinson's disease and thus provide greater opportunity to create animal models (Mhyre et al. 2012). Toxic and genetic models are evaluated effectively for Parkinson's disease and are currently available for research. Toxic models imply the use of toxic compounds that are known to offer selective toxicity for dopaminergic neurons. 1-methyl-4-phenyl-1,2,3,6-tetrahydropyridine (MPTP), 6-hydroxydopamine(6-OHDA), rotenone and paraquat are the widely employed neurotoxins and they act by destroying dopaminergic neurons in the substantia nigra of the brain. Genetic models of Parkinson's disease created by the successful mutation in α-synuclein gene and the manipulation of dopaminergic transcription factors are also employed (Konnova and Swanberg 2018).

10.3.3.1 Toxin Based Models
Toxin based models based on neurotoxins allow degeneration of dopamine neurons. Figure 10.3 summarizes the major toxins used the creation of toxin based models.

10.3.3.1.1 1-Methyl-4-Phenyl-1,2,3,6-Tetrahydropyridine (MPTP) Induced Parkinson's Disease
1-methyl-4-phenyl-1,2,3,6-tetrahydropyridine (MPTP) gets converted to an intermediate, 1-methyl-4-phenyl-2,3-dihydropyridinium by monoamine oxidase B in glial cells. 1-methyl-4-phenyl-2,3-dihydropyridinium further gets oxidized to 1-methyl-4-phenylpyridinium (MPP^+). Upon entry into the dopaminergic neurons, MPP^+ gets either sequestrated into synaptosomal vesicles or concentrated within the

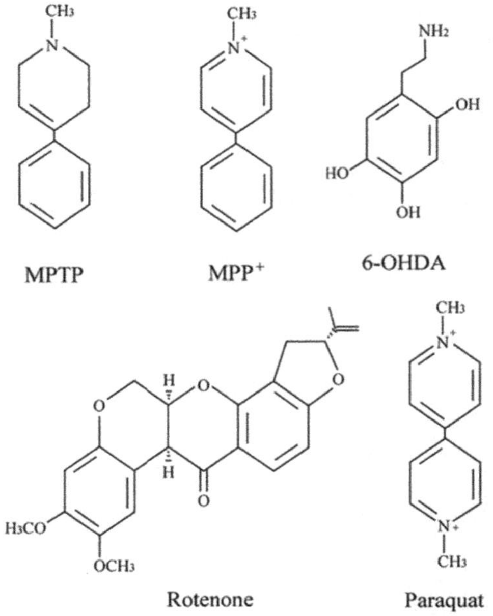

Fig. 10.3 Chemical structures of main neurotoxins used to reproduce the symptoms of Parkinson's disease in animal models. (Image curtesy: Zeng et al. 2018). *MPTP* 1-methyl-4-phenyl-1,2,3,6-tetrahydropyridine, *MPP⁺* 1-methyl-4-phenylpyridinium ion, *6-OHDA* 6-hydroxydopamine

mitochondria by utilizing mitochondrial transmembrane potential. In the mitochondria, MPP^+ blocks the electron transport chain by inhibiting Mitochondrial Complex I, thereby causing neuronal damage. However, the mechanism of conversion intermediate to MPP^+ is not clearly known. In addition, rats have proven resistant to MPTP-induced toxicity and typical Parkinson's disease behavior is not fully observed. Taking into account practical considerations, the MPTP mouse model is more beneficial and popular and it is reported to cause greater damage to dopaminergic neurons in substantia nigra pars compacta (Blesa et al. 2017).

10.3.3.1.2 6-Hydroxydopamine (6-OHDA) Induced Parkinson's Disease

6-hydroxydopamine (6-OHDA) was the first selective neurotoxin to be reported to cause lesions in dopaminergic neurons in rats (Ungerstedt 1968). It has also subsequently shown to work in other animal models, viz., such as mice (Thiele et al. 2012). 6-OHDA accumulates in the cytosol and promotes formation of hydrogen peroxide, other reactive oxygen species and quinines by auto-oxidation (Cohen 1984; Simola et al. 2007). 6-OHDA being hydrophilic fails to cross the blood brain barrier and thus requires the direct administration to substantia nigra pars compacta. Injection into substantia nigra pars compacta causes dopaminergic neuronal death in less than 24 h (Jeon et al. 1995).

10.3.3.1.3 Rotenone Induced Parkinson's Disease

Rotenone is used as a broad spectrum insecticide and pesticide. It functions by blocking the mitochondrial electron transport chain through inhibition of complex I, similar to MPTP. Rotenone also blocks mitosis and inhibits cell proliferation by perturbation of microtubule assembly and decreasing the GTP hydrolysis rate (Srivastava and Panda 2007). Chronic systemic exposure to rotenone in rats causes many features of Parkinson's disease, including nigrostriatal dopamine degeneration. Using this model, one can reproduce almost all the features of Parkinson's disease, including the formation of intracellular inclusion (Sherer et al. 2003). Rotenone can be injected intraperitoneally, intravenously or subcutaneously for systemic treatment. It has also been directly injected into the brain stereotaxically (Xiong et al. 2009). Rotenone is highly lipophilic and easily crosses the blood–brain barrier (Talpade et al. 2000). However, despite demonstrating the slow and specific loss of dopaminergic neurons, this model is difficult to replicate due to the high mortality observed in rats, when treated with rotenone (Fleming et al. 2004).

10.3.3.1.4 Paraquat Induced Parkinson's Disease

Paraquat, N,N'-dimethyl-4,4'-bipyridinium dichloride is one of the most widely used herbicides and structurally similar to MPP^+. It causes oxidative stress in the cell through the generation of reactive oxygen species. It is also reported to cause dopaminergic neuronal degeneration in substantia nigra pars compacta and induces the formation of lewy body in dopamine neurons in mice and rats (McCormack et al. 2002; Cicchetti et al. 2005).

10.3.3.2 Genetic Models of Parkinson's Disease

Genetic models are created by mutating the genes that are known to be involved in the progression of Parkinson's disease. Genes, viz., α-synuclein, LRRK 2, Parkin, DJ1, and PINK1 are regularly targeted for the development of Parkinson's disease models. However, single gene models (monogenic models) are less likely to be successful because several gene functions are expected to be altered in more common forms of Parkinson's disease.

10.3.3.2.1 α-Synuclein Induced Parkinson's Disease

α-synuclein makes up the major part of Lewy body protein in the brains of Parkinson's disease patients. Injection of mutant α-synucelin protein is reported to induce loss of dopaminergic neurons and motor impairment in rodent models (Oliveras-Salvá et al. 2013). Many α-synuclein mutant lines have been developed in mice that demonstrate the decrease in dopamine levels and increase in inclusion bodies. However, these mutant models have failed to show significant degeneration of nigrostriatal Parkinson's disease neurons. In similar lines, mutant lines are also developed in rats with significant dopamine loss and formation of inclusion bodies (Oueslati et al. 2012).

10.3.3.2.2 LRRK 2 Induced Parkinson's Disease

LRRK2 gene codes for a protein (2527 amino acid) with multiple domains. Mutations in this gene is known to cause an autosomal familial form of Parkinson's disease. LRRK 2 knockout mice have been reported to express abnormal aggregation and accumulation of proteins including α-synuclein.

10.3.3.2.3 Parkin Induced Parkinson's Disease

Parkin gene mutation have been associated for an autosomal recessive form of Parkinson's disease. Parkin genetic rodent models have developed and most of them do not exhibit loss of dopaminergic neurons in substantia nigra (Liu et al. 2013; Bian et al. 2012). However, few genetic rodent models have demonstrated modest loss of dopaminergic neurons (Rompuy et al. 2014).

10.3.3.2.4 DJl Induced Parkinson's Disease

DJ1 is a molecular chaperone that play role in inhibiting α-synuclein aggregate formation under redox reductions. DJ1 mutations are associated with early onset of Parkinson's disease. Mice DJ1 genetic models are widely employed, wherein Rousseaux et al. (2012) proved the usage of DJ1-C57 mouse as a tool to study the preclinical aspects of neurodegeneration. If reproduced, this model would be extremely beneficial to study the early onset of Parkinson's disease (Jagmag et al. 2016). Furthermore, a DJ1 rat model has also been developed, which exhibits loss in dopaminergic neuron and motor abnormalities (Dave et al. 2014).

10.3.3.2.5 PINK1 Induced Parkinson's Disease

PINK1 codes for a protein, mitochondrial kinase, which functions in recruiting Parkin from the cytosol to the mitochondria, thereby increasing the ubiquitination activity of Parkin, and induces Parkin-mediated mitophagy (Lazarou et al. 2013). Mutations in the PARK6 locus of PINK1 gene result in the early onset of Parkinson's disease. Because PINK1 and the Parkin function in the same pathway, the phenotypes of PINK1 and Parkin knockout mice are very similar. Use of PINK1 knockout mice is shown to affect mitochondrial functional defects and increases the sensitivity to oxidative stress. However, no significant dopaminergic neuron abnormalities or Lewy body formation was observed (Kitada et al. 2009). Over expression in PINK1 knockout mice has resulted in significant dopamine loss and increased levels of α-synuclein but no degeneration in substantia nigra (Oliveras-Salvá et al. 2013). In similar lines, PINK1 knockout rats exhibited dopamine loss and motor impairment which more closely mimic Parkinson's disease phenotype (Dave et al. 2014).

10.3.3.3 Experimental Evidences in Rodent Models

Many psychobiotic strains have been evaluated clinically for the treatment of Parkinson's disease. *L. acidophilus* and *B. infantis* at a dose of 120 mg/day given twice daily for 12 weeks to the patients suffering from Parkinson's disease was found to alleviate abdominal pain and bloating (Georgescu et al. 2016). Supplementation of 10^9 CFU of *L. acidophilus, B. bifidum, L. reuteri, and L. fermentum* per day for 12 weeks decreased movement disorders society-unified Parkinson's disease rating scale, C-reactive protein, and malondialdehyde levels with significant increase in glutathione levels (Tamtaji et al. 2019). However, very few scientific evidences have demonstrated the effectiveness of psychobiotics in the treatment of Parkinson's disease using above mentioned animal models. The effectiveness of psychobiotic strains, viz., *L. acidophilus* NCFM and *B. lactis* HN019 was evaluated using

MPTP-induced C57BL/6 N male mice mouse model of Parkinson's disease for 4 weeks. Psychobiotics significantly attenuated gastrointestinal and motor symptoms by decreasing pro-inflammatory cytokines, thereby offering neuroprotective effects in the pathogenesis of Parkinson's disease (Qian et al. 2018). Furthermore, transplantation of fecal flora to MPTP-induced murine Parkinson's disease model from normal C57BL/6 mice for a week significantly inhibited activation of glial cell activation and inflammation of neurons thereby offering neuroprotection (Sun et al. 2018). Much research evidences need to be gathered to establish the role of probiotics as an effective remedial measure to combat Parkinson's disease.

10.4   Conclusion

The probiotic industry is an ever-growing entity with the continual expansion of products being taken to market. This has driven scientific research with the aspirations to uncover probiotic strains that provide conclusive evidence of improvements in health and disease outcomes. These opportunistic endpoints have not currently been met, evidenced by the fact that no certified health claims credited to probiotic products are currently in place. This is likely owing to the wide interpersonal variations in commensal bacteria as well as fundamental differences between probiotic strains (Day et al. 2019). Dietary interventions comprising psychobiotics (Perez-Pardo et al. 2017) may influence gut–brain axis by altering GIT composition or by affecting neuronal functioning in CNS (Parashar and Udayabanu 2017). Therefore, these interventions might provide opportunities to complement traditional therapies treating neurodegenerative disorders (Day et al. 2019).

Animal models play a crucial role in defining disease associated mechanisms and have been of prime importance in evaluating the effectiveness of novel therapeutic agents, owing their initial studies performed in rodent models (Drummond and Wisniewski 2017). Many lines of evidence have shown that probiotics (psychobiotics) modulation of the gut microbiota could improve neurodegenerative disorders such as Alzheimer's disease, Parkinson's disease in animal models. However, the current animal models of human neurodegenerative disorders may not faithfully recapitulate the key aspects of disease pathology. Taken together, these animal studies show that psychobiotics may play an important role in the bidirectional communication between the gut and the brain, and support the notion that psychobiotics modulation could ameliorate the onset of disease pathology (Wong et al. 2018). However, clinical trials investigating psychobiotic effect on neurodegenerative disorders are comparatively small in number at present are less compelling than the animal model data (Wallace and Milev 2017).

References

Athari Nik Azm S, Djazayeri A, Safa M, Azami K, Djalali M, Sharifzadeh M, Vafa M (2017) Probiotics improve insulin resistance status in an experimental model of Alzheimer's disease. Med J Islam Repub Iran 31:103

Athari Nik Azm S, Djazayeri A, Safa M, Azami K, Ahmadvand B, Sabbaghziarani F, Sharifzadeh M, Vafa M (2018) Lactobacilli and bifidobacteria ameliorate memory and learning deficits and oxidative stress in β-amyloid (1–42) injected rats. Appl Physiol Nutr Metab 43(7): 718–726

Azad M, Kalam A, Sarker M, Li T, Yin J (2018) Probiotic species in the modulation of gut microbiota: an overview. Biomed Res Int 2018:9478630

Bercik P, Verdu EF, Foster JA, Macri J, Potter M, Huang X, Malinowski P, Jackson W, Blennerhassett P, Neufeld KA, Lu J (2010) Chronic gastrointestinal inflammation induces anxiety-like behavior and alters central nervous system biochemistry in mice. Gastroenterology 139(6):2102–2112

Bian M, Liu J, Hong X, Yu M, Huang Y, Sheng Z, Fei J, Huang F (2012) Overexpression of parkin ameliorates dopaminergic neurodegeneration induced by 1-methyl-4-phenyl-1, 2, 3, 6-tetrahydropyridine in mice. PLoS One 7(6):e39953

Blesa J, Trigo-Damas I, Dileone M, Del Rey NL, Hernandez LF, Obeso JA (2017) Compensatory mechanisms in Parkinson's disease: circuits adaptations and role in disease modification. Exp Neurol 1(298):148–161

Bonaz B, Bazin T, Pellissier S (2018) The vagus nerve at the interface of the microbiota-gut-brain axis. Front Neurosci 12:49

Bonfili L, Cecarini V, Cuccioloni M, Angeletti M, Berardi S, Scarpona S, Rossi G, Eleuteri AM (2018) SLAB51 probiotic formulation activates SIRT1 pathway promoting antioxidant and neuroprotective effects in an AD mouse model. Mol Neurobiol 55(10):7987–8000

Bravo JA, Forsythe P, Chew MV, Escaravage E, Savignac HM, Dinan TG, Bienenstock J, Cryan JF (2011) Ingestion of lactobacillus strain regulates emotional behavior and central GABA receptor expression in a mouse via the vagus nerve. Proc Natl Acad Sci 108(38):16050–16055

Briguglio M, Dell'Osso B, Panzica G, Malgaroli A, Banfi G, Zanaboni Dina C, Galentino R, Porta M (2018) Dietary neurotransmitters: a narrative review on current knowledge. Nutrients 10(5): 591

Burokas A, Moloney RD, Dinan TG, Cryan JF (2015) Microbiota regulation of the mammalian gut–brain axis. In: Advances in applied microbiology, vol 91. Academic Press, San Diego, CA, pp 1–62

Cheng LH, Liu YW, Wu CC, Wang S, Tsai YC (2019) Psychobiotics in mental health, neurode-generative and neurodevelopmental disorders. J Food Drug Anal 27(3):632–648

Cicchetti F, Lapointe N, Roberge-Tremblay A, Saint-Pierre M, Jimenez L, Ficke BW, Gross RE (2005) Systemic exposure to paraquat and maneb models early Parkinson's disease in young adult rats. Neurobiol Dis 20(2):360–371

Cohen G (1984) Oxy-radical toxicity in catecholamine neurons. Neurotoxicology 5(1):77–82

Dave KD, De Silva S, Sheth NP, Ramboz S, Beck MJ, Quang C, Switzer RC III, Ahmad SO, Sunkin SM, Walker D, Cui X (2014) Phenotypic characterization of recessive gene knockout rat models of Parkinson's disease. Neurobiol Dis 70:190–203

Dawson TM, Golde TE, Lagier-Tourenne C (2018) Animal models of neurodegenerative diseases. Nat Neurosci 21(10):1370–1379

Day RL, Harper AJ, Woods RM, Davies OG, Heaney LM (2019) Probiotics: current landscape and future horizons. Fut Sci OA 5(4):FSO391

De Palma G, Blennerhassett P, Lu J, Deng Y, Park AJ, Green W, Denou E, Silva MA, Santacruz A, Sanz Y, Surette MG (2015) Microbiota and host determinants of behavioural phenotype in maternally separated mice. Nat Commun 6(1):1–3

Desbonnet L, Garrett L, Clarke G, Kiely B, Cryan JF, Dinan TG (2010) Effects of the probiotic Bifidobacterium infantis in the maternal separation model of depression. Neuroscience 170(4): 1179–1188

Desbonnet L, Clarke G, Shanahan F, Dinan TG, Cryan JF (2014) Microbiota is essential for social development in the mouse. Mol Psychiatry 19(2):146–148

Dinan TG, Stanton C, Cryan JF (2013) Psychobiotics: a novel class of psychotropic. Biol Psychiatry 74(10):720–726

Distrutti E, O'Reilly JA, McDonald C, Cipriani S, Renga B, Lynch MA, Fiorucci S (2014) Modulation of intestinal microbiota by the probiotic VSL# 3 resets brain gene expression and ameliorates the age-related deficit in LTP. PLoS One 9(9):e106503

Divyashri G, Prapulla SG (2015) An insight into kinetics and thermodynamics of gamma-aminobutyric acid production by enterococcus faecium CFR 3003 in batch fermentation. Ann Microbiol 65(2):1109–1118

Divyashri G, Krishna G, Prapulla SG (2015) Probiotic attributes, antioxidant, anti-inflammatory and neuromodulatory effects of enterococcus faecium CFR 3003: in vitro and in vivo evidence. J Med Microbiol 64(12):1527–1540

Drummond E, Wisniewski T (2017) Alzheimer's disease: experimental models and reality. Acta Neuropathol 133(2):155–175

Finegold SM, Downes J, Summanen PH (2012) Microbiology of regressive autism. Anaerobe 18(2):260–262

Fleming SM, Zhu C, Fernagut PO, Mehta A, DiCarlo CD, Seaman RL, Chesselet MF (2004) Behavioral and immunohistochemical effects of chronic intravenous and subcutaneous infusions of varying doses of rotenone. Exp Neurol 187(2):418–429

Fung TC, Olson CA, Hsiao EY (2017) Interactions between the microbiota, immune and nervous systems in health and disease. Nat Neurosci 20(2):145

Georgescu D, Ancusa OE, Georgescu LA, Ionita I, Reisz D (2016) Nonmotor gastrointestinal disorders in older patients with Parkinson's disease: is there hope? Clin Interv Aging 11:1601

Giau VV, Wu SY, Jamerlan A, An SS, Kim S, Hulme J (2018) Gut microbiota and their neuroinflammatory implications in Alzheimer's disease. Nutrients 10(11):1765

Gourie-Devi M (2014) Epidemiology of neurological disorders in India: review of background, prevalence and incidence of epilepsy, stroke, Parkinson's disease and tremors. Neurol India 62(6):588

Ilie OD, Ciobica A, McKenna J, Doroftei B, Mavroudis I (2020) Minireview on the relations between gut microflora and Parkinson's disease: further biochemical (oxidative stress), inflammatory, and neurological particularities. Oxid Med Cell Longev 2020:4518023

Jagmag SA, Tripathi N, Shukla SD, Maiti S, Khurana S (2016) Evaluation of models of Parkinson's disease. Front Neurosci 9:503

Jeon BS, Jackson-Lewis V, Burke RE (1995) 6-Hydroxydopamine lesion of the rat substantia nigra: time course and morphology of cell death. Neurodegeneration 4(2):131–137

Kaushal A, Wani WY, Anand R, Gill KD (2013) Spontaneous and induced nontransgenic animal models of AD: modeling AD using combinatorial approach. Am J Alzheimers Dis Other Demen 28(4):318–326

Kitada T, Tong Y, Gautier CA, Shen J (2009) Absence of nigral degeneration in aged parkin/DJ-1/PINK1 triple knockout mice. J Neurochem 3:696–702

Kobayashi Y, Sugahara H, Shimada K, Mitsuyama E, Kuhara T, Yasuoka A, Kondo T, Abe K, Xiao JZ (2017) Therapeutic potential of Bifidobacterium breve strain A1 for preventing cognitive impairment in Alzheimer's disease. Sci Rep 7(1):1

Konnova EA, Swanberg M (2018) Animal models of Parkinson's disease. Exon Publ:83–106

Kumar A, Singh A (2015) A review on Alzheimer's disease pathophysiology and its management: an update. Pharmacol Rep 67(2):195–203

Lazarou M, Narendra DP, Jin SM, Tekle E, Banerjee S, Youle RJ (2013) PINK1 drives Parkin self-association and HECT-like E3 activity upstream of mitochondrial binding. J Cell Biol 200(2): 163–172

Liang S, Wang T, Hu X, Luo J, Li W, Wu X, Duan Y, Jin F (2015) Administration of Lactobacillus helveticus NS8 improves behavioral, cognitive, and biochemical aberrations caused by chronic restraint stress. Neuroscience 310:561–577

Liang S, Wu X, Hu X, Wang T, Jin F (2018) Recognizing depression from the microbiota–gut–brain axis. Int J Mol Sci 19(6):1592

Liu B, Traini R, Killinger B, Schneider B, Moszczynska A (2013) Overexpression of parkin in the rat nigrostriatal dopamine system protects against methamphetamine neurotoxicity. Exp Neurol 247:359–372

Liu WH, Yang CH, Lin CT, Li SW, Cheng WS, Jiang YP, Wu CC, Chang CH, Tsai YC (2015a) Genome architecture of lactobacillus plantarum PS128, a probiotic strain with potential immunomodulatory activity. Gut Pathog 7(1):22

Liu X, Cao S, Zhang X (2015b) Modulation of gut microbiota–brain axis by probiotics, prebiotics, and diet. J Agric Food Chem 63(36):7885–7895

Liu YW, Liu WH, Wu CC, Juan YC, Wu YC, Tsai HP, Wang S, Tsai YC (2016) Psychotropic effects of lactobacillus plantarum PS128 in early life-stressed and naïve adult mice. Brain Res 1631:1–2

Lyte M, Villageliú DN, Crooker BA, Brown DR (2018) Symposium review: microbial endocrinology—why the integration of microbes, epithelial cells, and neurochemical signals in the digestive tract matters to ruminant health. J Dairy Sci 101(6):5619–5628

Ma Q, Xing C, Long W, Wang HY, Liu Q, Wang RF (2019) Impact of microbiota on central nervous system and neurological diseases: the gut-brain axis. J Neuroinflammation 16(1):53

Magistrelli L, Amoruso A, Mogna L, Cantello R, Pane M, Comi C (2019) Probiotics may have beneficial effects in Parkinson's disease: in vitro evidence. Front Immunol 10:969

McCormack AL, Thiruchelvam M, Manning-Bog AB, Thiffault C, Langston JW, Cory-Slechta DA, Di Monte DA (2002) Environmental risk factors and Parkinson's disease: selective degeneration of nigral dopaminergic neurons caused by the herbicide paraquat. Neurobiol Dis 10(2):119–127

Mhyre TR, Boyd JT, Hamill RW, Maguire-Zeiss KA (2012) Protein aggregation and fibrillogenesis in cerebral and systemic amyloid disease, vol 65, p 389

Musa NH, Mani V, Lim SM, Vidyadaran S, Majeed AB, Ramasamy K (2017) Lactobacilli-fermented cow's milk attenuated lipopolysaccharide-induced neuroinflammation and memory impairment in vitro and in vivo. J Dairy Res 84(4):488–495

Nimgampalle M, Kuna Y (2017) Anti-Alzheimer properties of probiotic, lactobacillus plantarum MTCC 1325 in Alzheimer's disease induced albino rats. J Clin Diagn Res 11(8):KC01

O'Mahony SM, Clarke G, Borre YE, Dinan TG, Cryan JF (2015) Serotonin, tryptophan metabolism and the brain-gut-microbiome axis. Behav Brain Res 277:32–48

Oliveras-Salvá M, Van der Perren A, Casadei N, Stroobants S, Nuber S, D'Hooge R, Van den Haute C, Baekelandt V (2013) rAAV2/7 vector-mediated overexpression of alpha-synuclein in mouse substantia nigra induces protein aggregation and progressive dose-dependent neurodegeneration. Mol Neurodegener 8(1):44

Oueslati A, Paleologou KE, Schneider BL, Aebischer P, Lashuel HA (2012) Mimicking phosphorylation at serine 87 inhibits the aggregation of human α-synuclein and protects against its toxicity in a rat model of Parkinson's disease. J Neurosci 32(5):1536–1544

Parashar A, Udayabanu M (2017) Gut microbiota: implications in Parkinson's disease. Parkinsonism Relat Disord 38:1–7

Perez-Pardo P, Kliest T, Dodiya HB, Broersen LM, Garssen J, Keshavarzian A, Kraneveld AD (2017) The gut-brain axis in Parkinson's disease: possibilities for food-based therapies. Eur J Pharmacol 817:86–95

Qian Y, Yang X, Xu S, Xiao Q (2018) Neuroprotection effects of probiotics strains on a chronic MPTP-induced mouse model of Parkinson's disease. In: Movement disorders, vol 33. Wiley, Hoboken, NJ, p S800

Rezaei Asl Z, Sepehri G, Salami M (2019) Probiotic treatment improves the impaired spatial cognitive performance and restores synaptic plasticity in an animal model of Alzheimer's disease. Behav Brain Res 376:112183

Ribeiro FM, Camargos ER, de Souza LC, Teixeira AL (2013) Animal models of neurodegenerative diseases. Braz J Psychiatry 35(Suppl 2):S82–S91

Rompuy AS, Lobbestael E, Perren AV, Haute CV, Baekelandt V (2014) Long-term overexpression of human wild-type and T240R mutant Parkin in rat substantia nigra induces progressive dopaminergic neurodegeneration. J Neuropathol Exp Neurol 73(2):159–174

Roshchina VV (2016) New trends and perspectives in the evolution of neurotransmitters in microbial, plant, and animal cells. In: Microbial endocrinology: Interkingdom signaling in infectious disease and health. Springer, Champions, pp 25–77

Rousseaux MW, Marcogliese PC, Qu D, Hewitt SJ, Seang S, Kim RH, Slack RS, Schlossmacher MG, Lagace DC, Mak TW, Park DS (2012) Progressive dopaminergic cell loss with unilateral-to-bilateral progression in a genetic model of Parkinson disease. Proc Natl Acad Sci 109(39): 15918–15923

Sanchez B, Delgado S, Blanco-Míguez A, Lourenço A, Gueimonde M, Margolles A (2017) Probiotics, gut microbiota, and their influence on host health and disease. Mol Nutr Food Res 61(1):1600240

Sarkar A, Lehto SM, Harty S, Dinan TG, Cryan JF, Burnet PW (2016) Psychobiotics and the manipulation of bacteria–gut–brain signals. Trends Neurosci 39(11):763–781

Savignac HM, Kiely B, Dinan TG, Cryan JF (2014) Bifidobacteria exert strain-specific effects on stress-related behavior and physiology in BALB/c mice. Neurogastroenterol Motil 26(11): 1615–1627

Sherer TB, Betarbet R, Testa CM, Seo BB, Richardson JR, Kim JH, Miller GW, Yagi T, Matsuno-Yagi A, Greenamyre JT (2003) Mechanism of toxicity in rotenone models of Parkinson's disease. J Neurosci 23(34):10756–10764

Simola N, Morelli M, Carta AR (2007) The 6-hydroxydopamine model of Parkinson's disease. Neurotox Res 11(3–4):151–167

Srivastava P, Panda D (2007) Rotenone inhibits mammalian cell proliferation by inhibiting microtubule assembly through tubulin binding. FEBS J 274(18):4788–4801

Stenman LK, Patterson E, Meunier J, Roman FJ, Lehtinen MJ (2020) Strain specific stress-modulating effects of candidate probiotics: a systematic screening in a mouse model of chronic restraint stress. Behav Brain Res 379:112376

Sudo N (2019) Role of gut microbiota in brain function and stress-related pathology. In: Bioscience of microbiota, food and health, pp 19–006

Sun MF, Zhu YL, Zhou ZL, Jia XB, Xu YD, Yang Q, Cui C, Shen YQ (2018) Neuroprotective effects of fecal microbiota transplantation on MPTP-induced Parkinson's disease mice: gut microbiota, glial reaction and TLR4/TNF-α signaling pathway. Brain Behav Immun 70:48–60

Takahashi K, Nakagawasai O, Nemoto W, Odaira T, Sakuma W, Onogi H, Nishijima H, Furihata R, Nemoto Y, Iwasa H, Tan-No K (2019) Effect of enterococcus faecalis 2001 on colitis and depressive-like behavior in dextran sulfate sodium-treated mice: involvement of the brain–gut axis. J Neuroinflammation 16(1):1–6

Talpade DJ, Greene JG, Higgins DS Jr, Greenamyre JT (2000) In vivo labeling of mitochondrial complex I (NADH: UbiquinoneOxidoreductase) in rat brain using [3H] Dihydrorotenone. J Neurochem 75(6):2611–2621

Tamtaji OR, Taghizadeh M, Kakhaki RD, Kouchaki E, Bahmani F, Borzabadi S et al (2019) Clinical and metabolic response to probiotic administration in people with Parkinson's disease: a randomized, double-blind, placebo-controlled trial. Clin Nutr 38(3):1031–1035

Thiele SL, Warre R, Nash JE (2012) Development of a unilaterally-lesioned 6-OHDA mouse model of Parkinson's disease. J Vis Exp 60:e3234

Umbrello G, Esposito S (2016) Microbiota and neurologic diseases: potential effects of probiotics. J Transl Med 14(1):298

Ungerstedt U (1968) 6-Hydroxy-dopamine induced degeneration of central monoamine neurons. Eur J Pharmacol 5(1):107–110

Wallace CJ, Milev R (2017) The effects of probiotics on depressive symptoms in humans: a systematic review. Ann Gen Psychiatry 16(1):14

Wong CB, Kobayashi Y, Xiao JZ (2018) Probiotics for preventing cognitive impairment in Alzheimer's disease. In: Gut microbiota-brain axis. IntechOpen, London

Xiong N, Huang J, Zhang Z, Zhang Z, Xiong J, Liu X, Jia M, Wang F, Chen C, Cao X, Liang Z (2009) Stereotaxical infusion of rotenone: a reliable rodent model for Parkinson's disease. PLoS One 4(11):e7878

Yano JM, Yu K, Donaldson GP, Shastri GG, Ann P, Ma L, Nagler CR, Ismagilov RF, Mazmanian SK, Hsiao EY (2015) Indigenous bacteria from the gut microbiota regulate host serotonin biosynthesis. Cell 161(2):264–276

Zeng XS, Geng WS, Jia JJ (2018) Neurotoxin-induced animal models of Parkinson disease: pathogenic mechanism and assessment. ASN Neuro 10:1759091418777438

Zhu S, Jiang Y, Xu K, Cui M, Ye W, Zhao G, Jin L, Chen X (2020) The progress of gut microbiome research related to brain disorders. J Neuroinflammation 17(1):25

CPSIA information can be obtained
at www.ICGtesting.com
Printed in the USA
BVHW051846250123
657140BV00003B/55